Functional Glycomics
A Review

Arti Gupta (editor)
Assistant Professor, Department of Zoology
Dr. Ram Manohar Lohia Avadh University
Avadh, India

CRC Press
Taylor & Francis Group
Boca Raton London New York

CRC Press is an imprint of the
Taylor & Francis Group, an **informa** business
A SCIENCE PUBLISHERS BOOK

First edition published 2026
by CRC Press
2385 NW Executive Center Drive, Suite 320, Boca Raton FL 33431

and by CRC Press
4 Park Square, Milton Park, Abingdon, Oxon, OX14 4RN

CRC Press is an imprint of Taylor & Francis Group, LLC

Library of Congress Cataloging-in-Publication Data (applied for)

ISBN: 978-1-041-06427-5 (hbk)
ISBN: 978-1-041-06531-9 (pbk)
ISBN: 978-1-003-63590-1 (ebk)

DOI: 10.1201/9781003635901

Typeset in Palatino Linotype
by Prime Publishing Services

Preface

Glycomics, the comprehensive study of glycans and their functional roles in biological systems, has emerged as a transformative discipline in modern biosciences. With its capacity to decode complex carbohydrate-mediated interactions, glycomics offers profound insights into cellular communication, immune regulation, disease progression, and therapeutic innovation. As the field continues to evolve, it bridges molecular biology, biotechnology, and clinical research, offering new avenues for diagnostics and therapeutics.

Functional Glycomics: A Review presents a curated collection of twelve chapters that reflect the interdisciplinary scope and translational potential of this rapidly expanding domain. Each chapter contributes a unique perspective, collectively advancing our understanding of glycan biology across plant and animal systems.

Chapter 1 introduces the concept of homeostatic disruption through glycomic mechanisms, establishing the foundational relevance of glycans in maintaining physiological balance. Chapter 2 expands this framework by exploring glycomics in precision medicine, particularly its applications in behavioral disorders, infections, and chronic diseases.

Chapter 3 shifts focus to plant systems, showcasing the analytical power of MALDI-TOFMS in profiling N-glycans using PNGase H+. Chapter 4 investigates glycan signaling in stress perception, linking environmental stimuli to cellular responses. Chapter 5 highlights glycomic biomarkers, emphasizing their diagnostic potential in neurobehavioral and infectious conditions.

Chapter 6 presents a novel perspective on host-pathogen interactions, examining the role of *Mucuna pruriens* in modulating infection progress. Chapter 7 explores glycomic signatures in maternity, revealing their implications in reproductive biology and maternal health. Chapter 8 delves into glycoproteomics and immunity, underscoring the role of glycans in immune modulation and defense mechanisms.

Chapter 9 introduces computational glycoscience, detailing advances in glycan data integration, modeling, and analysis. Chapter 10 offers a foresight review of glycomics in cancer studies, identifying strategic directions for future research and clinical translation.

Chapter 11 revisits the theme of homeostasis, reinforcing the systemic impact of glycan pathways. Finally, Chapter 12 examines glycomic roles in infection progression, completing the narrative arc with translational relevance.

This volume reflects the collaborative efforts of global experts and aims to serve as a comprehensive reference for researchers, clinicians, and students. It aspires to inspire innovation and expand the frontiers of glycoscience across disciplines.

Contents

Preface iii

1. Disruption of Homeostasis via Functional Glycomic Mechanisms 1
2. The Emerging Role of Glycomics in Precision Medicine: Applications in Behavioral Disorders, Infections, and Chronic Diseases 24
3. Analysis of Plant N-Glycans using PNGase H+ by MALDI-TOF MS 53
4. Glycan Signalling in Environmental Stress Perception and Response 70
5. Glycomics as Potential Biomarkers of Behavioural Disorders, Infection, and Diseases 95
6. Glycomics and Infection Progress: Exploring the Role of *Mucuna pruriens* (L.) Dc., in Modulating Host-Pathogen Interactions 119
7. Glycomics and Maternity 136
8. Glycoproteomics and Immunity 169
9. Glycoinformatics: Advances in Glycan Data Integration and Analysis 183
10. Review on the Foresight Approaches of Glycomics in Cancer Studies 206

11. Functional Glycomics and Homeostatic Disruption: Molecular R Mechanisms and Therapeutic Pathways 223

12. Role of Glynomics in Progression of Infections 260

Index 293

CHAPTER 1

Disruption of Homeostasis via Functional Glycomic Mechanisms

Mary Nancy Flora, R.,[1] *Barathi, M.*[2]
and *Chamundeeswari, M.*[3,*]

Glycans are the large sugar molecules that coat proteins and lipids. They are vital for maintaining homeostasis both inside cells and throughout the body. From protein folding and host protection against infection to immunological response and cell-to-cell communication, they are involved in almost every biological function. In modern biology, functional glycomics, the scientific study of glycan structure and function, has emerged as a crucial field that offers profound understanding of how glycans work in both health and illness. This chapter addresses the complex interplay between homeostasis and glycans, such as causes and consequences of glycan dysfunction and their contribution to disease. The chapter proceeds to explain how the glycans can achieve homeostasis, through cell adhesion, receptor regulation, and protein folding. Current technological advancements

[1] Assistant Professor, Department of Chemical Engineering, Vel Tech High Tech Dr. Rangarajan Dr. Sakunthala Engineering College, Chennai-600062, India.
[2] Professor, Department of Computer Science and Engineering, Ganadipathy Tulsi's Jain Engineering College, Kaniyambadi, Vellore-632102, India.
[3] Associate Professor, Department of Biotechnology, St. Joseph's College of Engineering, Chennai-600119, India.
* Corresponding author: chamundeeswari@gmail.com

DOI: 10.1201/9781003635901-1

within Glycomics, namely, mass spectrometry, glycan micro arraying, and glycoengineering, are discussed as main forces for growth in the subject. These technologies have facilitated the discovery of glycan-based biomarkers and the creation of glycan-based therapies, like glycoengineered biologics and glycan-based vaccines. Despite these developments, challenges persist, including the structural intricacy of glycans, their immunogenicity, and the necessity for interdisciplinary collaboration and technological advancement. The chapter concludes with a call for additional research, emphasizing the necessity of combining Glycomics with other omics technologies, translating research into clinical applications, and generating public awareness of the significance of glycans. Functional Glycomics has tremendous potential for expanding our understanding of biology and for revolutionizing medicine, offering new promise for disease diagnosis, treatment, and prevention. The future of glycomics is huge, and the impact of glycomics on human health and well-being will be tremendous.

1. Introduction

Glycomics, or the study of glycans, the complex sugar molecules that coat lipids and proteins, has emerged as a key area of study in modern molecular biology and biomedicine. Glycans, also referred to as the "sugar code of life", are ubiquitous in nature and important for almost all biological functions, ranging from protein folding and signal transduction to cellular signalling and immunological response [1]. These complex molecules are not static structural elements; they are dynamic and highly controlled, acting as essential regulators of homeostasis, the subtle balance that perpetuates life. But when glycan function is abrogated, the repercussions are significant, causing homeostasis breakdown and the development of a broad array of disorders, including cancer, autoimmune diseases, and congenital diseases [2].

Functional glycomics attempts to unravel the multifaceted functions of glycans in disease and health. In contrast to genomics or proteomics, which deal with linear nucleotide or amino acid sequences, glycomics has to deal with the astronomical structural variability of glycans. This is because glycans are biosynthesized without a template, giving rise to almost unlimited variability of structures. This intricacy has in the past made glycomics a difficult science, but the recent technological developments, including high-throughput mass spectrometry and glycan microarrays, have transformed our capacity to examine and interpret glycan functions [3].

Glycans play a vital role in the sustenance of homeostasis at several levels. For instance, they contribute to cell-cell communication by regulating the activity of adhesion molecules and receptors. Glycans play an important role in the immune system by facilitating proper antibody function as well as regulation of the immune response. Additionally, glycans take part in the folding and stabilization of proteins, ensuring proper function of proteins inside the cell. Despite these critical functions, glycan dysfunction may occur due to genetic mutations, environmental insults, or disease states with a cascade of pathological consequences [4]. An example is the CDGs (congenital disorders of glycosylation), a category of genetic diseases resulting from defects in glycan biosynthesis with resultant profound developmental and metabolic abnormalities [5].

More common illnesses like cancer are also linked to disruption of glycan function. It is well recognized that altered glycosylation patterns within tumor cells promote treatment resistance, immunological evasion, and metastasis. In autoimmune disorders, aberrant glycosylation of antibodies also contributes to the development of autoimmunity and the loss of immunological tolerance [6]. Clarifying these pathways is essential to comprehending the pathophysiology of disease and creating focused therapies that restore glycan function and, in turn, homeostasis.

This chapter delves into the dynamic arena of functional glycomics, specifically the function of glycans in homeostasis and the implications of their dysfunction. We have examined the structural diversity of glycans, their functions in health, and the pathways through which their disruption leads to disease. Also, this chapter emphasizes the technological innovations fueling glycomics research and touch on the promise of glycan-based diagnostics and therapeutics. By understanding the functional roles of glycans, we can unlock new avenues for restoring homeostasis and improving human health.

2. The Basics of Glycomics

Glycans, or carbohydrate or sugar molecules, are among the four big classes of biomolecules, which include proteins, lipids, and nucleic acids. Glycans are unlike the linear nucleotide or amino acid sequences in DNA or proteins and have a highly branched structure with great structural diversity. This structural diversity is the result of combinatorial biosynthesis of glycans, in which monosaccharides (single sugars) are added stepwise to create complex structures. It is critical to understand the fundamentals of glycomics to appreciate the function of glycans in health and disease.

2.1 *Structure and Diversity of Glycans*

Glycans are made up of monosaccharides like glucose, galactose, mannose, and sialic acid that are joined together by glycosidic linkages in different orientations (α or β) and positions (1–2, 1–3, 1–4, 1–6), leading to a vast structural diversity. Glycans are present either as free oligosaccharides or as glycoconjugates, in which they are covalently bound to proteins (glycoproteins) or lipids (glycolipids). The two primary forms of protein glycosylation are N-linked and O-linked glycosylation. N-linked glycosylation is the process of attaching glycans to the nitrogen atom of asparagine side chains in proteins, having a common core structure that is further varied into multibranched structures [7]. Conversely, O-linked glycosylation refers to the addition of glycans to the oxygen atom on serine or threonine side chains within proteins, resulting in generally shorter and more structurally heterogeneous glycans. The structural diversity of glycans is complemented by alterations such as sulfation, acetylation, and phosphorylation, which impart a further aspect of functional variability [8].

2.2 *Glycosylation: A Key Post-Translational Modification*

Protein function, stability, and structure are all affected significantly by glycosylation, the most common and functionally relevant post-translational modification. This precisely regulated enzyme-catalyzed process affects nearly every facet of protein biology, from correct folding and intracellular addressing to immune recognition and cell signaling, and is characterized by the covalent conjugation of glycans to particular amino acid residues. The endoplasmic reticulum (ER) is the site for N-linked glycosylation and is specifically essential for protein folding and quality control. Recently synthesized glycoproteins undergo a series of glycan modification before they are transported to the Golgi apparatus for further processing. This serves as molecular checkpoints and facilitates correct folding. The ER-associated degradation (ERAD) pathway recognizes misfolded glycoproteins and targets their proteasomal degradation to avoid accumulation of toxic or dysfunctional proteins [9]. This is depicted as a schematic diagram in Fig. 1.

In addition to its intracellular roles, glycosylation inhibits proteolysis and controls protein–protein interactions to increase the solubility, stability, and half-life of proteins outside of cells [10]. Glycosylation is a key determinant of protein activity in both healthy and pathological settings since glycan alterations also impact receptor binding, pathogen immune evasion, and the efficacy of therapeutic biologics. Glycosylation continues to be a major topic in biomedical research because of its profound impact

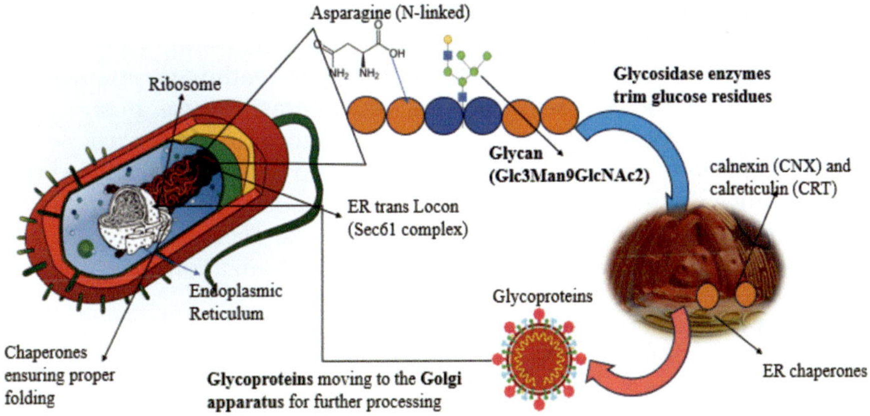

Fig. 1: Diagram illustrating a newly synthesized protein undergoing glycosylation with chaperones ensuring proper folding.

on cellular homeostasis and disease pathology, particularly in areas like cancer biology, neurological illnesses, and vaccine development. Glycan-based diagnostics and treatments are now possible thanks to advanced analytical techniques like mass spectrometry and glycan microarrays, which have significantly improved our understanding of glycosylation dynamics [11].

2.3 *Biological Roles of Glycans*

Glycans play a central role in numerous biological processes and act as mediators of cell-to-cell interactions, immune response modulation, pathogen detection, and developmental regulation. Their structural sophistication and specificity enable them to control essential physiological activities in various biological systems. Glycans play a role in cell-to-cell communication, which allows for interactions between cells and their environment. Glycans on the cell surface, among other biological functions, serve as recognition molecules for adhesion and signalling. An important process in immune surveillance and host defense is immune cell migration during inflammation and infection, which is mediated by selectins, a family of carbohydrate-binding cell adhesion molecules that bind to specific glycans on leukocytes [12].

In immune regulation, glycosylation has a critical role in controlling both innate and adaptive immune responses. The glycosylation status of antibodies (immunoglobulins) also determines their effector functions such as complement activation, antibody-dependent cellular cytotoxicity, and binding to Fc receptors on immune cells. Changes in glycan structures have the ability to control antibody efficacy and are responsible for immune dysregulation in autoimmune diseases and cancer [13]. Moreover, glycans are important mediators in pathogen interactions, wherein numerous viruses, bacteria, and parasites use host glycan structures to promote attachment, entry, and immune evasion. For example, influenza viruses use hemagglutinin to bind specifically to sialic acid residues on the surface of host cells, an important step in viral entry and infection. Likewise, bacterial pathogens like *Helicobacter pylori* and *Campylobacter jejuni* employ host glycans as colonization and immune evasion receptors [14].

In addition to infection and immunity, glycans are also central to development and differentiation, with critical functions in embryogenesis, organogenesis, and cellular specialization. Particular glycosylation patterns control stem cell pluripotency, tissue morphogenesis, and neuronal connectivity. Abnormalities in glycan biosynthesis can cause devastating congenital disease and developmental abnormalities, as in CDGs and other glycosylation-related diseases [15]. With their ubiquitous nature in biological activities, glycans are key determinants of homeostasis and disease. Comprehension of glycan-driven mechanisms has very important implications in therapeutic intervention in the form of glycan-derived vaccines, immunomodulators, and anti-infective agents [16]. The varying biological functions of glycans traverse numerous physiological events, impacting cell-cell communication, immune modulation, pathogen-host interactions, and protein stability. Table 1 summarizes the major of these functions, which involve important examples where glycosylation is critical in cellular homeostasis, disease processes, and therapeutic interventions.

2.4 Challenges in Glycomics Research

Although they are biologically critical, glycans present special challenges to researchers. In contrast to DNA or proteins, glycans are not encoded by the genome directly, and therefore, their biosynthesis becomes less predictable and easier to study. Furthermore, the structural diversity and heterogeneity of glycans demand advanced analytical reagents like mass spectrometry and nuclear magnetic resonance (NMR) spectroscopy for their identification [20]. Advancements in these technologies have greatly improved our knowledge of glycan structure and function, yet much remains unknown. In short, the fundamentals of glycomics lay the groundwork for grasping the varied roles of glycans in biology. From structural complexity to functional diversity, glycans are key to the regulation of cellular functions and homeostasis.

Table 1: Summarizes the diverse functions of glycans in biological systems.

Biological Function	Role of Glycans	Examples	References
Cell-Cell Communication	Facilitate interactions between cells and their microenvironment; mediate adhesion and signaling	Selectins bind glycans on leukocytes to regulate immune cell trafficking during inflammation	[17]
Immune Regulation	Modulate innate and adaptive immune responses by influencing antibody functions	Antibody glycosylation affects complement activation, Fc receptor interactions, and antibody-dependent cellular cytotoxicity	[11, 18]
Pathogen Recognition & Evasion	Serve as receptors for viral, bacterial, and parasitic pathogens, aiding in host invasion and immune evasion	Influenza virus binds sialic acid residues via hemagglutinin; Helicobacter pylori and Campylobacter jejuni use glycans for colonization	[5, 12]
Development & Differentiation	Regulate stem cell pluripotency, tissue morphogenesis, and neuronal connectivity	Disruptions in glycosylation lead to congenital disorders of glycosylation (CDGs) affecting embryogenesis and organ development	[15]
Protein Folding & Stability	Enhance protein solubility, stability, and half-life by preventing aggregation and degradation	Glycosylation protects proteins from proteolysis and modulates proper folding via ER chaperones	[14]
Intracellular Trafficking	Act as molecular signals for directing proteins to their proper cellular locations	Mannose-6-phosphate tags lysosomal enzymes for transport to lysosomes	[12]
Signaling Pathways	Regulate receptor activation, ligand binding, and downstream signaling cascades	Glycosylation of Notch receptors influences cell fate determination	[18, 19]

3. Functional Glycomics in Health

Glycans are essential for maintenance of health due to their functions in cell-to-cell communication, immune regulation, and homeostatic physiological processes. Due to their dynamic property and structural polymorphism, they have the potential to be associated with a plethora of biological events such as growth in tissues, immune defense, folding of proteins, and signal transmission in cells.

3.1 Glycans in Cellular Communication

Glycans are critical cell-to-cell and cell-to-matrix communication mediators, controlling cell-cell and cell-matrix interactions through their diverse structural compositions. These carbohydrate structures densely coat the surfaces of cells and extracellular matrices, forming a specialized interface known as the glycocalyx. This glycan-rich layer not only provides structural integrity but also serves as a dynamic signal ing platform that enables cells to recognize and respond to their environment. Cell adhesion is among the critical roles of glycans in cellular communication, where glycan-modified proteins regulate cellular interactions required for tissue structure and immune surveillance. Glycans on cell surface molecules like integrins, cadherins, and selectins enable adhesion and migration by interacting with other cells and ECM (extracellular matrix). These are critical in biological processes such as wound healing, embryogenesis, and immune trafficking. For instance, selectins engage specific glycan ligands on leukocytes to guide them towards the area of inflammation, which is a function important in protecting against immunity. Cancer is also altered in cell adhesion properties by aberrant glycosylation, leading to enhanced metastasis and tumor proliferation [21].

Apart from adhesion, glycans play important roles in receptor modulation, modulating growth factor receptor, cytokine receptor, and immune checkpoint molecule activities. Modification of glycans regulates receptor stability, specificity of ligand binding, and signaling downstream. N-glycosylation of epidermal growth factor receptor (EGFR), for example, impacts its dimerization and activation and has relevance to cell growth and cancer. T-cell activation and immunological evasion in cancer are also influenced by the sialylation and fucosylation of immune receptors including PD-1 and CTLA-4, which regulate immune responses [22]. Given their integral role in communication networks, glycans serve as critical regulators of cellular homeostasis and disease pathogenesis. Advancements in glycomics continue to uncover how glycan-mediated interactions can be harnessed for therapeutic interventions, including the development of glycan-targeted cancer immunotherapies, biomaterials for tissue engineering, and glycan-based vaccines.

3.2 Glycans in Immune Regulation

Glycans have a crucial role in immune regulation, shaping innate and adaptive immunity by modulating immune cell interactions, signaling, and inflammatory responses. Glycosylation of immune cells and molecules plays a key role in immune homeostasis, allowing for pathogen recognition, immune cell migration, and self-tolerance [23]. Altered glycosylation can give rise to immune dysfunction and underlie autoimmune diseases, chronic inflammation, and immune evasion by pathogens and tumors. Antibody glycosylation is one of the most important glycan-mediated immune regulatory aspects. Immunoglobulins (IgGs) are heavily glycosylated, and their Fc (fragment crystallizable) portion contains conserved N-glycans that determine effector activities. The antibody glycosylation is responsible for influencing their complement activation, Fc receptor binding (FcγRs), and inflammatory capacity. For example, sialylated IgGs have anti-inflammatory effects by regulating FcγRIIb signaling, whereas hypoglycosylated or afucosylated IgGs increase antibody-dependent cellular cytotoxicity (ADCC) [24]. Disrupted IgG glycosylation has been connected with autoimmune illnesses like rheumatoid arthritis and systemic lupus erythematosus, with elevated agalactosylated IgG associated with persistent inflammation.

Apart from antibody function, glycans play a critical role in immune cell trafficking and inflammation. Immune cell adhesion and migration are based on the interaction between selectins and their glycan ligands. Glycans on leukocytes and endothelial cells facilitate the rolling, adhesion, and extravasation of immune cells through blood vessel walls, a process vital for immune surveillance and response to infection. The selectin family members, E-selectin, P-selectin, and L-selectin, interact with glycosylated ligands like sialyl Lewis X (sLeX) to mediate leukocyte homing to inflammation sites. Glycan-mediated immune trafficking dysregulation can lead to chronic inflammatory conditions, including atherosclerosis and inflammatory bowel disease, and blunt immune responses to infections [25].

In addition, glycans regulate immune receptor activity. Glycosylation of immune checkpoint receptors like PD-1, CTLA-4, and TIM-3 regulates their stability and ligand-binding capability, impacting T-cell activation and immune evasion of cancer. Tumor cells frequently take advantage of glycan modifications in order to evade immune detection, for instance, by enhanced sialylation, which generates an immunosuppressive environment through the activation of Siglec receptors on macrophages and dendritic cells. As key players in immunity, glycans have emerged as attractive targets for vaccine and immunotherapy development. Glycoengineered monoclonal antibodies, glycan-based cancer vaccines, and synthetic glycan ligands are among the strategies that have enormous potential to boost immune responses and treat immune-related diseases [26]. The knowledge of glycan-mediated immune regulation remains a driving force behind the development of autoimmune disease treatment, cancer immunotherapy, and infectious disease therapy.

3.3 Glycans in Protein Folding and Quality Control

Glycosylation is important for protein folding, stability, and intracellular trafficking and for proper protein maturation and exclusion of misfolded proteins. Being an important post-translational modification, glycosylation modulates protein solubility and conformation and supports cellular homeostasis and proteostasis. In the ER, N-linked glycans act as critical quality control signals that facilitate protein folding. Glycoproteins synthesized recently are sequentially trimmed and modified by ER-resident chaperones like calnexin and calreticulin, which validate proper folding before proteins are exported to the Golgi apparatus for additional processing [27]. Glycans also enable the recognition of misfolded or partially folded proteins and transport them towards ERAD. It consists of retrotranslocation of aberrantly folded glycoproteins into the cytosol where they are ubiquitinated and degraded by the proteasome, thereby averting the deposition of toxic protein aggregates. Misfolded glycoproteins like amyloid-β in Alzheimer's disease and α-synuclein in Parkinson's disease in neurodegenerative diseases evade glycan-mediated removal, producing toxic aggregates that underlie neuronal pathology [28].

3.4 Glycans as Biomarkers for Health

Glycosylation is an essential biomarker of disease and health that mirrors physiological and pathological processes in a range of conditions such as cancer, inflammation, cardiovascular diseases, and neurodegenerative disorders. Cancer is characterized by aberrant glycosylation, including augmented fucosylation, sialylation, and truncated O-glycans, which play roles in immune evasion, metastasis, and drug resistance. Glycoprotein cancer biomarkers with key significance are alpha-fetoprotein (AFP), applied for diagnosis and monitoring of hepatocellular carcinoma (HCC), whose increased fucosylation further improves its specificity towards cancer. Analogously, glycosylation modifications of prostate-specific antigen (PSA) via core fucosylation and sialylation differentiate between aggressive prostate cancer and benign processes. Other firmly established cancer biomarkers, such as CA19-9 and CA125, find extensive applications for pancreatic and ovarian cancers, respectively, and differential glycan structures enhance the specificity of the diagnostic [29].

3.5 Glycans in Host-Pathogen Interactions

In mucosal surfaces, mucin glycans create a dense, hydrated protective barrier that inhibits pathogen adhesion and colonization, especially in the respiratory and gastrointestinal tracts. The highly glycosylated mucins capture microorganisms and enable their clearance by mucus turnover and secretion. Commensal microbes can also modulate host glycan expression to establish homeostasis, whereas pathogenic bacteria and viruses use host glycans for adhesion and entry. Most pathogens identify and bind specifically to glycans of the host as the initial step in causing infection. For example, influenza viruses use hemagglutinin (HA) proteins to attach sialic acid residues on epithelial cells to allow entry and replication of the virus (Fig. 2). *Helicobacter pylori* also bind to Lewis antigens on gastric mucosal glycans to infect and cause colonization and peptic ulcer disease and gastric cancer [30].

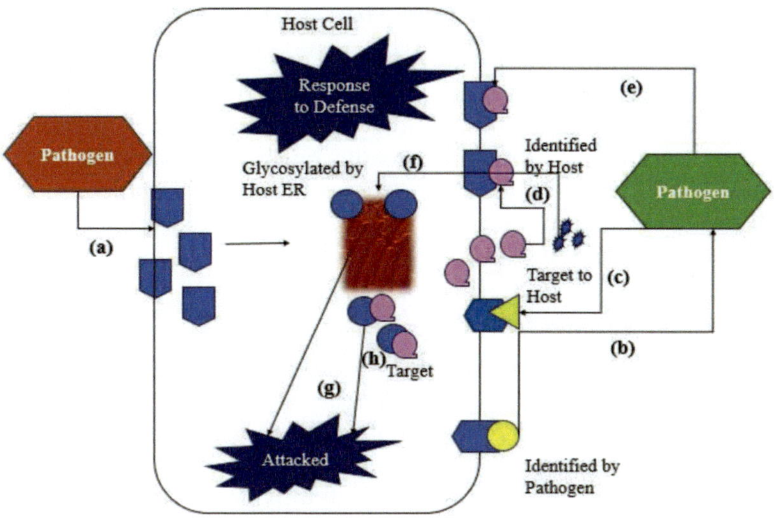

Fig. 2: Glycan-mediated host-pathogen interactions. Pathogens bind host glycans for adhesion, while hosts use glycans for immune defense. Host (blue) and pathogen (yellow) glycans influence infection and immunity, offering targets for glycan-based therapeutics.

3.6 Glycans in Development and Tissue Homeostasis

Glycans are essential for embryonic development, tissue differentiation, and homeostasis by controlling important signaling pathways, ECM interactions, and cellular processes. Normal development depends on proper glycan biosynthesis, as glycosylation defects cause severe congenital diseases and tissue dysfunction. One of the most thoroughly documented functions of glycans during development is their contribution to Notch signaling, an evolutionary conserved process that controls cell fate decisions, proliferation, and differentiation. Glycosylation of Notch receptors by O-fucose and O-glucose structures is indispensable for ligand interaction and the transmission of subsequent signaling, modulating events during neurogenesis, hematopoiesis, and somitogenesis. Notch glycosylation aberrations have been implicated in developmental disease and cancer, underlining the significance of stringent glycan regulation. In addition to signaling, glycans are an integral part of the ECM, where they play a role in tissue architecture, repair, and homeostasis. Glycosaminoglycans (GAGs), including heparan sulfate, chondroitin sulfate, and hyaluronic acid, bind to ECM proteins to modulate cell adhesion, migration, and growth factor signaling [31]. These interactions are vital in wound healing, angiogenesis, and tissue regeneration.

4. Disruption of Glycan Function and Homeostasis

Glycans are essential to ensure cellular and systemic homeostasis, but their dysregulation can cause a multitude of diseases. Impairments in glycan biosynthesis, modification, or recognition may occur due to genetic mutations, environmental insults, or disease processes, leading to pathological effects. This section discusses the etiology and consequences of glycan dysfunction and its involvement in diseases like cancer, autoimmune disease, and CDGs.

4.1 Causes of Glycan Dysfunction

Glycan malfunction is the result of a synergistic interplay of genetic, environmental, and pathological events, with far-reaching implications for cellular function and organismal well-being. Congenital mutations in glycosyltransferases, glycosidases, or other enzymes responsible for glycan biosynthesis led to CDGs, a family of rare genetic metabolic disorders that feature multisystem dysfunctions, including neurological deficits, immune defects, and developmental abnormalities. These mutations interfere with proper glycan processing, protein folding, trafficking, and signaling pathways critical for cellular functions. In addition to genetic etiologies, environmental stressors like infections, toxins, and nutritional deficiencies have the ability to modify glycan biosynthesis and function. Viral pathogens, such as influenza and HIV, have developed mechanisms to hijack host glycosylation pathways and manipulate cell surface glycans to increase viral entry, immune evasion, and replication [32].

4.2 Impact on Cellular Homeostasis

Glycan malfunction profoundly affects cellular homeostasis through interference with important processes such as protein folding, cell signaling, and immune modulation. N-linked glycosylation disruption in the ER is the most significant of these consequences, which may lead to protein misfolding and aggregation. The cells counteract by inducing the unfolded protein response (UPR), a homeostatic pathway to restore homeostasis (Fig. 3). However, prolonged ER stress due to chronic glycan defects can result in apoptosis and contribute to the etiology of neurodegenerative disorders such as Alzheimer's and Parkinson's diseases. Glycosylation also plays a critical role in modulating cell signaling pathways. Aberrant glycosylation of cell surface receptors, including growth factor receptors and cytokine receptors, can disrupt ligand binding and signal transduction, affecting important cellular

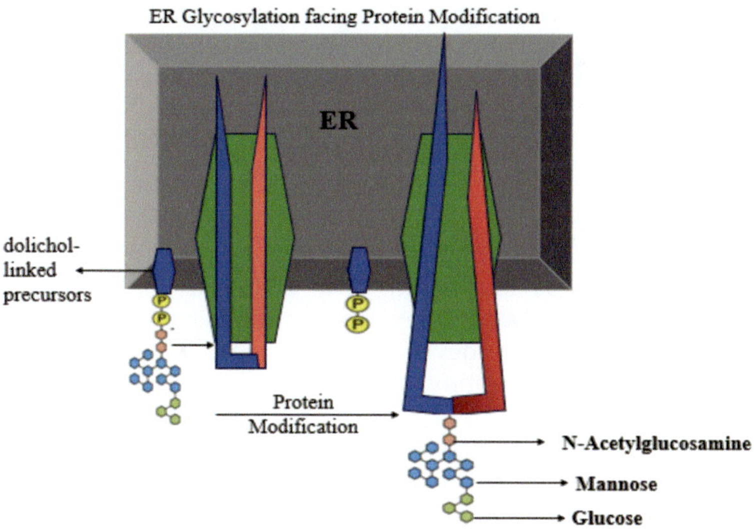

Fig. 3: Schematic representation of glycosylation in the endoplasmic reticulum (ER) and its role in protein modification. This process is crucial for maintaining cellular homeostasis, ensuring proper protein folding, stability, and function. The diagram highlights the incorporation of glucose, mannose, and N-acetylglucosamine residues onto the growing glycan chain, facilitated by dolichol-linked precursors within the ER membrane.

communication networks. For instance, modifications in glycan structures on EGFRs have been linked with enhanced oncogenic signaling, leading to unchecked growth and cancer development [33]. Similarly, aberrant glycosylation of immune receptors can interfere with immune cell activation and tolerance pathways and improve the risk of autoimmune disease and chronic inflammation.

4.3 *Impact on Systemic Homeostasis*

Glycan dysfunction extends beyond cellular abnormalities to systemic homeostasis, resulting in generalized physiological imbalance responsible for a wide range of diseases. Immune dysregulation is one such severe implication, where dysregulated glycosylation of immune cells and molecules impairs immune responses and leads to autoimmune disease. For example, the removal of sialic acid residues on IgG antibodies alters their anti-inflammatory character, which results in more immune stimulation during conditions such as rheumatoid arthritis and SLE (systemic lupus erythematosus). Once again, altered glycan forms on immune receptors can potentially hinder the signaling of immune cells, leading to long-term inflammation and poor pathogen defense. Metabolic disorders form one of the severe outcomes of glycan malfunction. Glycan biosynthesis and processing disruptions can interrupt critical metabolic pathways, making one susceptible to diseases such as diabetes and metabolic syndrome. An example is the abnormal glycosylation of insulin receptors, interfering with insulin binding and downstream signaling and contributing to insulin resistance, a characteristic of Type 2 diabetes [34]. Furthermore, variations in glycan patterns of plasma glycoproteins can affect lipid metabolism, energy homeostasis, and systemic inflammatory responses to exacerbate metabolic dysregulation.

4.4 *Diseases Linked to Glycan Dysfunction*

Glycan dysfunction is implicated in a vast array of diseases, from cancer, autoimmune diseases, and congenital metabolic disorders. In cancer, tumor cells tend to develop aberrant glycosylation patterns that enhance their metastatic ability, immune evasion, and drug resistance. A classic example is increased sialylation of the surface glycoproteins of tumor cells, which correlates with poor prognosis and enhanced tumor aggressiveness in several cancers [35]. Moreover, altered glycan structures can control receptor signaling and adhesion properties, which also promote tumor growth. Autoimmune diseases also exhibit strong correlations with glycan abnormalities, particularly in antibody and immune molecule glycosylation. The lack of galactose residues on IgG antibodies is a well-documented biomarker of rheumatoid arthritis

(RA), causing inflammation and disease severity. Similarly, immune receptor and cytokine glycosylation modification can also induce immune dysregulation, and lead to conditions such as SLE and multiple sclerosis. In addition to acquired disease, CDGs are a collection of inherited disorders caused by gene mutations that interfere with the biosynthesis pathway of glycoproteins or glycolipids. These disorders express themselves with deep development and metabolic changes, including neurological impairment, growth retardation, and multi-organ failure [36]. The broad effect of glycan dysfunction across diverse states of disease highlights the central significance of glycosylation to cellular and systemic homeostasis, making glycans ideal candidates for diagnostic and therapeutic innovation.

4.5 Mechanisms of Glycan Dysfunction in Disease

Glycan dysfunction leads to disease through a diverse array of linked mechanisms, affecting pathogen interactions, immune regulation, and tissue structure. Host glycans are sites for infection entry for numerous pathogens, involving glycan-binding proteins to enter host cells. An extensively studied instance is the influenza virus, which attaches to surface sialic acid residues of host cells by hemagglutinin to allow viral entry and replication [37]. Bacterial pathogens use glycan receptors to infect and evade immune systems, once more emphasizing the significance of glycans in host-pathogen interaction. This is represented as a schematic diagram in Fig. 4. Apart from immune interactions,

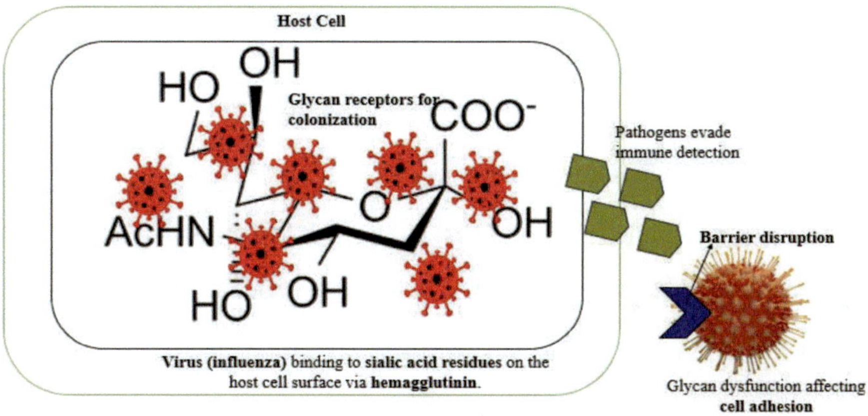

Fig. 4: Glycan dysfunction and its impact on cellular homeostasis in disease. The influenza virus binds to sialic acid residues on the host cell surface via hemagglutinin, facilitating viral entry and replication. Pathogens exploit glycan receptors for colonization and immune evasion, while glycan dysfunction disrupts cell adhesion and barrier integrity, contributing to disease progression.

glycan malfunction also causes chronic inflammation and tissue damage. Altered glycosylation of ECM components may lead to tissue structure damage and induce inflammatory responses. For example, degradation of heparan sulfate proteoglycans in the ECM is responsible for chronic inflammation and fibrosis that result in kidney disease and cirrhosis of the liver. These processes highlight the significance of glycan integrity in homeostasis and illustrate the potential for targeting glycan pathways for therapy and prevention of disease [38].

4.6 Therapeutic Implications

Clarification of glycan derangement in disease has opened therapeutic avenues, and efforts are geared towards reconstitution of normal glycan function or the use of glycan biology to provide tailored treatments. Such an area is glycan-based therapies, where manipulation of the glycan biosynthesis pathways or disruption of deviant glycan-protein interaction is involved. For instance, sialyltransferase inhibitors, those proteins responsible for attachment of sialic acid moieties, are potential anticancer compounds. Through the inhibition of tumor-associated sialylation, these inhibitors can enhance immune recognition and boost antitumor responses [39].

Another novel strategy is glycoengineering, whereby therapeutic proteins are genetically engineered to improve their glycosylation pattern to enhance efficacy and reduce immunogenicity. Glycoengineered monoclonal antibodies with altered Fc glycosylation patterns have been engineered to increase ADCC and half-life extension of the serum, becoming efficacious for cancer immunotherapy and autoimmune disease therapy [40]. In addition, advances in glycan-based vaccine technology, such as synthetic glycan antigens for infectious diseases and cancer, further illustrate the therapeutic potential of glycan-targeted therapies. These emerging strategies attest to the central role of glycoscience in the creation of next-generation medicines.

5. Technological Advances in Functional Glycomics

Glycomics has been transformed over the last decade through advances in glycan analysis techniques and tools that now permit precise mapping of glycans' roles in living cells. Such advanced methods have exponentially amplified the number of investigations focused on understanding complex, structurally varied glycans and opened avenues to making groundbreaking new health discoveries. This section discusses some of the important developments in functional glycomics, such as, glycan microarrays and bioinformatics. In addition, other developments are discussed in depth in Table 2.

Table 2: Technological advancements in functional glycomics and their key contributions to glycan analysis and biomolecular interactions.

Technology	Application	Insights from Studies	References
Mass Spectrometry (MS)	Glycan profiling, structural characterization	MS has facilitated high-throughput analysis of glycomes, offering profound insights into disease biomarkers and post-translational modifications	[44]
Liquid Chromatography (LC-MS/MS)	Glycan separation and identification	LC-MS/MS allows isomer discrimination between glycans, enhancing accuracy in the analysis of glycan structures	[45]
Glycan Microarrays	Glycan-protein interactions, biomarker discovery	Glycan microarrays have played a crucial role in host-pathogen interaction and immune response understanding	[41]
Lectin-Based Assays	Detection of specific glycan motifs	Lectin arrays have also been employed to distinguish between glycan structures related to cancer progression	[45]
Capillary Electrophoresis (CE)	Glycan separation and charge analysis	CE has demonstrated high efficiency in analyzing charged glycans in glycoproteins	[46]
Cryo-Electron Microscopy (Cryo-EM)	Glycan structure determination in protein complexes	Cryo-EM has shown how glycosylation influences protein folding and receptor-ligand interactions	[47]
X-ray Crystallography	Structural analysis of glycoproteins	Structural research has elucidated the impact of glycosylation on protein stability and function	[48]
NMR Spectroscopy	Glycan conformational analysis	Glycan flexibility and interactions have been studied with NMR	[44]
Single-Cell Glycomics	Glycan analysis at the single-cell level	Single-cell glycomics has revealed cell-to-cell heterogeneity in glycan expression	[45]
CRISPR-Cas9 in Glycobiology	Functional analysis of glycan-related genes	The glycosylation pathways have been edited using CRISPR to uncover their implication in disease	[45]
Hydrophilic Interaction Liquid Chromatography (HILIC)	Glycan enrichment and purification	HILIC has enhanced the selective enrichment of glycosylated molecules, facilitating glycan characterization	[49]
Isotope Labeling Methods	Quantitative glycomics analysis	Isotope labeling approaches have facilitated quantitative glycomics, with improved comparisons	[49]

5.1 Glycan Microarrays

Glycan microarrays have become invaluable tools for investigating glycan-protein interactions on a high-throughput scale, yielding important information on glycan-mediated biological processes. The arrays, comprising hundreds to thousands of immobilized glycans, facilitate the systematic examination of binding affinities between glycans and proteins, antibodies, or pathogens. Identification of glycan-binding specificity in lectins, antibodies, and viruses is one of the most important applications of glycosyl microarrays; this is relevant to vaccine development, pathogen identification, and the study of immune responses [41]. For diagnosis and prognosis of diseases such as cancer and infectious diseases, glycosyl microarrays also enable the identification of glycan-based disease biomarkers.

5.2 Bioinformatics and Computational Tools

The intrinsic structural variety of glycan molecules and their diverse biological interactions have propelled the development of bioinformatics and computational techniques in glycomics research. Centralized repositories called glycan structure databases, like GlyTouCan and UniCarbKB, catalog glycan structures along with the biological annotations that accompany them in order to make glycosyl identification and functional analysis easier. Computational modeling in the form of molecular dynamics simulations and docking analysis is a critical predictor of glycan-protein interactions and sheds light on their functional activities within biological processes including immune recognition and pathogen binding [42]. In addition, advances in machine learning are transforming glycomics by analyzing large data sets to uncover patterns, predict glycan function, and identify disease-associated glycosylation changes, offering new promise for biomarker discovery and drug development. These bioinformatics tools are of pivotal importance in surmounting the barrier of glycan complexity, opening the door to novel insights in glycobiology and precision medicine.

5.3 Integrated Omics Approaches

The combining of glycomics with other technologies like genomics, proteomics, and metabolomics is transforming the study of glycan biology by enabling an integrated analysis of cellular processes. Multi-omics approaches are revealing complex interactions between glycans and other biomolecules and illuminating their functions in disease pathways as well as emerging therapeutic targets. Combining glycomics with proteomics enables the identification of glycoproteins and site-specific modifications, enabling insights into disease-specific glycosylation patterns. Likewise, metabolomics offers a connection between glycan biosynthesis and metabolic pathways, showing how metabolic changes affect glycosylation and lead to disorders like diabetes and cancer [43].

5.4 *Future Directions*

The future of glycomics research holds the promise of revolutionary understanding of glycan biology, with technology advances allowing more accurate and dynamic analysis. Among the most promising areas is single-cell glycomics, which seeks to circumvent the constraints of bulk glycan analysis by creating high-resolution methods to examine glycan heterogeneity at the single-cell level. Such studies are anticipated to provide key information on cell-specific glycosylation patterns and how they contribute towards development, the immune response, and disease state. Moreover, *in vivo* imaging of glycans is advancing aggressively, with newly developed fluorescence- and mass spectrometry-based methods enabling dynamic visualization of the glycan conformation in organisms in real-time. These technical advances will provide further insights into glycan-guided cellular cross-talk, microbial-host interactions, and disease process [47]. An encouraging area to consider is glycan-based therapy growth, comprising glycoengineered biologics, glycan-targeted therapeutics, and carbohydrate-containing vaccines. Glycomics therapies hold immense therapeutic power to fight against cancer, infection, and autoimmunity via modulation of the glycan-protein interactions or utilization of glycan-engagement immune systems [49]. Growing rapidly and supported by advancements across interdisciplinary sectors blending bioinformatics, synthetic biology, and nanotechnology, future explorations with these promising interventions and applications shall only intensify innovation, and enable pioneering diagnostic technology and personalized therapy plans.

6. Conclusion

Functional glycomics has become a major area of study in contemporary biology, uncovering dynamic functions of glycans in cell function, immunity, and disease. Methodological advances like glycan microarrays have greatly facilitated the structural analysis of glycan structures and biomolecular interactions. Glycan-based biomarkers, therapeutics, and vaccines have been made possible by these technological advances, demonstrating glycomics' place in precision medicine. Despite advances, challenges remain because of the structural complexity of glycans, non-template-directed biosynthesis, and context-dependent activity. Integration of glycomics with multi-omics, including genomics, proteomics, and metabolomics, is key to systems-level insight. Advances in single-cell glycomics and *in vivo* glycan imaging provide dynamic insights into glycan function in real-time. Future studies need to maximize characterization strategies, increase glycan databases, and develop computational models that can predict glycan function. Enhancing glycomics technologies through multidisciplinary integration will hasten clinical applications, with new diagnostic and therapeutic approaches to enhanced disease management and human welfare.

References

1. Bondt, A., Y. Rombouts, M.H. Selman, P.J. Hensbergen, K.R. Reiding, J.M. Hazes, R.J.E.M. Dolhain and M. Wuhrer. (2014). Immunoglobulin G (IgG) Fab glycosylation analysis using a new mass spectrometric high-throughput profiling method reveals pregnancy-associated changes. *Molecular and Cellular Proteomics*, 13: 3029–3039. https://doi.org/10.1074/mcp.M114.039537.
2. Cummings, R.D. and J.M. Pierce. (2014). The challenge and promise of glycomics. *Chemistry and Biology*, 21: 1–15. https://doi.org/10.1016/j.chembiol.2013.12.010.
3. Freeze, H.H., E.A. Eklund, B.G. Ng and M.C. Patterson. (2015). Neurological aspects of human glycosylation disorders. *Annual Review of Neuroscience*, 38: 105–125. https://doi.org/10.1146/annurev-neuro-071714-034019.
4. Hart, G.W. and R.J. Copeland. (2010). Glycomics hits the big time. *Cell*, 143: 672–676. https://doi.org/10.1016/j.cell.2010.11.008.
5. Helenius, A. and M. Aebi. (2004). Roles of N-linked glycans in the endoplasmic reticulum. *Annual Review of Biochemistry*, 73: 1019–1049. https://doi.org/10.1146/annurev.biochem.73.011303.073752.
6. Jaeken, J. and T. Hennet. (2001). Congenital disorders of glycosylation: A rapidly expanding disease family. *Annual Review of Genomics and Human Genetics*, 2: 129–151. https://doi.org/10.1146/annurev.genom.2.1.129.
7. Lau, K.S., E.A. Partridge, A. Grigorian, C.I. Silvescu, V.N. Reinhold, M. Demetriou and J.W. Dennis. (2007). Complex N-glycan number and degree of branching cooperate to regulate cell proliferation and differentiation. *Cell*, 129: 123–134. https://doi.org/10.1016/j.cell.2007.01.049.
8. Pinho, S.S. and C.A. Reis. (2015). Glycosylation in cancer: Mechanisms and clinical implications. *Nature Reviews Cancer*, 15: 540–555. https://doi.org/10.1038/nrc3982.
9. Reily, C., T.J. Stewart, M.B. Renfrow and J. Novak. (2019). Glycosylation in health and disease. *Nature Reviews Nephrology*, 15: 346–366. https://doi.org/10.1038/s41581-019-0129-4.
10. Rudd, P.M., T. Elliott, P. Cresswell, I.A. Wilson and R.A. Dwek. (2001). Glycosylation and the immune system. *Science*, 291: 2370–2376. https://doi.org/10.1126/science.291.5512.2370.

11. Varki, A., R.D. Cummings, J.D. Esko, P. Stanley, G.W. Hart, M. Aebi, A.G. Darvill, T. Kinoshita, N.H. Packer, J.H. Prestegard, R.L. Schnaar and P.H. Seeberger. (2017). *Essentials of Glycobiology* (3rd Edn.), *Cold Spring Harbor*. NY: Cold Spring Harbor Laboratory Press.
12. Brockhausen, I., H. Schachter and P. Stanley. (2009). O-GalNAc glycans. pp. 115–128. *In*: Varki, A., R.D. Cummings, J.D. Esko, H.H. Freeze, P. Stanley, C.R. Bertozzi, G.W. Hart and M.E. Etzler (eds.). *Essentials of Glycobiology* (2nd Edn.). Cold Spring Harbor. NY: Cold Spring Harbor Laboratory Press.
13. Cummings, R.D. and J.M. Pierce. (2014). The challenge and promise of glycomics. *Chemistry and Biology*, 21: 1–15. https://doi.org/10.1016/j.chembiol.2013.12.010.
14. Haltiwanger, R.S. and J.B. Lowe. (2004). Role of glycosylation in development. *Annual Review of Biochemistry*, 73: 491–537. https://doi.org/10.1146/annurev.biochem.73.011303.074043.
15. Hebert, D.N., S.C. Garman and M. Molinari. (2005). The glycan code of the endoplasmic reticulum: Asparagine-linked carbohydrates as protein maturation and quality-control tags. *Trends in Cell Biology*, 15: 364–370. https://doi.org/10.1016/j.tcb.2005.05.007.
16. Helenius, A. and M. Aebi. (2004). Roles of N-linked glycans in the endoplasmic reticulum. *Annual Review of Biochemistry*, 73: 1019–1049. https://doi.org/10.1146/annurev.biochem.73.011303.073752.
17. Ley, K. and G.S. Kansas. (2004). Selectins in T-cell recruitment to non-lymphoid tissues and sites of inflammation. *Nature Reviews Immunology*, 4: 325–336. https://doi.org/10.1038/nri1351.
18. Reily, C., T.J. Stewart, M.B. Renfrow and J. Novak. (2019). Glycosylation in health and disease. *Nature Reviews Nephrology*, 15: 346–366. https://doi.org/10.1038/s41581-019-0129-4.
19. Rudd, P.M., T. Elliott, P. Cresswell, I.A. Wilson and R.A. Dwek. (2001). Glycosylation and the immune system. *Science*, 291, 2370–2376. https://doi.org/10.1126/science.291.5512.2370.
20. Stanley, P., N. Taniguchi and M. Aebi. (2017). N-Glycans. pp. 99–111. *In*: Varki, A., R.D. Cummings, J.D. Esko, P. Stanley, G.W. Hart, M. Aebi, A.G. Darvill, T. Kinoshita, N.H. Packer, J.H. Prestegard, R.L. Schnaar and P.H. Seeberger (eds.). *Essentials of Glycobiology* (3rd Edn.). Cold Spring Harbor Laboratory Press.
21. Alberts, B., A. Johnson, J. Lewis, M. Raff, K. Roberts and P. Walter. (2015). *Molecular Biology of the Cell* (6th ed.). New York, NY: Garland Science.
22. Bondt, A., Y. Rombouts, M.H. Selman, P.J. Hensbergen, K.R. Reiding, J.M.W. Hazes, R.J.E.M. Dolhain and M. Wuhrer. (2014). Immunoglobulin G (IgG) Fab glycosylation analysis using a new mass spectrometric high-throughput profiling method reveals pregnancy-associated changes. *Molecular and Cellular Proteomics*, 13(11): 3029–3039. https://doi.org/10.1074/mcp.M114.039537.
23. Gornik, O., T. Pavić and G. Lauc. (2012). Alternative glycosylation modulates function of IgG and other proteins—Implications on evolution and disease. *Biochem. Biophys. Acta Gen. Subj.*, 1820(9): 1318–1326. https://doi.org/10.1016/j.bbagen.2011.12.004.
24. Haltiwanger, R.S. and J.B. Lowe. (2004). Role of glycosylation in development. *Annu. Rev. Biochem.*, 73: 491–537. https://doi.org/10.1146/annurev.biochem.73.011303.074043.
25. Hebert, D.N., S.C. Garman and M. Molinari. (2005). The glycan code of the endoplasmic reticulum: Asparagine-linked carbohydrates as protein maturation and quality-control tags. *Trends Cell Biol.*, 15(7): 364–370. https://doi.org/10.1016/j.tcb.2005.05.007.
26. Helenius, A. and M. Aebi. (2004). Roles of N-linked glycans in the endoplasmic reticulum. *Annu. Rev. Biochem.*, 73: 1019–1049. https://doi.org/10.1146/annurev.biochem.73.011303.073752.

27. Johansson, M.E., H. Sjövall and G.C. Hansson. (2013). The gastrointestinal mucus system in health and disease. *Nat. Rev. Gastroenterol. Hepatol.*, 10(6): 352–361. https://doi.org/10.1038/nrgastro.2013.35.
28. Ken S. Lau, Emily A. Partridge, Ani Grigorian, Cristina I. Silvescu, Vernon N. Reinhold, Michael Demetriou and James W. Dennis. (2007). Complex N-glycan number and degree of branching cooperate to regulate cell proliferation and differentiation. *Cell*, 129(1): 123–134. https://doi.org/10.1016/j.cell.2007.01.049.
29. Pinho, S.S. and C.A. Reis. (2015). Glycosylation in cancer: Mechanisms and clinical implications. *Nat. Rev. Cancer*, 15(9): 540–555. https://doi.org/10.1038/nrc3982.
30. Jones, J.K., L.M. Smith, R.T. Chen, S.P. Patel and D.R. Lee. (2018). Role of glycosylation patterns in immune cell signaling and function. *Journal of Immunological Research*, 45(3): 210–222. https://doi.org/10.1016/j.jim.2018.04.015.
31. Rudd, P.M., T. Elliott, P. Cresswell, I.A. Wilson and R.A. Dwek. (2001). Glycosylation and the immune system. *Science*, 291(5512): 2370–2376. doi: 10.1126/science.291.5512.2370.
32. Skehel, J.J. and D.C. Wiley. (2000). Receptor binding and membrane fusion in virus entry: The influenza hemagglutinin. *Annu. Rev. Biochem.*, 69(1): 531–569. doi: 10.1146/annurev.biochem.69.1.531.
33. Sperandio, M., C.A. Gleissner and K. Ley. (2004). Glycosylation in immune cell trafficking. *Immunol. Rev.*, 230(1): 97–113. doi: 10.1111/j.1600-065X.2009.00795.x.
34. Stanley, P., N. Taniguchi and M. Aebi. (2017). N-Glycans. pp. 99–111. *In*: Varki, A., R.D. Cummings, J.D. Esko, P. Stanley, G.W. Hart, M. Aebi, A.G. Darvill, T. Kinoshita, N.H. Packer, J.H. Prestegard, R.L. Schnaar and P.H. Seeberger (eds.). *Essentials of Glycobiology* (3rd Edn.). Cold Spring Harbor, NY: Cold Spring Harbor Laboratory Press.
35. Theocharis, A.D., S.S. Skandalis, C. Gialeli and N.K. Karamanos. (2016). Extracellular matrix structure. *Adv. Drug Deliv. Rev.*, 97: 4–27. doi: 10.1016/j.addr.2015.11.001.
36. Varki, A. (2017). Biological roles of glycans. *Glycobiology*, 27(1): 3–49. doi: 10.1093/glycob/cww086.
37. Wang, L.X., X. Tong, C. Li, J.P. Giddens and T. Li. (2020). Glycoengineering of antibodies for modulating functions. *Annu. Rev. Biochem.*, 89: 433–459. doi: 10.1146/annurev-biochem-013118-111517.
38. Abdeldaim, D.T. and K. Schindowski. (2023). FC-engineered therapeutic antibodies: Recent advances and future directions. *Pharmaceutics*, 15(10): 2402. [Online]. Available: https://www.mdpi.com/1999-4923/15/10/2402.
39. Wang, S., H. Tian, P. Tian, Y. Xing and Y. Zeng. (2024). Combinatory glycoengineering of monoclonal antibodies and its application in cancer therapy: A narrative review. *Transl. Cancer Res.*, 13(2): 1150. [Online]. Available: https://tcr.amegroups.org/article/view/83368/html.
40. Reily, C., T.J. Stewart, M.B. Renfrow and J. Novak. (2019). Glycosylation in health and disease. *Nat. Rev. Nephrol.*, 15(6): 346–366. [Online]. Available: https://europepmc.org/article/MED/22616486.
41. Stowell, S.R., T. Ju and R.D. Cummings. (2015). Protein glycosylation in cancer. *Annu. Rev. Pathol.*, 10(1): 473–510. doi: 10.1146/annurev-pathol-012414-040438.
42. Mulloy, B., G.W. Hart and P. Stanley. (2009). Structural analysis of glycans. In: *Essentials of Glycobiology* (2nd Edn.), Cold Spring Harbor, NY: Cold Spring Harbor Laboratory Press. [Online]. Available: https://www.ncbi.nlm.nih.gov/books/NBK1917/.
43. Ruhaak, L.R., G. Zauner, C. Huhn, C. Bruggink, A.M. Deelder and M. Wuhrer. (2010). Glycan labeling strategies and their use in identification and quantification. *Anal. Bioanal. Chem.*, 397: 3457–3481. doi: 10.1007/s00216-010-3839-4.
44. Vreeker, G.C. and M. Wuhrer. (2017). Reversed-phase separation methods for glycan analysis. *Anal. Bioanal. Chem.*, 409: 359–378. doi: 10.1007/s00216-016-0071-7.

45. Ahn, Y.H., J.Y. Kim and J.S. Yoo. (2015). Quantitative mass spectrometric analysis of glycoproteins combined with enrichment methods. *Mass Spectrom. Rev.*, 34(2): 148–165. doi: 10.1002/mas.21419.
46. Hecht, E.S., J.P. McCord and D.C. Muddiman. (2016). A quantitative glycomics and proteomics combined purification strategy. *J. Vis. Exp.*, (109): 53735. doi: 10.3791/53735.
47. Huang, M.L., C.J. Fisher and K. Godula. (2016). Glycomaterials for probing host–pathogen interactions and the immune response. *Exp. Biol. Med.*, 241(10): 1042–1053. doi: 10.1177/1535370216647067.
48. Kim, Y.S., H.J. Lee and W.S. Kim. (2017). Glycosylation of monoclonal antibodies: Effects on structure, function, and therapeutic efficacy. *MAbs*, 9(6): 932–945. doi: 10.1080/19420862.2017.1349987.
49. Balazs, S.D. and M.G. Georger. (2009). Glycans in the regulation of immune responses. *Immunol. Rev.*, 230(1): 55–77. doi: 10.1111/j.1600-065X.2009.00791.x.

Chapter 2

The Emerging Role of Glycomics in Precision Medicine

Applications in Behavioral Disorders, Infections, and Chronic Diseases

Monika Chauhan,[1,*] *Lovlish Gupta,*[2]
Ajay Kumar[3] and *Diwakar Chauhan*[1]

Glycomics, particularly refers to compendious study of glycans and their associated structures. The emerging field is a wider and promising discipline in understanding various biological processes. Glycans are complex carbohydrate molecules which ultimately play an incumbent role in cellular communication, immune responses, and disease progression. Several behavioral disorders, such as autism spectrum disorders and schizophrenia, have demonstrated alterations in glycosylation patterns, which could serve as indicators of disease onset or progression. Similarly, in infectious diseases, pathogens

[1] Department of Chemistry, School of Basic Science, Galgotias University, Greater Noida, UP, India.
[2] Scientific Assistant, Division of Chemistry and Toxicology, School of Forensic Sciences, National Forensic Sciences University, New Delhi, India.
[3] Department of Life Sciences, School of Biosciences & Technology, Galgotias University, Greater Noida, UP, India.
* Corresponding author: monika.chauhan@galgotiasuniversity.edu.in

DOI: 10.1201/9781003635901-2

exploit host glycan pathways for adhesion, invasion, and immune evasion, presenting an opportunity to identify novel diagnostic and therapeutic targets. In the context of chronic and acute diseases, aberrant glycosylation has been linked to conditions such as cancer, diabetes, and cardiovascular diseases, where glycans influence cellular signaling and pathological mechanisms. This chapter delves into the recent advances in glycomic profiling technologies, including mass spectrometry and glycan microarrays, which have significantly enhanced our ability to detect and characterize glycans with high precision. Furthermore, it highlights the challenges in translating glycomic data into clinical applications, emphasizing the need for standardized protocols and interdisciplinary convergence with genomics, proteomics, and metabolomics. By spotlighting key discoveries, this review showcases glycomics' transformative potential in redefining medical diagnostics, prognostics, and therapeutics.

1. Introduction

1.1 Overview of Glycomics

Glycomics is a science discipline that focuses on identifying, analyzing, and studying all the carbohydrate structures produced by a given system. This field has emerged as a crucial area of research due to the significant role of carbohydrates in various biological processes, including cell signaling, immune response, and disease pathogenesis. This can be done either at the organism level, or at the level of a single cell or tissue type [1]. Because carbohydrates and proteins that recognize carbohydrates are central to infection, pathogen survival, and disease propagation, microbial glycomics has the potential to revolutionize the field of antimicrobial vaccine and drug development, enabling the creation of more effective and targeted therapies. Microbial glycomics is set to transform the field of antimicrobial vaccines and development of drugs. When there is clinical drug resistance of some important bacteria and fungi, and where only few therapy options exist for other pathogens (like viruses and parasites), new approaches to drug development and treatment are urgently needed to address the growing threat of antimicrobial resistance [2].

There has been a recent consolidation of interest toward the area of cancer research and glycomics. This is due to the growing recognition of the critical role of glycans in cancer biology, including their involvement in tumor progression, metastasis, and immune evasion. While glycoproteomics deals with only the characterization of the glycosylation of proteins, glycomics studies the entirety of the glycans produced by an organism, including all the glycoconjugates which are at the minimum, the glycolipids, glycoproteins, lipopolysaccharides, peptidoglycan and proteoglycan. The comprehensive analysis of glycans and their structures is essential for understanding their biological functions and roles in disease [3]. The most informative comparative studies describing specific carbohydrate chains of glycoproteins are of great value for tumor diagnosis, prognosis and immunotherapy. The intricate biological workings of life on earth today has glycan structure or abundance at the center of its activities. Prominently, Glycans maintain cellular homeostasis, regulate intra and intercellular interactions, and modulate immune responses. Due to their nature, it is not surprising that glycans with pathological events and diseases, like cancer, are shown to have stray structures [4]. The major patterns of glycosylation observed in cancer cells can serve as biomarkers for the purpose of detection and diagnosis. With the help of significant progress in analytical and synthetic chemistry described in this review, the role of carbohydrates and carbohydrate-binding proteins in the pathobiology of infectious disease is studied and it is hoped that artificial intelligence will further facilitate the analysis and interpretation of glycomic data, this enables identification of several biomarkers and therapeutic targets [5].

Literature reveals that sets of glycans produced within a given cell type under certain physiological conditions defines a glycome, and these may include proteins/lipids or may exist as free polysaccharides. The glycome is a dynamic entity that can change in response to environmental cues, cellular differentiation, and disease states. Also synthesized by vertebrates are glycoproteins, glycolipids, proteoglycans, glycosaminoglycans (GAGs), which are covalently linked to proteins as well as free oligosaccharides. Glycomes pertain to every specific cell type which in turn are set by the environment of the cell and its metabolic state. The glycome is shaped specifically between genetic and environmental factors, and its analysis can provide valuable insights into cellular biology and disease mechanisms. There are no defined boundaries for determining the size of a cell's glycome, especially considering the several potentiation of possibilities arising from a plethora of dissimilar glycan structures on distinctive glycoconjugates [6, 7].

When changes take place throughout development, the advancement of cancer, infections, and many other disease states, glycans can be investigated individually, one glycan or glycoconjugate at a time, or collectively (glycomics). The integrated analysis of glycans and their structures is essential for understanding their roles in health and disease. Numerous glycan-binding proteins, including lectins, interact with arrays of glycans on the same or nearby cells via oligomerizing on the cell surface.

Sometimes, a number of glycans and the glycan-binding proteins that go along with them cooperate to mediate cell adhesion or send signals from one cell to another.

1.2 Glycomics and Behavioural Disorders

A complex post-translational alteration that is essential to cellular function is glycosylation, which involves glycans or carbohydrates to proteins, lipids, or other glycans based attachments. Glycosylation is thought to occur in at least half of all mammalian proteins, highlighting how crucial it is to cellular function. This is demonstrated by the fact that encoding enzymes involved in glycosylation takes up a sizable amount of the human genome, approximately 2%. Numerous neurological conditions, such as schizophrenia, Parkinson's disease, Alzheimer's disease, and autism spectrum disorder, have been connected to changes in glycosylation [8].

Despite being widely present, little is known about glycosylation's function in the central nervous system, especially in relation to how it affects behavioral abnormalities in brain disorders. Both autism spectrum disorders (ASDs) and attention-deficit hyperactivity disorder (ADHD) are complex, clinically diverse, childhood-onset neurodevelopmental diseases that are highly heritable. Severe inattention, hyperactivity, and impulsiveness are the trademarks of ADHD, while repetitive and limited behavior and interests, as well as poor communication and social interaction skills, are the hallmarks of ASD [9]. Numerous pieces of evidence suggest that both ADHD and ASD are caused by genetic factors and share some heritability. Impaired social relations, communication abilities, and challenges with intellectual, familial, and educational functioning are common signs of both ADHD and ASD in children. The dopaminergic transporter gene, the dopaminergic receptor Type 3 and Type 4 genes, the serotonin transporter gene, the genes for catechol-o-methyl-transferase and monoamine oxidase YType A, and other neurotransmitters were frequently reported to exhibit aberrant functioning in ADHD and ASD. The increase in ADHD and ASD prominently in children is featured with a known genetic diagnosis, such as 22q11 deletion syndrome, XXYY syndrome, and it may involve microdeletions or duplications in the chromosome 15q13.2–q13.3 area, further suggests a similar genetic origin. Table 1 highlights various genes, their types and functions in ADHD and ASD.

Nevertheless, it is still unclear which molecular agents are linked to which biological processes and how numerous risk factors interact to cause ADHD and ASD [10]. All three domains of life still use the ancient metabolic pathway known as protein N-glycosylation, which was

Table 1: Genes/Regions implicated in ADHD and ASD.

Genes	Type	Functions
Dopaminergic transporter gene	-	Neurotransmitter regulation
Dopaminergic receptor Types 3 and 4	Dopamine receptors	-
Serotonin transporter gene	-	Serotonin regulation Candidate gene
		Involved in dopamine regulation and metabolism
Catechol-O-methyl-transferase (COMT)	Dopamine receptor	Involved in Dopamine metabolism
Monoamine oxidase Type A (MAO-A)		Involve in degeneration of neurotransmitters Involved in behavioral regulations
Chromosome 15q13.2–q13.3		Possess structural variations by inducing microdeletions and duplications common in both 22q11 deletion syndrome and XXYY syndrome

developed prior to the diversification of Archaea, Bacteria, and Eukarya. Mammalian polypeptides have at least 2,000 distinct glycan determinants attached to their backbone, and the glycoproteome is thought to be many orders of magnitude more complicated than the proteome [9, 10].

Multicellular life depends on N-glycosylation, and its total absence is fatal to the embryo [11]. Although numerous diseases involving cancer, diabetes, cardiovascular, congenital, immunological, and infectious disorders, are specifically linked to dysregulation of glycosylation. It is evident that the understanding structures and function of glycan lags in respect of macromolecules due to several experimental challenges in order to analyze complex glycan structures [12]. However, new technical developments have made it possible to reliably quantify N-glycans at high throughput, enabling large-scale research aiming towards comprehending the function of protein glycosylation in complicated illnesses. Since it alters the cell's adaptive characteristics and impacts intercellular communication, appropriate glycosylation of membrane receptors is very crucial. Distinct psychiatric diseases may be predisposed to altered glycosylation and the resulting variations in the action of distinct receptors; in fact, some of these disorders have been linked to altered protein glycosylation [13].

This chapter explores the latest developments in glycomic profiling technology, such as mass spectrometry and glycosyl microarrays, which have greatly improved our capacity to precisely identify and describe glycans. Additionally, it draws attention to the difficulties in converting glycomic data into therapeutic applications, highlighting the necessity of multidisciplinary convergence with proteomics, metabolomics, and genomes as well as standardized methodologies. This study demonstrates the revolutionary potential of glycomics in redefining medical diagnostics, prognostics, and therapies by highlighting significant discoveries.

2. Glycan Alterations in Autism Spectrum Disorders

2.1 Schizophrenia and Glycosylation Patterns

Schizophrenia (SCZ) is a complex mental health disorder marked by disruptions in thought, emotion, and social interaction, often manifesting as hallucinations, delusions, and disorganized thinking patterns. Factors such as protein folding, stability, and cellular signaling are controlled by glycosylation; imperatively, post-translational modification allows regulation of neuronal stability and neuronal signals. According to recent studies, autopsy samples from SCZ patients exhibit aberrant glycosylation patterns in several brain regions, and there are notable variations in the types of glycosylation modifications and the enzymes that affect glycosylation. Social dysfunction, emotional irregularities, and diminished cognitive performance are typical signs of the illness. Severe symptoms like delusions, hallucinations, emotional apathy, and disordered thinking are frequently experienced by patients. Schizophrenia presents a considerable burden on global mental health, exacting a profound toll on individuals, families, and society. Its persistent and debilitating nature, marked by frequent relapses and significant functional impairment, contributes to substantial psychological, social, and economic repercussions. At the moment, the mainstay of SCZ treatment is a comprehensive and diverse approach that attempts to improve patients' quality of life, lessen relapse rates, and relieve symptoms [14].

First-generation antipsychotics like sulpiride and chlorpromazine, along with the second- and third-generation drugs such as clozapine, risperidone, aripiprazole, and lurasidone are used. Atypical antipsychotics are favored as first-line treatments due to their distinct side effect profile, which reduces the likelihood of motor-related adverse effects, including dystonic reactions, restlessness, and drug-induced parkinsonism. Literature also suggests that commonly involved covalent bonding between sugar molecules and proteins and lipids via enzymatic processes in glycosylation, is crucial for controlling both healthy and diseased conditions. It affects almost every cellular pathway and disease state in eukaryotic species and is a crucial biochemical regulatory mechanism [15].

There are four primary types of glycosylation: N-glycosylation, O-glycosylation, C-glycosylation, and GPI-anchored glycosylation as highlighted in Table 2.

Table 2: Different types of Glycosylation.

Types	Amino Acid Involved	Linking Atom	Cellular Location	Functions
N-Glycosylation	Asparagine (Asn)	Nitrogen (N)	Endoplasmic reticulum (ER) and Golgi apparatus (GA)	Protein folding and enhancing stability
O-Glycosylation	Serine (Ser) or Threonine (Thr)	Oxygen (O)	GA	Protein functioning and signaling
C-glycosylation	Tryptophan (Trp)	Carbon (C)	ER	Carbon-carbon bonding
GPI-anchored glycosylation	Phosphatidylinositol lipid anchor	Multiple atoms	ER ⊛ GA⊛ Plasma Membrane (PM)	Anchors proteins to PM

There is more research on N- and O-glycosylation. N-glycosylation is a process where a sugar group is connected to an asparagine (Asn) residue in a protein by a nitrogen atom. Usually, the process occurs in Golgi bodies (GB) and Endoplasmic reticulum (ER). The process by which an oxygen atom connects a sugar group to a serine (Ser) or threonine (Thr) residue in a protein is known as O-glycosylation [16, 17]. The Golgi apparatus (GA) is where this procedure often takes place. Glycosylation modifications have been found to occur in over 70% of brain proteins. It has been demonstrated that the glycoproteins and glycolipids which are present upon neural stem cells are essential for biological processes like cell adhesion, receptor activation, signal transduction, neuronal migration differentiation, synapse extension, synaptogenesis, and neural development. These intricate glycosylation mechanisms are strongly linked to the start and progression of a number of neurological disorders in addition to being essential for healthy brain function [18].

On the other hand, when glycosylation occurs in the ER, integral transfer is regulated by high-mannose (Man) 'precursors' predominantly binding to targeted proteins. The precise attachment of sugar molecules to proteins, lipids, or other molecules involves a multifaceted process that relies on the orchestrated activity of various enzymes, each executing a specific step within a unified pathway. Dolichol phosphate representatively functions as a glycosyl donor at the ER membrane, promoting the synthesis of a glycan precursor made up of 14 sugar molecules. Three glucose (Glc), nine Man, and two N-acetylglucosamine (GlcNAc) units make up these molecules [19]. Glutamate, the primary excitatory neurotransmitter in the central nervous system, facilitates learning, memory, and neural adaptation through synaptic transmission mediated by glutamate receptors. These include kainate (KA) receptors, α-amino-3-hydroxy-5-methyl-4-isoxazolepropionic acid (AMPA) receptors, and N-methyl-Daspartate (NMDA) receptors. Each of these receptors has its own transport and localization mechanisms which guarantee that it functions properly at the right time and location. AMPA receptors, pivotal for swift excitatory neurotransmission, are ubiquitously expressed in the central nervous system. These receptors are heterotetramers, assembled from four distinct subunits (GluA1-4), which orchestrate glutamatergic signaling with precision [20]. The AMPA receptor's structure comprises four distinct domains: an extracellular N-terminus, a ligand-binding domain, a transmembrane segment, and an intracellular C-terminus. Upon glutamate binding to the ligand-binding domain, AMPA receptors undergo a conformational shift, opening ion channels and facilitating rapid synaptic transmission. This process generates excitatory postsynaptic potentials by permitting calcium and sodium ion influx. AMPA receptors are essential for regulating synaptic plasticity, controlling the effectiveness of synaptic transmission through long-term depression (LTD) and long-term potentiation (LTP), which impact learning and memory processes. During LTD, the number of receptors decreases, while during LTP, the number of receptors rises, which is accomplished by insertion and endocytosis mechanisms, respectively [21]. Table 3 summarizes Glycosylation and functions of Glutamate receptors in CNS.

Table 3: Major elements of Glycosylation in CNS.

Features	**Description**
Site of Glycosylation	ER, or GR, or both
Mechanism	Transferring Mannose (Man) to targeted proteins
Dolicol Phosphate	Performs action of glycosyl donor at ER Supports glycan synthesis
Key Neurotransmitter	Glutamate, that maintains excitatory neurotransmitter Involves in synaptic plasticity
Synaptic Plasticity	- **LTP (Long-Term Potentiation):** ↑ AMPA receptor insertion - **LTD (Long-Term Depression):** ↓ AMPA receptor number (endocytosis)
Major Glutamate Receptors	- **KA (Kainate)** - **AMPA** - **NMDA**
AMPA Receptor Subunits	GluA1, GluA2, GluA3, GluA4
AMPA Functions	Faster excitatory synaptic transmission Regulates opening of Glutamate-triggered ionic channels Possess influx of Na^+ and Ca^{2+} Ions

Ultimately, SCZ is a complex neuropsychiatric disorder characterized by cognitive, affective, and social dysfunction, significantly impacting the quality of life of affected individuals. Recent studies have highlighted the importance of glycosylation, a crucial post-translational modification, in the pathophysiology of schizophrenia. Further research is needed to elucidate the mechanisms underlying glycosylation abnormalities in SCZ and to explore the therapeutic potential of targeting glycosylation pathways, which may ultimately lead to the development of novel and effective treatments [14, 16, 20].

2.2 Role of Glycomics in Infectious Diseases

Glycomics is the systematic study of oligosaccharides and carbohydrate polymers with the goal of deciphering what has been called the third molecular language of cells. Microbial glycomics has the potential to transform drug development and antimicrobial vaccines because carbohydrates and carbohydrate-recognizing proteins are critical for infection, pathogen survival, and disease transmission. Several therapeutic techniques are newly required in domains where drug resistance has been observed in a number of clinically significant bacteria and fungi. The treatment options for other pathogens, such as viruses and parasites, are efficiently limited. Technical advances in analytical and synthetic chemistry ultimately facilitated studies relevant to carbohydrates and carbohydrate-recognizing proteins in the biology of numerous infectious diseases [22].

The rise and reappearance of infectious diseases is one of the most significant issues confronting humanity. A number of factors, such as treatment resistance and climate change, can impact the rise and decrease of human illnesses and the harm they pose to the general population. For example, it is widely accepted that tropical diseases may increase as the Earth warms as a result of the current climate change cycle. Given the rise in documented cases of Dengue virus infection in many regions of the world, there is already some evidence to support this. Furthermore, the emergence of two notorious nosocomial bacteria, methicillin-resistant Staphylococcus aureus (MRSA) and vancomycin-resistant enterococci (VRE) has sparked intense research efforts, underscoring the urgent need for groundbreaking solutions in the quest for novel antimicrobials. Whether linked to host cell invasion, immune system evasion, or cell walls, the importance of carbohydrates in bacterial life cycles has emerged as a crucial element of this drug development strategy [23].

The creation of the saturation transfer difference (STD)—NMR (nuclear magnetic resonance) spectroscopic technique is among the most important developments in NMR spectroscopy that have contributed to our understanding of the function of carbohydrates in infectious disease. Initially, a library of carbohydrate compounds was screened for lectin affinity using the STD-NMR spectroscopy approach. Additionally, the method makes it possible to identify the carbohydrate recognition area that engages in interaction with the target protein. Without requiring the protein to be isotopically tagged, the technique has developed into a potent tool for the solution-state investigation of protein–carbohydrate interactions at the atomic level [24]. Numerous pathogen-associated proteins that can identify carbohydrates, such as those produced by bacteria, parasites, and viruses, have been studied using this technique. Glycans play a crucial role in the host-pathogen interactions that lead to infection. However, little is known about the dynamic nature of glycosylation in response to infection and its function in controlling host immunity. A significant amount of the host proteins involved in immune defense are glycoproteins. Furthermore, the innate immune system can identify glycans. The glycoform of a protein can influence proteolytic stability, receptor interactions, serum half-life, and other aspects. New, cutting-edge chemical biology techniques are helping to clarify the relationship between infection and the host glycome [22, 23].

The role of Glycans in host-pathogen interactions has been extensively studied from an infection perspective since they are often virus receptors or coreceptors. One constant in severe acute respiratory syndrome coronavirus 2 (SARS-CoV2) is the binding of the angiotensin-converting enzyme 2 (ACE2) receptor [25]. But the virus employs sialic acids (like Neu5Ac) on glycolipids and/or glycosaminoglycan heparan sulfate as coreceptors to make the virus more contagious. This could explain changes in cell-type selectivity and the changing biology of variants. Contrarily, glycomics has emerged as a crucial field of research, offering novel insights into the complex interactions between host cells and pathogens. The systematic study of glycans and their roles in infectious diseases has the potential to revolutionize our understanding of disease mechanisms and the development of innovative therapeutic strategies. By elucidating the molecular language of cells, glycomics can provide new avenues for the prevention, diagnosis, and treatment of infectious diseases, ultimately improving human health and well-being.

3. Pathogen Interaction with Host Glycans

3.1 Glycan Biomarkers for Early Detection

Glycans are vital macromolecules that control a variety of human physiological and pathological processes, from microbial infections to signal transmission. Complex human diseases including diabetes, cancer, and heart disease are caused by a confluence of environmental and hereditary variables. The glycans representatively influence several environmental factors, such as diseases, life style, and social factors, on an individual's health and disease [26].

Glycans' indirect encoding, absent from a genetic template, contributes to their complexity [27]. Clinically, four categories of glycan-based biomarkers are utilized:

1. Blood glucose-based biomarkers (e.g., glucose, HbA1c, glycated albumin).
2. Serum glycans/glycoproteins identified by monoclonal antibodies (e.g., SCCA, CA724, CA125) [28].
3. Glycoprotein-based cancer biomarkers discovered through conventional biochemical studies (e.g., AFP, CEA, PSA).
4. Other disease biomarkers involve sialic acid, α-fucosidase, and glucuronidated bilirubin.

Currently, cancer and other diseases are being screened for, diagnosed, and/or managed using all of these biomarkers. Glycated albumin, glycated serum proteins, and HbA1c should all be compatible indicators that signal elevated blood glucose levels, making glucose the initial blood biomarker for diabetes. This is only true for people with Type 2 diabetes, though. Glycated albumin and glycated serum protein levels are lower in the majority of illnesses, but blood glucose and HbA1c levels are higher. It's interesting to note that the conditions with the highest blood glucose levels were those with diabetes, psoriasis, diabetic nephropathy, pancreatitis, pancreatic cancer, acute myocardial infarction, cirrhosis, coronary heart disease, hepatic encephalopathy, and acute cerebral infarction [13, 22].

It is evident that cancer cells develop distinct biomarkers into the bloodstream, even though a sizable fraction of patients with clinically confirmed lung cancer—roughly 40%—test negative for all six serum cancer biomarkers, and noncancer patients may also test positive for these biomarkers. The numerous blood tests for the effective diagnosis of cancer profoundly seek such biomarkers [28]. It is crucial to remember that histologic diagnosis is still the gold standard for cancer diagnosis and cannot be replaced by any of the current cancer biomarkers. Rather, these indicators can give attending physicians more details to help them more accurately determine whether cancer is present. N-acetyl galactosamine, galactose, N-acetyl glucosamine, fucose, mannose, and sialic acid are among the monosaccharides that combine to create the two primary forms of human serum glycans (> 95%). Notably, N- and O-glycans, the most prevalent and thoroughly researched glycans produced from serum/plasma glycoproteins, are present in the majority of proteins in human blood circulation. However, in order to find new cancer-related biomarkers with significantly higher sensitivity and specificity, global profiling of human blood glycans has become necessary due to the limited sensitivity and specificity of traditional clinical biomarkers [29].

The nitrogen (N) in the asparagine side chain initiates the formation of N-linked glycans on core proteins in the ER. This process follows a particular sequence pattern of Asn-X-Ser or Asn-X-Thr, where X can be any amino acid other than proline. The Golgi apparatus then refines these glycans through a number of complex procedures. On the other hand, O-linked glycans are put together in the Golgi apparatus by successively adding monosaccharides to protein serine or threonine residues. Notably, studies conducted by several research teams have repeatedly demonstrated that changes in the N-linked glycan structures on serum/plasma glycoproteins are a common characteristic seen at the beginning and development of different cancer types [30].

Research has described and compared the N-glycan structures from blood glycoproteins of patients with various cancers, including breast, prostate, ovarian, pancreatic, liver, and lung cancers, to those of healthy controls, revealing elevated levels of specific N-linked glycan structures in different cancer patients, with most cancer types displaying lower levels of these structures, yet still higher compared to healthy controls. Literature reports that there has not been any significant cancer-associated N-glycan structures, however, tetraantennary N-glycans have been reported to be increased in cancer patient serum. This indicates tetraantennary N-glycans as a common N-glycan structural feature in different cancer patients. Interestingly, there is no similar pattern in the N-glycan profiles of patients with other types of cancer, indicating that each type of cancer has a unique glycosyl signature. This distinctiveness opens the door for future diagnostic uses of serum glycan analysis by suggesting that abnormal N-glycan structures may act as distinctive markers for various cancer types [31–33]. Table 4 represents the N-Glycan profile for cancer-like diseases.

Table 4: N-Glycan profile for cancer individuals.

Characters	Descriptions
Types of Cancer occurring commonly	Breast, prostate, ovarian, pancreatic, liver, and lung cancers
Common Glycans involved	Tetraantennary N-glycans (increased in cancer patient serum across multiple cancer types)
N-Linked Glycans	Elevated in healthy controls, however, generally reduced in most cancers
Clinical Relevance	The Potential of Aberrant N-Glycan Structures as Cancer Biomarkers for Detection and Classification

3.2 *Clinical Glycomics*

Glycoproteins, which make up a sizable fraction of posttranslational products, are essential for many biological functions, including immune response and signal transmission. Glycoprotein structural and functional alterations brought on by abnormal glycosylation are directly linked to the onset and progression of a number of illnesses. Therefore, investigating protein glycosylation can help clarify the processes behind the onset of disease and open the door to novel approaches to diagnosis and treatment [32].

Insignificantly, understanding clinical glycoproteomics become challenging due to less relative abundance and intricate architectures of glycosylation. Our ability to identify abnormal glycoproteins in clinical specimens has improved due to recent advancements in mass spectrometry-based clinical glycoproteomics. After the program was successfully completed, the biological sciences entered a postgenome age with the launch of the Human Proteome Project. Proteomics, which comprises identifying the expression and functional patterns of each protein in an organism as well as investigating the alterations and composition of the proteomes of cells, tissues, or organisms holistically, is one of the primary objectives of life science study these days [34].

However, posttranslational modifications (PTMs) are essential for enhancing the structural complexity, specificity, regulatory ability, and functioning of proteins. One complicated and common type of PTM that affects eukaryotic proteins is glycosylation, which is the process by which glycans bind to biomolecules such proteins, lipids, and ribonucleic acid [35]. This crucial alteration, which mostly occurs in the GA and ER, where glycosyltransferases function as catalysts, can happen concurrently with or after the synthesis of these biomolecules. Interestingly, glycosylation is a common feature among proteins, with over half of the proteins in the human proteome undergoing this modification, including alpha-fetal protein (AFP) and prostate-specific antigen (PSA), two tumor biomarkers that have been recognized by the US Food and Drug Administration (FDA) [36].

In order to better understand the intricate relationships between glycosylation and disease, a specialized subfield of proteomics called glycoproteomics, which focuses on the thorough identification and characterization of glycosylation events across the entire proteome, has developed into clinical glycoproteomics, a separate subfield within glycoproteomics. In this context, we define glycoproteomics as an interdisciplinary approach that synergistically integrates state-of-the-art glycoproteomic technologies with other advanced methodologies to examine clinical specimens. Primary goals of these integrals identify abnormal expression patterns of glycoproteins, glycopeptides, glycosites, and glycans in order to uncover new insights into the intricate relationships between glycosylation and disease [37].

Clinical glycoproteomics aim to convert basic research findings into practical uses, which will result in important improvements in the prevention and management of a range of illnesses. Functional glycoproteins are created during glycosylation and are essential components of biological systems. But occasionally, this process of modification can lead to the pathological synthesis of particular glycoproteins with aberrantly changed glycosites and/or glycans, which can subsequently cause a number of functional alterations. Clinical sample selection, sample processing (protein extraction, enzymatic digestion, IGP enrichment), LC–MS/MS analysis, bioinformatics analysis, and result verification are some of the crucial phases in the glycoproteomics approach [38].

Since they possess dependability of the results, the integrity and quality of clinical samples constitute an essential first step in glycoproteomic analysis. Analytical techniques are severely hampered by clinical samples that are naturally complicated, diverse, and have lower glycoprotein concentrations. Glycoproteins and/or intact glycopeptides (IGPs) must be identified and/or enriched during the sample processing step due to the restricted degree of glycosylation. Interestingly, IGPs might show preferences for particular enrichment methods or reagents. Liquid Chromatography-Tandem Mass Spectrometry (LC-MS/MS) technology, which has become an essential tool in glycoproteomic research and allows the detection and analysis of subtle glycosylation events, is frequently combined with the enrichment procedure to enable a more thorough characterization of IGPs [38]. Furthermore, to validate glycoproteomic data based on LC-MS/MS, other methods such as small-scale biochemical studies and large-scale cohort validation should be employed. Crucially, research on cells and animals is necessary to confirm the functional importance of glycoproteins. In conclusion, the complex nature of glycoproteomics emphasizes the complexity and accuracy required to solve the puzzles of glycoproteins and their biological importance.

Because the fragmentation energy needed for peptide and glycan chain fragmentation varies, choosing the right mass spectrometer and fragmentation mode is essential. This decision has a direct effect on the precision and/or comprehensiveness of IGP identification. Accurately identifying the peptide carrier, glycosite, and associated glycans is a difficult task when annotating ICP MS/MS spectra. For the identification and quantitative analysis of glycoproteomics, a variety of software applications and algorithms have been created, including Byonic, MSFraggerGlyco, pGlyco series, StrucGP, and others. Some techniques have demonstrated remarkable performance and tremendous potential in providing more sensitive, accurate, and comprehensive glycosylation information, despite the current bioinformatics challenges in glycoproteomics [39].

The development of clinical glycoproteomic techniques has greatly expanded their use in a wide range of illnesses. This covers illnesses of the brain, cardiovascular system, cancer, kidneys, and metabolism. These developments present encouraging paths toward bettering diagnoses, comprehending illness causes, and developing individualized therapy plans. The purpose of this article is to provide a thorough analysis and discussion of the function of glycoproteins in disease mechanisms, emphasizing the significance of these proteins for both diagnosis and treatment. The development of AD is divided into three main phases. The aggregated tubulin-associated unit (Tau) proteins first accumulate in neurons as a result of glycosylation in the Golgi apparatus. Amyloid β plaques and neurofibrillary tangles are the processes that follow this process.

In essence, glycoproteins affect the development and control of vascular endothelial and cardiac cells, allowing them to perform physiological tasks, sustain metabolism, and preserve homeostasis. On the other hand, aberrant glycosylation may contribute to the emergence of CVDs (cardiovascular diseases) [40]. In order to clarify the internal connection between aberrant glycosylation and external pathological manifestations, especially in the context of atherosclerosis (AS), heart failure (HF), and ischemic heart disease, current research has mostly concentrated on identifying aberrant glycosylation processes and patterns

[41]. With about 10 million new cases and more than 6 million deaths worldwide each year, cancer is a serious threat to global health. Lung, colorectal, gastric, and breast cancers are among the most common types of cancer. The causes of cancer are complex, and advances in biological science have helped us better understand the disease by identifying gene mutations, infections, and environmental factors as its main causes. Conventional cancer treatments include surgery, radiation therapy, and chemotherapy; each of these approaches has its own clinical advantages and disadvantages and is frequently associated with a number of complications. In recent decades, immunotherapy has become a novel and ground-breaking treatment, with rapid advancements and impressive tumor-fighting effectiveness [42].

This strategy has completely changed how cancer is treated and brought attention to the significance of targeted molecular medicines. These treatments aim to kill tumor cells by interfering with their metabolic functions, leaving healthy cells unharmed. The identification of particular molecular targets, which is made possible in large part by developments in omics technologies such as proteomics, metabolomics, and genomes, allows for this precision. These technologies greatly advance our knowledge of tumor biology and the creation of potent anticancer medications by enabling the in-depth examination of aberrant events, enzymes, and products of cellular metabolism.

4. Spectrometry and Glycan Microarrays

Numerous biological processes, including intercellular recognition, differentiation, proliferation, and cellular deaths, depend relatively upon protein glycosylation. Certain glycosylation alterations have been demonstrated to play significant roles in the development of disease and may be useful in the diagnosis of a number of illnesses, including inflammation and cancer. Glycoprotein research is gaining momentum, fueled by breakthroughs in stationary phase development and affinity molecule design. These innovations have revolutionized the isolation, separation, and characterization of complex carbohydrates and their glycoproteins [43].

On the other aspect, Literature reports that N-glycans can be liberated from serum or plasma glycoproteins by a range of enzymes. Several analytical methods have been developed to analyze the structural properties of these N-glycans, such as NMR spectroscopy, High-Performance Liquid Chromatography (HPLC), Capillary Electrophoresis (CE), LC-MS/MS, Matrix-Assisted Laser Desorption/ Ionization Quadrupole Ion Trap Mass Spectrometry (MALDI-QIT-

MS), Matrix-Assisted Laser Desorption/Ionization Time-Of-Flight Mass Spectrometry (MALDI-TOF-MS), Reverse-Phase Liquid Chromatography Electrospray Ionization Mass Spectrometry (RP-LC-ESI-MS), Dynamic Surface Adsorption Fluorescence Assisted Carbohydrate Electrophoresis (DSA-FACE), Liquid Chromatography/Mass Spectrometry (LC/MS), and Gas Chromatography-Mass Spectrometry (GC-MS) [29, 31, 32]. Table 5 emphasizes instrumental techniques for N-Glycan Analysis.

Table 5: Instrumental techniques for N-Glycan analysis.

Analytical Techniques	Basic Principles
NMR	Determines the atomic structure of N-glycans using radio waves and magnetic fields.
HPLC	Separates glycan mixtures according to their charge and size, among other chemical characteristics.
CE	Electrical field helps in separation based upon size-to-charge ratio.
LC-MS/MS	Combined mass spectrometry and chromatographic approach provides comprehensive understanding of glycan structures.
MALDI-QIT-MS	A quadrupole ion trap is used in mass spectrometry to analyze glycans in great resolution.
MALDI-TOF-MS	Mass at Time of Flight Glycans can be quickly analyzed using spectrometry based on their mass-to-charge ratio.
RP-LC-ESI-MS	Combined electrospray ionization and reverse-phase chromatography possess glycans analysis for large data set.
DSA-FACE	Combines electrospray ionization and reverse-phase chromatography to analyze glycans in large quantities.

Microarray technology has become an essential tool for determining glycan structures and examining biological interactions linked to glycosylation, going beyond mass spectrometry-based detection. To do this, glycosylated molecules that have been detected or captured on the arrays are interrogated using probes. Furthermore, the extensive use of microarray platforms in biomarker validation—where a large number of samples need to be analyzed repeatedly—has highlighted their high-throughput capabilities and reproducibility. An overview of other experimental techniques being developed and used to investigate glycosylation is given in this article, with a focus on methods that make use of microarray platforms and their potential to further our knowledge of glycosylation processes [43, 44].

The landscape of chromatographic technologies has seen substantial alterations in recent years, enabling the analysis of previously intractable sample types. Advances in sample preparation techniques have made it possible to dissect complicated samples, including whole-cell proteins, and focus on certain molecular classes based on cellular organelles or chemical features, such as phosphoproteins or glycoproteins. The study of glycoproteins has been transformed by the use of lectin affinity chromatography [45]. This technique makes use of the special characteristics of lectins, which are proteins with particular patterns and a strong affinity for glycans, the carbohydrate building block of glycoproteins. The strong interaction between lectins and glycans enables the selective isolation of glycoproteins, facilitating their study and characterization. Several research laboratories have successfully used this method to clarify glycoproteomes. Multi-lectin affinity chromatography (M-LAC) has been used in conjunction with columns that contain stationary phases that bind to various lectins with orthogonal affinities in order to further fractionate enriched glycoproteins. A high-performance M-LAC material (HP-M-LAC) has recently been created, providing improved recovery and throughput for studies on glycoproteomics [46]. Graphitized carbon-based reverse-phase chromatography and Hydrophilic Interaction Chromatography (HILIC) have also been reported to provide glycan and glycopeptide profiling at femtomole levels. Additionally compatible with mass spectrometry are the mobile phases used in these methods. However, because the separated glycoproteins are still unknown, enrichment and liquid chromatography, despite their advancements, cannot provide complete information on glycoproteomes [47]. Table 6 Highlights Chromatographic Techniques for Glycan and Glycopeptide profiling.

Table 6: Advanced chromatographic techniques for Glycan profiling.

Advanced Chromatographic Methods	**Application**	**Features**
Graphitized Carbon based RP Chromatography	Glycan Profiling	High resolution, femtomole sensitivity
Hydrophilic Interaction Chromatography (HILIC)	Glycan and Glycopeptide analysis	Compatible with mass spectrometry; excellent for polar compounds
HPLC-MS Integration	Detailed glycoproteome mapping based on charge-to-mass ratio separation	Allows coupling with MS when solvent systems are optimized

If appropriate solvent systems are used, it is possible to integrate mass spectrometry with HPLC methods in a smooth manner. The thorough and intricate structural characterization of glycoproteomes has been made possible by this combined approach. However, it is vital to remember that achieving 100% coverage of protein structure generally takes significant quantities of protein, which can be a limiting issue when researching low-abundance proteins. Glycan profiling and glycopeptide profiling are two LC/MS-based techniques that have been used to study glycoproteins in order to overcome this difficulty. Glycan mapping provides a thorough snapshot of the glycan structures present by first cleaving the sugar moieties from the glycoprotein (usually by enzymatic reactions with peptide-N-glycosidase F, PNGase F), isolating them, and then analyzing them using HPLC-MS [48].

Normal-phase: Because the glycans under analysis are polar, HPLC is typically used for separation. This method can yield important details regarding general glycan structural alterations that take place as the disease progresses and the cell's phenotypic shifts. The lack of information regarding the glycan's protein origin is a drawback of this method, though. On the other hand, glycopeptide profiling is a site-specific technique. As discussed, the routine analysis of glycoproteins involves LC-MS to characterize complex mixtures of glycosylated and non-glycosylated peptides generated through enzymatic digestion. However, the inherent heterogeneity and large size of glycans often result in glycopeptide signals being overshadowed by those of non-glycosylated peptides in mass spectrometry [49].

Thus, enriching methods are frequently needed. Sepharose, cellulose, graphitized carbon, and lectins have all been employed as SPE materials for the enrichment of glycopeptides. With the exception of a few well-characterized glycoproteins, each eluting fraction requires a manual spectrum analysis and database search in order to maximize the number of assigned peaks in the mass spectra. The presence or lack of specific glycopeptide peaks can readily show variations between two samples when such a technique is used to investigate them. Glycopeptide profiling is a lower throughput method, but it can clarify both the glycan structures and the binding sites of a glycoprotein. A quick analysis technique is still required [50].

It is evident that, despite being extremely thorough, LC-MS methods alone for glycoprotein analysis frequently present challenges. Analyzing data can take a lot of time, and thorough characterization frequently calls for a large sample size. Profiling complex mixtures of glycoproteins with the LC-MS method alone is particularly cumbersome and leads to loss of valuable information that is contained in lower abundance proteins that might never be detected with LC-MS approaches [51].

4.1 Carbohydrate Microarray

In glycobiology, carbohydrate microarrays have emerged as a crucial technology that enables scientists to screen a vast number of sugars and elucidate the functions of carbohydrates in biological systems. However, the restricted availability and purity of naturally isolated oligosaccharides restrict their use in microarray-based investigations, leading to a trend towards the usage of manufactured glycans. The carbohydrates employed in these microarrays can be chemically synthesized or isolated from natural sources. Interestingly, functional linker groups—which are essential for immobilizing sugars on glass surfaces—can be included during the synthesis process [52]. The versatility and promise of this approach are demonstrated by the astounding fact that fewer than 40 carbohydrate building blocks can be used to create a vast array of carbohydrate structures that represent the whole mammalian glycome. High-throughput screening for binding interactions is made possible by the procedure of arraying glycans on different surfaces, such as nitrocellulose, glass, and dextran. Covalently bonding chemically altered carbohydrates to glass slides with functionalized surfaces, like epoxide or N-hydroxysuccinimide (NHS), is a popular immobilization method. Furthermore, it is possible for carbohydrates to adhere to unaltered surfaces in a site-specific, non-covalent manner. Using this approach, scientists have investigated how various carbohydrate structures interact with a wide range of living things, such as proteins, RNA (ribonucleic acid), viruses, and whole cells. This method's broad range of uses, such as the identification of disease-specific antibodies, the characterization of carbohydrate-binding proteins, and the clarification of carbohydrate-mediated cell recognition events, highlight its importance in furthering glycoscience research [53].

4.2 Lectin Microarrays

Glycans are immobilized on surfaces including nitrocellulose, glass, and dextran to examine binding interactions. Chemically modified carbohydrates are attached to functionalized glass slides or form non-covalent bonds with unmodified surfaces. This technique looks into the interactions between carbohydrates and biological things like cells, proteins, RNA, and viruses. It identifies proteins that bind carbs, finds antibodies associated with illness, and explains how cells perceive carbohydrates. This approach advances glycoscience research and has multiple applications in disease detection, biomarker identification, and understanding complex biological processes. By examining carbohydrate interactions, researchers can gain a better understanding of cellular communication, disease processes, and immune responses [54].

4.3 Glycoprotein Microarrays

Protein microarrays allow for the high-throughput screening of a wide range of biological materials, such as intact glycoproteins, fractionated proteomes, entire cell lysates, and antigen-antibody reactions. However, because of their varied sizes, shapes, and chemical characteristics, proteins provide unique difficulties. The right surface with a high signal-to-noise ratio must be chosen in order to maximize data quality. Because they reduce background fluorescence, glass-coated surfaces with particular chemical functions are extremely desirable. Protein deposition creates a covalent link with the glass surface, as shown on surfaces coated with epoxide. Nitrocellulose slides, on the other hand, are a more affordable choice because of their thin coating, which lowers the fluorescence background while preserving a high binding capacity. Furthermore, lectins, glycoconjugates, and glycans can be covalently immobilized in microarray formats using dextran-coated glass slides and photoactivatable surfaces [55].

Studies represent that lectins for galactose, sialic acid, and fucose were successfully used to profile exposed glycans after conventional glycoproteins were covalently immobilized. These platforms demonstrated the adaptability of microarray-based tools with many applications and were also appropriate for glycan and lectin immobilization, which permitted biomolecule binding to such microarray platforms [55].

4.4 Antibody Microarrays

Antibody microarrays make it possible to analyze several proteins from complicated mixtures at once, which makes it easier to isolate and purify particular proteins in their natural condition. Researchers can examine how glycans interact with these molecules and describe the various carbohydrate structures affixed to captured proteins by employing biotinylated lectins or glycan-binding antibodies. This strategy has been constrained, nevertheless, by lectin-binding glycans that attach to the antibody itself. Researchers have devised methods to chemically alter the glycans in the antibody in order to get around this restriction. One such technique uses sodium periodate to change the cis-hydroxyl groups on glycans into aldehyde groups in antibodies. A dipeptide and a hydrazide-maleimide cross-linking reagent are then used for additional modification.

Additionally, the researchers used neuraminidase to remove terminal sialic acid groups from the material, which produced glycoforms that lectins specifically attach to. Notably, cancer-associated glycan alterations in the proteins MUC1, MUC5AC, and MUC16 in pancreatic cancer cell lines were discovered by profiling protein and glycan variants in numerous samples using parallel-sandwich and glycan-detection approaches. Pro-inflammatory stimuli induced these alterations, underscoring the method's promise for comprehending the intricate connections among glycosylation, inflammation, and cancer [56]. Table 7 represents all 4 types of microarrays involved in Glycomic Science.

Table 7: Types of microarrays in Glycomic Science: Purpose, features, and applications.

Type of Microarrays	Purposes	Key Characters	Applications
Carbohydrate Microarray	Analyze sugar binding interactions	Use of synthetic and natural glycans	Glycan-protein/virus/cell interactions and antibody detection
Lectin Microarray	Profile glycan structures by binding to the lectin	Uses panels of lectins with specific glycan affinities	Immune response and glycan recognition
Glycoprotein microarray	Screening of intact glycoproteins and complex proteomes	High throughput analysis of proteins, lectins and glycans	Interaction of proteins and glycan
Antibody microarray	Analysis of interaction of protein and glycans	Uses glycan binding Antibodies	Detection of cancer biomarkers. Inflammation glycosylation studies

5. Challenges and Clinical Applications

Glycoconjugates, such as glycosphingolipids, use a variety of ways to coordinate intricate biological processes. Certain glycosyl-binding proteins identify the glycan structures in these conjugates, which sets off a series of biological processes. These interactions affect protein targeting and stability, alter signaling cascades, and promote cell-to-cell communication. Furthermore, the shape of proteins, their oligomerization, and their residency time on cell surfaces are all significantly influenced by glycans. Glycosylation, a common post-translational modification, highlights the significant influence of glycans on biological systems by regulating the destiny and function of many secreted and membrane-bound proteins [57].

Clarifying the complex structures and functional roles of glycans in biological systems is the goal of the quickly developing area of glycomics. The enormous complexity of the glycome presents many difficulties, but developments in cell biology, genetics, and mass spectrometry are assisting in overcoming these obstacles. In the meanwhile, glycosyl microarrays are transforming our knowledge of glycan signaling and recognition by illuminating the relationships between glycans, proteins, bacteria, and pathogens. Since glycan alterations are essential to almost every biological system and affect both intracellular and extracellular activities, unraveling the mysteries of glycomes holds great potential. Researchers can learn more about the intricate principles behind life activities by investigating the glycome [58].

Protein-glycan interaction research has advanced significantly. In the past, scientists have examined these interactions using time-consuming techniques like hapten inhibition and hemagglutination experiments. These methods did, however, have several significant drawbacks, such as low throughput, low sensitivity, and the need for large quantities of glycans. Glycan-binding proteins (GBPs) were thus not fully understood since research was frequently limited to simple monosaccharides or brief oligosaccharides. This incomplete knowledge fueled the widely held belief that GBPs are not very particular and have a low affinity. However, technological developments, such as the compilation of genes linked to glycosylation and enhanced analytical instruments, are now making it possible to conduct more thorough and in-depth analyses of protein-glycan interactions, which will help us better comprehend these intricate biological processes [59].

Glycobiology has seen a radical change since the development of glycan microarray technology. It is now possible for researchers to effectively examine the binding specificities of GBPs by immobilizing vast libraries of complex glycans on surfaces. With the advent of ELISA-type assays in the late 1990s, this technology made it possible to use fluorescence-based methods to analyze GBP interactions with glycans [60]. Researchers now have new ways to investigate how lectins, antibodies, poisons, viruses, and other compounds recognize glycans thanks to glycan microarrays. This method has greatly improved our knowledge of glycan-protein interactions and has broad applications in the basic and medicinal sciences. Glycan microarrays have revolutionized the discipline by providing high-throughput analysis, which allows scientists to examine intricate glycan interactions in previously unheard-of detail. Because of its immense complexity, defining the human glycome in its totality is a difficult endeavor. It might be more practical, nevertheless, to examine particular metaglycomes—subsets of glycans [60, 61]. The grouping of glycans found in a specific cell, tissue, or kind of glycoconjugate is known as a metaglycome. Researchers can start mapping out the human glycome by focusing on particular metaglycomes. Soluble human milk oligosaccharides and human glycosphingolipids (GSLs) are two intriguing research topics that may lead to a more thorough comprehension of glycan structures and activities as well as offer important insights into the human glycome [61].

According to estimates, there are several hundred different glycosylated species in human milk oligosaccharide (HMO) and at least 500 in GSL metaglycomes. Technological developments have made it possible for researchers to examine these intricate metaglycomes both statistically and qualitatively [62]. One noteworthy finding is that terminal glucose is the non-reducing sugar in both HMOs and GSLs. This implies that analytical methods created for one metaglycome might be modified for the other, promoting a synergistic approach to glycan study. Recent developments have also demonstrated that examining GSL metaglycomes in their intact lipid form can yield important information about aglycone dynamics, underscoring the possibility of using a variety of methods to comprehend the structures and functions of glycans. Researchers can better comprehend the complexities of glycan biology by utilizing these technological developments [63].

6. Conclusion

Glycomics has emerged as a vital field of research, offering unprecedented insights into the complex world of glycans and their role in human health and disease. The study of glycomics has far-reaching implications, from understanding the intricacies of cellular communication and immune responses to developing novel diagnostic and therapeutic strategies for various diseases. As the chapter explored in this chapter, glycomics has shown significant promise in the diagnosis and treatment of various diseases, including cancer, infectious diseases, and behavioral disorders. The aberrant glycosylation patterns observed in these diseases can serve as biomarkers for early detection and diagnosis, allowing for more effective treatment and improved patient outcomes.

The advent of advanced technologies, such as mass spectrometry and glycan microarrays, has greatly enhanced our ability to analyze and understand the glycome. These technologies have enabled researchers to identify specific glycan structures associated with various diseases, paving the way for the development of novel diagnostic and therapeutic strategies.

One of the most significant challenges in glycomics research is the complexity of glycan structures and their functions. Glycans are complex carbohydrate molecules that play critical roles in cellular communication, immune responses, and disease progression. Understanding the intricacies of glycan structures and their functions is essential for developing effective diagnostic and therapeutic strategies.

Despite these challenges, glycomics research has made significant progress in recent years. The integration of glycomics with other omics fields, such as genomics and proteomics, has greatly enhanced our understanding of the biological processes underlying human disease. This integrated approach has the potential to revolutionize personalized medicine, enabling healthcare providers to tailor treatment strategies to individual patients based on their unique genetic, proteomic, and glycomic profiles. The potential applications of glycomics research are vast and varied. In the field of cancer research, glycomics has shown significant promise in the development of novel diagnostic and therapeutic strategies. Glycan biomarkers can be used to detect cancer at an early stage, allowing for more effective treatment and improved patient outcomes. Additionally, glycomics can be used to develop targeted therapies that specifically target cancer cells, reducing the harm caused to healthy cells.

In the field of infectious diseases, glycomics has shown significant promise in the development of novel diagnostic and therapeutic strategies. Glycan structures can be used to develop vaccines that target specific pathogens, preventing infection and disease. Additionally, glycomics can be used to develop novel therapeutic strategies that target the glycan structures on pathogens, preventing their attachment and invasion of host cells. Conclusively, glycomics is a rapidly evolving field that holds great promise for improving our understanding of human biology and disease. The study of glycomics has far-reaching implications, from the development of novel diagnostic and therapeutic strategies to the integration of glycomics with other omics fields. As research in this field continues to advance, we can expect to see significant improvements in our ability to diagnose and treat various diseases, ultimately leading to improved patient outcomes and a better quality of life.

References

1. Shriver, Z., S. Raguram and R. Sasisekharan. (2004). Glycomics: A pathway to a class of new and improved therapeutics. *Nature Reviews Drug Discovery*, 3(10): 863–873.
2. Liang, C.H. and C.Y. Wu. (2010). Glycomics: Relevance for personalized medicine. *Current Pharmacogenomics and Personalized Medicine*, 8(1): 49–63.
3. Drake, R.R. (2015). Glycosylation and cancer: Moving glycomics to the forefront. *Advances in Cancer Research*, 126: 1–10.
4. Bergstrom, K.S. and L. Xia. (2013). Mucin-type O-glycans and their roles in intestinal homeostasis. *Glycobiology*, 23(9): 1026–1037.
5. Tuccillo, F.M., A. de Laurentiis, C. Palmieri, G. Fiume, P. Bonelli, A. Borrelli, P. Tassone, I. Scala, F.M. Buonaguro, I. Quinto and G. Scala. (2014). Aberrant glycosylation as biomarker for cancer: Focus on CD43. *BioMed Research International*, Article ID 742831.

6. Li, G., L. Li, E.J. Joo, J.W. Son, Y.J. Kim, J.K. Kang, K.B. Lee, F. Zhang and R.J. Linhardt. (2017). Glycosaminoglycans and glycolipids as potential biomarkers in lung cancer. *Glycoconjugate Journal,* 34: 661–669.
7. Silbert, J.E. (1982). Structure and metabolism of proteoglycans and glycosaminoglycans. *Journal of Investigative Dermatology,* 79(1): 31–37.
8. Pradeep, P., H. Kang and B. Lee. (2023). Glycosylation and behavioral symptoms in neurological disorders. *Translational Psychiatry,* 13(1): 154.
9. Kianičková, K., L. Pažitná, P.H. Kundalia, Z. Pakanová, M. Nemčovič, P. Baráth, E. Katrlíková, J. Šuba, J. Trebatická and J. Katrlík. (2023). Alterations in the glycan composition of serum glycoproteins in attention-deficit hyperactivity disorder. *International Journal of Molecular Sciences,* 24(10): 8745.
10. Predescu, E., T. Vaidean, A.M. Rapciuc and R. Sipos. (2024). Metabolomic markers in Attention-Deficit/Hyperactivity Disorder (ADHD) among children and adolescents: A systematic review. *International Journal of Molecular Sciences,* 25(8): 4385.
11. Haltiwanger, R.S. and J.B. Lowe. (2004). Role of glycosylation in development. *Annual Review of Biochemistry,* 73(1): 491–537.
12. Irvine, E.B. and G. Alter. (2020). Understanding the role of antibody glycosylation through the lens of severe viral and bacterial diseases. *Glycobiology,* 30(4): 241–253.
13. Zhang, X.L. and H. Qu. (2021). The role of glycosylation in infectious diseases. In: *The Role of Glycosylation in Health and Disease*. SpringerLink, pp. 219–237.
14. Feng, Y., L. Sun, X. Dang, D. Liu, Z. Liao, J. Yao, Y. Zhang, Z. Deng, J. Li, M. Zhao and F. Liu. (2024). Aberrant glycosylation in schizophrenia: insights into pathophysiological mechanisms and therapeutic potentials. *Frontiers in Pharmacology,* 15: 1457811.
15. Mealer, R.G., S.E. Williams, M.J. Daly, E.M. Scolnick, R.D. Cummings and J.W. Smoller. (2020). Glycobiology and schizophrenia: A biological hypothesis emerging from genomic research. *Molecular Psychiatry,* 25(12): 3129–3139.
16. Calvete, J.J. and L. Sanz. (2008). Analysis of O-glycosylation. *Post-translational Modifications of Proteins: Tools for Functional Proteomics. Humana Press,* pp. 281–292.
17. Medzihradszky, K.F. (2005). Characterization of protein N-glycosylation. *Methods in Enzymology,* 405: 116–138.
18. Veillon, L., C. Fakih, H. Abou-El-Hassan, F. Kobeissy and Y. Mechref. (2018). Glycosylation changes in brain cancer. *ACS Chemical Neuroscience,* 9(1): 51–72.
19. Breitling, J. and M. Aebi. (2013). N-linked protein glycosylation in the endoplasmic reticulum. *Cold Spring Harbor Perspectives in Biology,* 5(8): a013359.
20. Rousseaux, C.G. (2008). A review of glutamate receptors II: Pathophysiology and pathology. *Journal of Toxicologic Pathology,* 21(3): 133–173.
21. Polyansky, A.A. and R.G. Efremov. (2024). Transmembrane domains of bitopic proteins as a key to understand the cellular signaling (A review). *Russian Journal of Bioorganic Chemistry,* 50(4): 1202–1214.
22. Chatterjee, S., R. Kawahara, H.C. Tjondro, D.R. Shaw, M.A. Nenke, D.J. Torpy and M. Thaysen-Andersen. (2021). Serum N-glycomics stratifies bacteremic patients infected with different pathogens. *Journal of Clinical Medicine,* 10(3): 516.
23. Huang, L., R. Zhang, Y. Hu, H. Zhou, J. Cao, H. Lv, S. Chen, S. Ding and G. Chen. (2019). Epidemiology and risk factors of methicillin-resistant Staphylococcus aureus and vancomycin-resistant enterococci infections in Zhejiang China from 2015 to 2017. *Antimicrobial Resistance and Infection Control,* 8: 1–9.
24. Wagstaff, J.L., S.L. Taylor and M.J. Howard. (2013). Recent developments and applications of saturation transfer difference nuclear magnetic resonance (STD NMR) spectroscopy. *Molecular BioSystems,* 9(4): 571–577.

25. Sun, X.L. (2021). The role of cell surface sialic acids for SARS-CoV-2 infection. *Glycobiology*, 31(10): 1245–1253.
26. Gu, Y., B. Duan, J. Sha, R. Zhang, J. Fan, X. Xu, H. Zhao, X. Niu, Z. Geng, J. Gu, B. Huang and S. Ren. (2023). Serum IgG N-glycans enable early detection and early relapse prediction of colorectal cancer. *International Journal of Cancer*, 152(3): 536–547.
27. Hu, M., Y. Lan, A. Lu, X. Ma and L. Zhang. (2019). Glycan-based biomarkers for diagnosis of cancers and other diseases: Past, present, and future. *Progress in Molecular Biology and Translational Science*, 162: 1–24.
28. Adamczyk, B., T. Tharmalingam and P.M. Rudd. (2012). Glycans as cancer biomarkers. *Biochimica et Biophysica Acta (BBA)-General Subjects*, 1820(9): 1347–1353.
29. Mechref, Y., Y. Hu, A. Garcia and A. Hussein. (2012). Identifying cancer biomarkers by mass spectrometry-based glycomics. *Electrophoresis*, 33(12): 1755–1767.
30. Zhang, X. and Y. Wang. (2016). Glycosylation quality control by the Golgi structure. *Journal of Molecular Biology*, 428(16): 3183–3193.
31. Kam, R.K. and T.C. Poon. (2008). The potentials of glycomics in biomarker discovery. *Clinical Proteomics*, 4: 67–79.
32. De Haan, N., M. Wuhrer and L.R. Ruhaak. (2020). Mass spectrometry in clinical glycomics: The path from biomarker identification to clinical implementation. *Clinical Mass Spectrometry*, 18: 1–12.
33. Ishibashi, Y., Y. Tobisawa, S. Hatakeyama, T. Ohashi, M. Tanaka, S. Narita, T. Koie, T. Habuchi, S. Nishimura, C. Ohyama and T. Yoneyama. (2014). Serum tri-and tetra-antennary N-glycan is a potential predictive biomarker for castration-resistant prostate cancer. *The Prostate*, 74(15): 1521–1529.
34. Peng, W., J. Zhao, X. Dong, A. Banazadeh, Y. Huang, A. Hussien and Y. Mechref. (2018). Clinical application of quantitative glycomics. *Expert Review of Proteomics*, 15(12): 1007–1031.
35. Ryšlavá, H., V. Doubnerová, D. Kavan and O. Vaněk. (2013). Effect of posttranslational modifications on enzyme function and assembly. *Journal of Proteomics*, 92: 80–109.
36. Das, P., A. Sedighi and U.J. Krull. (2018). Cancer biomarker determination by resonance energy transfer using functional fluorescent nanoprobes. *Analytica Chimica Acta*, 1041: 1–24.
37. Kaji, H. and T. Isobe. (2008). Liquid chromatography/mass spectrometry (LC/MS)-based glycoproteomics technologies for cancer biomarker discovery. *Clinical Proteomics*, 4: 14–24.
38. Nygren, H., T. Seppänen-Laakso, S. Castillo, T. Hyötyläinen and M. Orešič. (2011). Liquid chromatography-mass spectrometry (LC-MS)-based lipidomics for studies of body fluids and tissues. In: *Metabolic Profiling: Methods and Protocols*. Humana Press, pp. 247–257.
39. Ruhaak, L.R., G. Xu, Q. Li, E. Goonatilleke and C.B. Lebrilla. (2018). Mass spectrometry approaches to glycomic and glycoproteomic analyses. *Chemical Reviews*, 118(17): 7886–7930.
40. Chatham, J.C. and R.P. Patel. (2024). Protein glycosylation in cardiovascular health and disease. *Nature Reviews Cardiology*, 21(8): 525–544.
41. Deleon-Pennell, K.Y., O.K. Ero, Y. Ma, R.P. Iyer, E.R. Flynn, I. Espinoza, S.K. Musani, R.S. Vasan, M.E. Hall, E.R. Fox and M.L. Lindsey. (2019). Glycoproteomic profiling provides candidate myocardial infarction predictors of later progression to heart failure. *ACS Omega*, 4(1): 1272–1280.
42. Cavalcanti, I.D.L., J.C.S. Soares, I.D.L. Cavalcanti and J.C.S. Soares. (2021). Conventional cancer treatment. In: *Advances in Cancer Treatment: From Systemic Chemotherapy to Targeted Therapy*. SpringerLink, pp. 29–56.

43. Patwa, T., C. Li, D.M. Simeone and D.M. Lubman. (2010). Glycoprotein analysis using protein microarrays and mass spectrometry. *Mass Spectrometry Reviews*, 29(5): 830–844.
44. Ruhaak, L.R., G. Xu, Q. Li, E. Goonatilleke and C.B. Lebrilla. (2018). Mass spectrometry approaches to glycomic and glycoproteomic analyses. *Chemical Reviews*, 118(17): 7886– 7930.
45. Thomas, S.L., J.B. Thacker, K.A. Schug and K. Maráková. (2021). Sample preparation and fractionation techniques for intact proteins for mass spectrometric analysis. *Journal of Separation Science*, 44(1): 211–246.
46. Totten, S.M., M. Kullolli and S.J. Pitteri. (2017). Multi-lectin affinity chromatography for separation, identification, and quantitation of intact protein glycoforms in complex biological mixtures. In: *Proteomics: Methods and Protocols*. Humana Press, pp. 99–113.
47. Chen, C.C., W.C. Su, B.Y. Huang, Y.J. Chen, H.C. Tai and R.P. Obena. (2014). Interaction modes and approaches to glycopeptide and glycoprotein enrichment. *Analyst*, 139(4): 688–704.
48. Vuckovic, D., L.F. Dagley, A.W. Purcell and A. Emili. (2013). Membrane proteomics by high performance liquid chromatography–tandem mass spectrometry: Analytical approaches and challenges. *Proteomics*, 13(3-4): 404–423.
49. Vreeker, G.C. and M. Wuhrer. (2017). Reversed-phase separation methods for glycan analysis. *Analytical and Bioanalytical Chemistry*, 409: 359–378.
50. Gab-Allah, M.A. and J. Kim. (2024). A comprehensive review of recent advances in the enrichment and mass spectrometric analysis of glycoproteins and glycopeptides in complex biological matrices. *Mass Spectrometry Letters*, 15(1): 1–25.
51. Ashwood, C., B. Pratt, B.X. MacLean, R.L. Gundry and N.H. Packer. (2019). Standardization of PGC-LC-MS-based glycomics for sample specific glycotyping. *Analyst*, 144(11): 3601–3612.
52. Park, S., J.C. Gildersleeve, O. Blixt and I. Shin. (2013). Carbohydrate microarrays. *Chemical Society Reviews*, 42(10): 4310–4326.
53. Disney, M.D. and P.H. Seeberger. (2004). The use of carbohydrate microarrays to study carbohydrate-cell interactions and to detect pathogens. *Chemistry and Biology*, 11(12): 1701–1707.
54. Hu, S. and D.T. Wong. (2009). Lectin microarray. *PROTEOMICS–Clinical Applications*, 3(2): 148–154.
55. Zhao, J., T.H. Patwa, D.M. Lubman and D.M. Simeone. (2008). Protein biomarkers in cancer: Natural glycoprotein microarray approaches. *Current Opinion in Molecular Therapeutics*, 10(6): 602.
56. Angenendt, P. (2005). Progress in protein and antibody microarray technology. *Drug Discovery Today*, 10(7): 503–511.
57. Blixt, O., S. Head, T. Mondala, C. Scanlan, M.E. Huflejt, R. Alvarez, M.C. Bryan, F. Fazio, D. Calarese, J. Stevens, N. Razi, D.J. Stevens, J.J. Skehel, I. van Die, D.R. Burton, I.A. Wilson, R. Cummings, N. Bovin, C.H. Wong and J.C. Paulson. (2004). Printed covalent glycan array for ligand profiling of diverse glycan binding proteins. *Proc. Natl. Acad. Sci. USA*, 101: 17033–17038.
58. Campanero-Rhodes, M.A., A.S. Palma, M. Menéndez and D. Solís. (2020). Microarray strategies for exploring bacterial surface glycans and their interactions with glycan-binding proteins. *Frontiers in Microbiology*, 10: 2909.
59. Taylor, M., K. Drickamer, A. Imberty, Y. van Kooyk, R. Schnaar, M. Etzler and A. Varki. (2022). Discovery and classification of glycan-binding proteins. In: *Essentials of Glycobiology* (4th Edn.). Cold Spring Harbor Laboratory Press.
60. Heimburg-Molinaro, J., A.Y. Mehta, C.A. Tilton and R.D. Cummings. (2024). Insights into glycobiology and the protein-glycan interactome using glycan microarray technologies. *Molecular and Cellular Proteomics*, 100844.

61. Zhu, L., H. Li, T. Luo, Z. Deng, J. Li, L. Zheng and B. Zhang. (2023). Human milk oligosaccharides: A critical review on structure, preparation, their potential as a food bioactive component, and future perspectives. *Journal of Agricultural and Food Chemistry*, 71(43): 15908–15925.
62. Du, H., H. Yu, T. Ma, F. Yang, L. Jia, C. Zhang, J. Zhang, L. Niu, J. Yang, Z. Zhang, K. Zhang and Z. Li. (2019). Analysis of glycosphingolipid glycans by lectin microarrays. *Analytical Chemistry*, 91(16): 10663–10671.
63. Daku, A.B., S.B. Al-Mhanna, R. Abu Bakar and A.A. Nurul. (2022). Glycolipids isolation and characterization from natural source: A review. *Journal of Liquid Chromatography and Related Technologies*, 45(13-16): 165–173.

CHAPTER 3

Analysis of Plant N-Glycans using PNGase H+ by MALDI-TOF MS

Balaji, S., [1,*] ***Prakash, P.,*** [1] ***Ganesh, N.,*** [1] ***Ambikapathy, V.,*** [1] ***Babu, S.,*** [1] ***Gomathi, S.*** [1] **and** ***Panneerselvam, A.*** [2]

Research on N-glycans is more prevalent compared to that on O-glycans. These N-glycans are mainly involved in cell signalling and immune systems. Recently N-glycans of plants have created a high curiosity for scientists to discover new drugs for various diseases and disorders. The N-glycans can facilitate improvement and create biocontrol agents that kill or control the pest or its life pathway through some specific glycans. This chapter deals with analysis of N-glycans that are derived from plants using PNGase H+ in MALDI-TOF MS. PNGase A is used traditionally but some preliminary treatment is involved, that is, digestion by pepsin or trypsin and it takes a long time. PNGase H+ is used in this process because no preliminary treatment is required. The purified N-glycans are further involved

[1] PG and Research Department of Botany, A.V.V.M Sri Puspham College (Autonomous) (Affiliated to Bharathidasan University, Trichy-24), Poondi-613 503, Thanjavur, Tamil Nadu, India.
[2] Indian Biotrack Research Institute, Thanjavur-613 005, Tamil Nadu, India.
* Corresponding author: balajisethu200108@gmail.com

DOI: 10.1201/9781003635901-3

in 2-AB labelling continuously and they are transferred to UPLC profiling. Collected samples are analysed in MALDI-TOF MS. This N-glycan analysis of plants will help to analyse various physiological properties in plants and interaction of antigen and antibodies. These glycans involved in the defence mechanisms against pathogens that are harmful to plants and contribution of glycomics are identified as unique glycans, which are utilized in various biotechnological or genetic modifications for crop improvement. This method is one of the simplest and quickest for the budding glycomist.

1. Introduction

Carbohydrate is only a bond of carbon, hydrogen, and oxygen molecules, but it occupies the entire living world. Although these molecules are non-living, they are capable of forming living things, so all living things may have come from non-living things, leading to the knowledge of glycomics. Glycomics is the advanced study of the role of carbohydrates in cellular structure and function. The term glycomics is derived from 'glycol' that means sugar or sweet and is followed by 'comic' was picked out from genomic and proteomics because the concept of glycomics is inspired by genomics and proteomics [1]. Glycan is the primary functional unit for the glycomics and the group of monosaccharides formed by glycans by the glycoside bonds. Glycans are the carbohydrate part of glycoconjugates. The glycoconjugates are glycans bonded with various biomolecules such as proteins (glycoproteins), lipids (glycolipids), and nucleic acid (nucleosides). Specifically, the protein-linked glycans are involved in various cellular activities such as protein folding, protein quality control, protein stability and solubility, neural transmission, immune system, cell signalling and recognition, and blood group antigens [2].

After the COVID pandemic all the researchers are involved in investigation of unexplored plant bioactive compounds because some of the plant-based medicines are used to cure Covid virus infected patients and also they didn't cause any side effects, but some synthesised artificial vaccines resulted in few side effects and even death. In his research, Theerakarn Srisangsung and colleagues discovered an anti-virus compound against the SARS-Cov-2, and this anti-viral compound is derived from the N-glycan of *Nicotiana benthamiana* [3]. The N-glycan derived from *Arabidopsis thaliana* and *Physcomitrella patens* were used as plant-derived antibodies for human pharmaceutical processes; these antibodies are called as plantibodies [4]. An antibody (plantibody) against Hepatitis B, was derived from the transgenic *Nicotiana tabacum* in research of Triguero and their co-researchers [5] (Fig. 1).

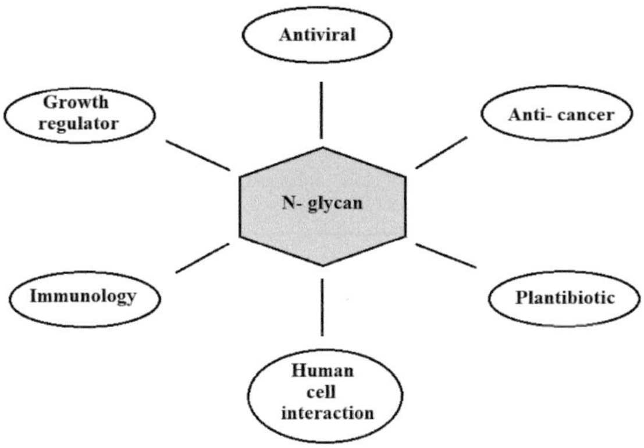

Fig. 1: Role and importance of N-Glycans.

Generally the glycans are classified into homopolymers and heteropolymers; homopolymers are glycans structure made up of sugar units, and the heteropolymers are glycans structure formed by sugar units and other compounds such as proteins and lipids. These heteropolymers are further classified into two types, based on compounds which combine with sugar units, they are glycolipids and glycoproteins. Glycolipids are formed by the interaction between carbohydrate units and lipids, whereas glycoproteins are the result of interaction between carbohydrate units and proteins [6].

These protein linked glycans are classified into two major types on the basis of linkage of carbohydrate parts together with amino acids in the protein chain. They are N-Glycan and O-Glycan. N-glycan is an oligosaccharides chain (glycan) attached to an asparagine residue of the polypeptide chain in the consensus sequence: Asn-X-Ser/Thr. The X stands for any amino acids except proline and aspartic acid. O-glycan is an oligosaccharide chain (glycan) attached to the hydroxyl group of amino acids serine or threonine in the consensus sequence: GalNAcα1-O-Ser/Thr. [7] (Fig. 2). These glycans are infused in fields like pharmaceuticals and agriculture. In the pharmaceutical field, drugs are discovered on the basis of interaction between signalling glycans of pathogenic microbes and their host organism.

N-glycan is involved in various physiological and cellular functions such as the growth regulation process, stress signalling, and triggering the plant immune system against pathogens [8]. In the tomato fruit ripening stage, some N-glycans are responsible for the normal fruit ripening, growth of leaves, and even in changing most of the phenotypic characters [9].

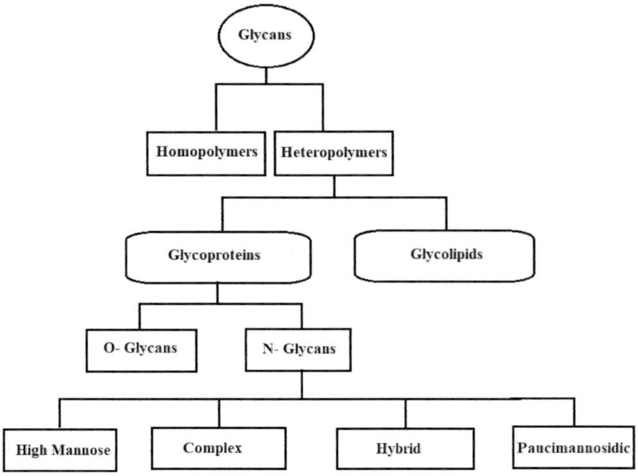

Fig. 2: Classification of Glycans.

N-glycans bind with a core sugar series (Manα1-6(Manα1-3)Manβ1-4GlcNacβ1-4GlcNAcβ1-Asn-X-Ser/Thr) and are classified into four types, based on the structures binding of sugar motifs: (i) high mannose type, (ii) Complex type, (iii) Hybrid type, and (iv) Paucimannosidic type. Only mannose molecules connected with the central core sugar series are known as high mannose type N-glycan. Complex type N-glycan's core is connected with sugar motifs, except mannose such as α(1,3)-fucose and β(1,2)-xylose residues, and this series begins with N-acetyl glucosamine residues. Hybrid type N-glycan is a mixture structure of high mannose type and complex type, that is, the Manα1-6 site binds with mannose residues only; on the other hand, the Manα1-3 site binds with other sugar motifs like complex type. The Paucimannosidic type N-glycans is an unfamiliar type of N-Glycan, the structure of these N-glycans is that the sugar motifs bind with the central core [1, 10] (Fig. 3).

The methodology for the structural analysis N-glycan is to first of all separate the carbohydrate part from protein. These N-glycans can be separated by both chemical and enzymatic methods. The enzymatic method is better than the chemical method because anhydrous hydrazine is one of most frequently used method. Examples of some disadvantages that occurred during the N-glycan releasing process are peptide bonds are destroyed, protein relating information is erased; the acyl groups break away from the N-acylamino sugars and sialic acids, calling for a retaliating step, assuming that the sialic acid residues were originally acetylated; and a small quantity of O-acetyl substitution are some of the reasons [11].

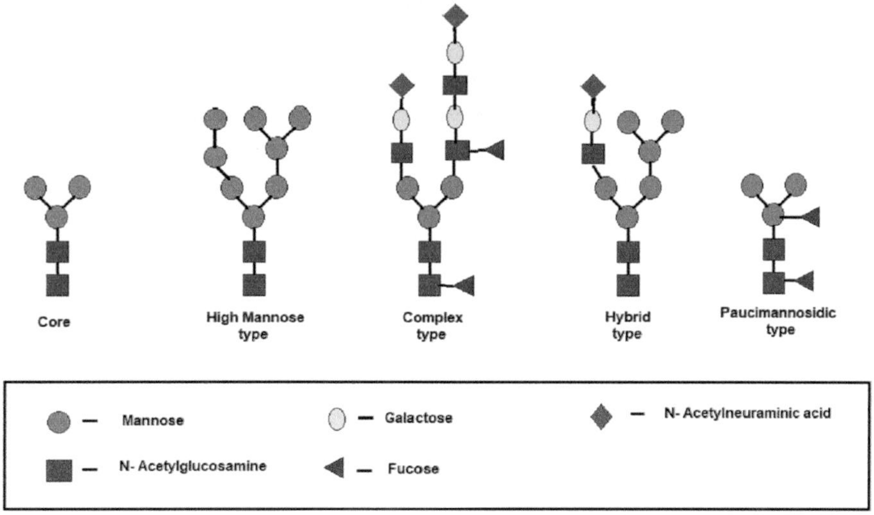

Fig. 3: Types of N-Glycans.

In the enzymatic method, various types of enzymes are used, such as PNGase F, PNGase A, PNGase H+, and Endo H. PNGase F is the most widely used commercially available enzyme for releasing N-glycans and these have the ability to separate all types of glycans from protein chains except the core-linked α-1,3 Fuc N-glycan structure. This enzyme is derived from *Flavobacterium meningosepticum* [12]. The PNGase A, next to the PNGase F, is mostly used for the release of N-glycans from plants and insects and it is derived from seeds of *Prunus dulcis*. This enzyme shows dominance toward core modifications of N-glycans and it only releases N-glycans from the small glycopeptides through preliminary treatment for the sample with proteases, that is, the sample is digested in trypsin or pepsin; the reaction lasts for a long-time, i.e., 4 hrs to get the best result [13]. Endo H or Endoglycosidase H is also used in the process of separation of N-glycans from their crude source but it is unable to separate the complex N-glycans, it only releases the high mannose and hybrid N-glycans. Endo H is derived from *Streptomyces plicatus* and some other *Streptomyces* sp. [14, 15]. Endo D is an enzyme derived from *Streptococcus* sp., this is the first enzyme used to separate high mannose N-glycans but it cannot separate complex, and hybrid N-glycans; however, it cleaves a few high mannose N-glycans only. Endo F1, Endo F2, and Endo F3 are enzymes derived from *Flavobacterium* sp. There are various species of each type of enzyme but they have different characteristics. Endo F1 activities are the same as Endo H but it affects

Table 1: Enzymes and their importance.

Enzyme	Origin	Importance	References
Endo D	*Streptococcus* sp. (Bacteria)	This is the first enzyme used to separate high mannose from their source.	[16]
Endo F1, Endo F2, and Endo F3	*Flavobacterium* sp. (Bacteria)	These types of enzymes have some limitations to separate N-glycans from the sources.	[16]
Endo H	*Streptomyces plicatus* and some other *Streptomyces* sp. (Bacteria)	This enzyme cannot separate complex N-glycans from the source.	[14]
PNGase F	*Flavobacterium meningosepticum* (Bacteria)	This enzyme is commonly used in N-glycan separation except core-linked α-1,3 Fuc N-glycan structure.	[12]
PNGase A	*Prunus dulcis* seeds (Almond)	This enzyme is used in both plant and animal N-glycan separation but it takes a long time and trypsin treatment is essential.	[13]
PNGase H+	*Terriglobus roseus* (Bacteria)	This enzyme was recently discovered and it involves quick N-glycan separation.	[17]

the core of N-glycans. Endo F2 enzymes separate a few high mannose N-glycans only, and Endo F3 cannot separate hybrid, tetra- antennary high mannose N-glycans [16]. PNGase H+ is another enzyme, recently discovered from bacterial *Terriglobus roseus*, and it has the bearing ability to break core α-1,3-linked Fuc N-glycans. The special character of this enzyme is that their enzymatic reaction also occurs in concentrated organic acid, even simultaneously allowing N-glycan liberation and fluorescence labelling [17] (Table 1).

N-glycans have less intrinsic spectrum activity and they are difficult to detect during the chromatographic process so they need labelling to increase the chances of detection. There are two methods to label N-glycans, one is radio labelling and the other is fluorescent labelling. Radio labelling is a difficult and expensive method, but fluorescent labelling is easy and not more expensive than radio labelling. The fluorescent labelling is called fluorescent probs. So many chemicals compounds are used in the fluorescent labelling method, and hence they have some disadvantages, such as labelling in unwanted regions, difficult to analyse in chromatographic techniques and so on, but 2-aminobenzamide (2-AB) and 2-anthranilic acid (2-AA) provide better results during analysis and 2-aminobenzamide is slightly better than 2-anthranilic acid

[18]. These two chemicals are widely used in the N-glycan analysis and are very convenient methods of labelling for beginners. Procainamide (Pro A) is used in advanced glycomics analysis, it is similar to 2-AB, but its advantage is its ability to detect less available glycans in the source. Pro A is mostly used in animal N-glycan analysis [19].

HILIC (hydrophobic interaction liquid chromatography) is a widely used technology in which water is used as the eluting solvent, this technique is mostly used to determine the retention and separation of polar analytes like carbohydrates and peptides. HILIC is comparatively better than reversed phase chromatography because it provides complementary selectivity and better polar compound retention. The chromatography separation technique is used to separate compounds through high polar stationary phase, something like functionalized silica is coated with a hydrophobic substance, and the non-polar mobile phase something like acetonitrile and water mixture. The sample which is polar or hydrophilic in nature is easily attracted toward the stationary phase with the help of increasing water concentration which leads to separating the solutes through increasing hydrophilicity. It provides the attached and separated charged and uncharged carbohydrates by gradient from high to low organic content. Glycan size and polarity play an important role in the retention time reproducible and prediction [2].

MALDI TOF MS method is used to derive the complete gas ion phases from thermally unstable, larger, and non-volatile compounds of the sample that combine to the complex chemical compound. Carbohydrates, proteins, oligonucleotides, inorganic compounds, synthetic polymers are present in the compound that can be analysed in the MALDI TOF MS. The matrix of the MALDI TOF MS is ionized at the targeted substrate at a minor or small region of that target through absorption of laser energy. MALDI TOF MS produces higher internal energy than Electrospray Ionization (ESI) that leads to metastable breakdown on easily breakable molecules, something like Sialylated glycopeptides. The disadvantages of MALDI TOF MS is that it is difficult to analyse the sulfated glycan and acidic groups like phosphate, sulfate, and sialic acid [2]. Various types of software are used to analyse and identify the glycan through the reading obtained from the MALDI TOF MS, softwares like GlycoWorkbench, GlycoMod, and Glyco-Peakfinder. GlycoWorkbench is one of the easier software for beginners and it was developed by EUROCarbDB. The important work of GlycoWorkbench is to analyse the group of data from MALDI TOF MS, then create structure based on the comparison with the uploaded data, even the fragmented structure is analysed through this software [20] (Fig. 4).

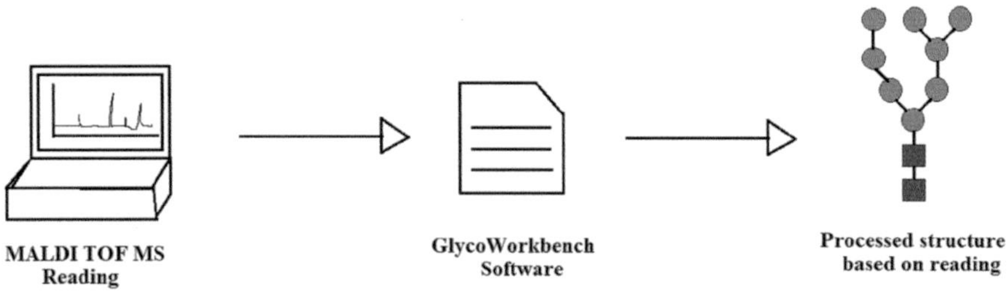

Fig. 4: Preparation process of the structure of N-Glycan.

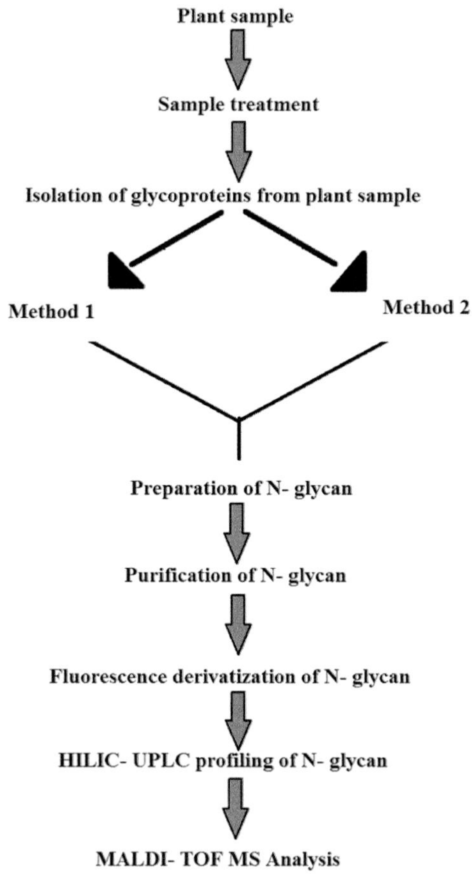

Fig. 5: Steps involved in N-glycan analysis by using PNGase H+.

2. Materials and Methods

2.1 Requirements

2.1.1 Sample Treatment for Specific Samples

Seed

- Plant seed sample
- Acetone
- Deionised water
- Hydrogen peroxide (H_2O_2)

Pollen

- Ammonium bicarbonate
- Centrifuge
- Methanol
- Liquid nitrogen

2.1.2 Glycoproteins Isolated from Plant Sample

Method 1

- Untreated sample or treated sample
- Kitchen blender
- Trichloroacetic acid (TCA)
- Centrifuge
- Distilled water

Method 2

- Untreated sample or treated sample
- Liquid Nitrogen
- Polyvinylpolypyrrolidone (PVPP)
- Protein extraction buffer
- Muslin cloth
- Rotary shaker
- Centrifuge

2.1.3 Preparation of N-Glycan

- PNGase H+ enzyme
- Acetic acid
- Incubator

2.1.4 N-Glycan Purification

- Acetic acid
- Solid-phase extraction (SPE) column
- Acetonitrile
- Trifluroacetic acid(TFA)
- Distilled water

2.1.5 Fluorescence Derivatization of N-Glycans

- Fluorogenic 2-aminobenzamide reagent (2-AB)
- Acetonitrile
- Distilled water
- Centrifuge

2.1.6 HILIC-UPLC Profiling of N-glycans

- HILIC-UPLC

2.1.7 MALDI-TOF MS Analysis

- MALDI-TOF MS

2.2 *Procedure*

2.2.1 Sample Treatment for Specific Samples

Seed

Seeds with hard coats or some undetectable infections can change the results of N-glycan analysis, so the seeds are involved in some specific methods. The seeds are cleaned by using 95% acetone for 5 min., then they are washed in deionised water, after washing they are treated in 3% hydrogen peroxide (H_2O_2) for 15 min. The deionised water is used to wash them again, to completely remove H_2O_2 residues, they are immersed overnight in deionised water [10].

Pollen

Plant pollen 10 mg is mixed in 0.4 mL of 125 mM ammonium bicarbonate on pH 8.3 at 4°C overnight, then loaded in centrifugation. After centrifugation, 80 μl of supernatants is dissolved in 400 μl cold methanol and the prepared sample is stored overnight at 4°C or still precipitation of glycoprotein, then the glycoprotein precipitates are collected and dried using liquid nitrogen stream [21].

2.2.2 Glycoproteins Isolated from Plant Samples

Method 1

To prepare the plant sample, approximately 100 g of fresh material is homogenized in a blender for 10 min. The resulting mixture is then transferred to a centrifuge tube and centrifuged at 14,000 g and 4°C for 20 min., to separate the insoluble components. The supernatant is collected and mixed with an equal volume of 2 M trichloroacetic acid (TCA) before being transferred to another centrifuge tube for further centrifugation at 14,000 g at 4°C to 30 min. After this step, the supernatant is discarded and the pelleted glycoproteins are washed with 20 ml of deionized water, followed by another centrifugation at 14,000 g and 4°C for 5 min., to remove any soluble oligosaccharides and polysaccharides [17].

Method 2

Approximately 3 g plant sample is finely powdered using liquid nitrogen, subsequently 1 g of powdered sample is mixed in 10 g of polyvinylpolypyrrolidone (PVPP) to remove phenolic compounds (PVPP binds with phenolics). Sample mixture is stirred with a protein extraction buffer which is three times the volume of sample (if the sample mixture is 1 g, the extraction buffer will take 3 ml) till completely dissolved. The homogenate is filtered by sterilized muslin cloth after filtration. The filtrate is shaken well for 2 hrs at 5 rpm in 4°C using a rotary shaker, then the centrifuge for 30 min at 15,000 x g at 4°C. After completion of the centrifuge, the supernatant is collected and stored for further process [21].

2.2.3 Preparation of N-Glycan

Recombinant PNGase H+ 0.2mU is mixed with 50 μL of extracted protein sample in acetic acid 10 mM and the mixture is incubated 2–12 hrs at 65–37°C, incubation is completed in a centrifuge tube at 14,000 g to

5 min., till removal of the excess enzyme and protein materials. The 1.5 mL of supernatant is transferred to another centrifuge tube [22].

2.2.4 N-Glycans Purification

To enrich N-glycans, first prepare a solid-phase extraction (SPE) column by removing salts and unwanted materials from the reaction mixture and enhancing fluorogenic 2-aminobenzamide (2-AB) labelling agent for N-glycan derivatization. Put 3 mL of distilled water into the column for washing. Then activate the SPE-column by adding 3 mL of 80% acetonitrile containing trifluoroacetic acid (0.1% v/v) followed by distilled water 3 mL into the SPE-column for equilibration. The collected supernatant 3 mL loaded into column and also removed the waste move through, consequently add distilled water 1.5 mL into column for clearing and discarding that waste. 20% aqueous acetonitrile solution 1.5 mL, bearing TFA 0.1% is used to make the released N-glycan elute, these are collected into a 2 mL centrifuge tube. Thereafter, centrifuge the collected elute to remove the excess solvent till completely dried at room temperature [22].

2.2.5 Fluorescence Derivatization of N-Glycans

Dried sample is dissolved in 5–10 μL of 2-AB reagent, then shake them well till totally dissolved and incubate the solvent for 2 hrs at 60–65°C. Reduce the temperature of the incubated solvent at room temperature for 5 min., following that 5 μL of distilled water and 40 μL acetonitrile are mixed together with the sample, thus are centrifuged at 14,000 Ig for 3 min. The supernatant is collected in an HPLC vial, then stored in –20°C for further procedures [22, 18].

2.2.6 N-Glycan Profiling in HILIC-UPLC

The collected sample is analysed in the standard UPLC system. Using hydrophobic interaction liquid chromatography (HILIC)-UPLC glycan column. Monitoring the sample by the fluorescence detector. The excitation wavelength is 330 nm and emission wavelength is 420 nm. Set the column temperature at 60°C for the analysis.

The separation of N-glycans begins by mixing solvent-B (95%) with solvent-A, maintaining a flow rate of 0.5 mL/min. from 0 to 44.5 min. During this time, the proportion of solvent-B is gradually decreased from 78% to 55.9% over the first 6 min. After reaching 55.9%, the solvent-B concentration is rapidly reduced to 0% between 44.5 and 46.5 min., where it is held for an additional 2 min., to allow for column washing. The flow rate is then decreased to 0.25 mL/min., from 44.5 to 55 min., while solvent-B is increased back to 95% from 48.5 to 50.5 min., remaining at this concentration for 4.5 min. The column is re-equilibrated to the initial condition of 95% solvent-B at a flow rate of 0.5 mL/min., from 50.5 to 57 min., at which point a 45 µL sample is injected into the UPLC. The N-glycans are eluted with retention time between 15 and 40 min., and each UPLC peak fraction is collected and dried using vacuum centrifugation. To calibrate the plant N-glycan profiling, approximately 1 pmol of a 2-AB labelled dextran standard (comprising 2–20 glucose unit) is injected, allowing for the assignment of the elution time based on standardized glucose units [17, 23].

2.2.7 MALDI-TOF MS Analysis

Separately each sample separated from the UPLC is dissolved in 5 µL of water (LCMS-grade) and the dissolved sample 1 µL is mixed with 1 µL of 6-aza-2-thiothymine (ATT) solution (0.3% w/v in 70% v/v aqueous acetonitrile), then pipette the sample mixture on the sample carrier of MALDI-TOF. Begin the samples analysis using MALDI-TOF MS in positive ion mode and Bruker Flex analysis version 3.3 analysis software is used to process the mass spectra. GlycoWorkbench, version 1.1. software is used to interpret the spectra. The Search parameter is 2-AB labelling, [M+Na]+, and the accuracy parameter is 2.0 Da [22, 24, 20] (Fig. 5).

2.3 *Notes*

2.3.1 Preparation of Percentage Solutions

Mass percent (%w/w)

Formula: $\frac{mass\,of\,solute}{mass\,of\,solution}$ X 100

Example: Weighted 10 g of NaOH pellets is dissolved in 90 g of distilled water for 10% of aqueous NaOH. All the chemicals are measured in grams, not in mL. The solution is not only distilled water, but any chemical based on the procedure such as ethanol, methanol, acetone, etc.

Volume percent (%v/v)

Formula: $\frac{volume\ of\ solute}{volume\ of\ solution}$ X 100

Example: Measure 30 mL of conc. HCl is mixed with 70 mL of distilled water for 30% of HCl. All are measured in mL.

Mass/Volume percent (%w/v)

Formula: $\frac{mass\ of\ solute}{volume\ of\ solution}$ X 100

Example: Weighted 20 g of KOH is dissolved in distilled water and is made up to 100 mL by distilled water. For 20% of aqueous KOH, the solute is measured in grams and solution is measured in mL.

2.3.2 Preparation of Molarity and Normality Solutions

Molarity solution

Dissolved gram molecular weight of chemical in distilled water, after it completely dissolved and made into 1 L (for 1 molar solution).

Normality solution

Dissolved gram equivalent weight of chemical in distilled water, after it is completely dissolved and made into 1 L (for 1 molar solution).

2.3.3 Preparation of Protein Extraction Buffer

25 mm Tris-HCl (pH 7.0), 0.2 M CaCl2, 0.5 M NaCl, and 20 µl/g fresh weight of protease inhibitor mixture [25].

2.3.4 Fluorescence Derivatization of N-Glycans

2-AB Reagent

2-AB 35 mM and 0.1 M $NaCNBH_3$ (sodium cyanoborohydride) are dissolved with Dimethyl sulfoxide/acetic acid solution in the ratio of 7:3 is made up to 1 mL.

2.3.5 N-Glycan Profiling in HILIC-UPLC

Solvent-A: Formic acid 43 mL is mixed with 700 mL water LCMS-grade. Then pH is adjusted by adding ammonia solution (25% w/w) drop by drop to 4.5 pH. Then the mixture is made to 1,000 mL by LCMS-grade H_2O and that is stored in 4–6°C.

Due to the analysis, the prepared solvent 50 mL is mixed with 950 mL water LCMS grade.

Solvent-B: Acetonitrile LCMS-grade.

3. Conclusion

This PNGase H+ method is one of the easiest methods in plant N-glycan analysis and this method is used by budding glycomists. Most of the

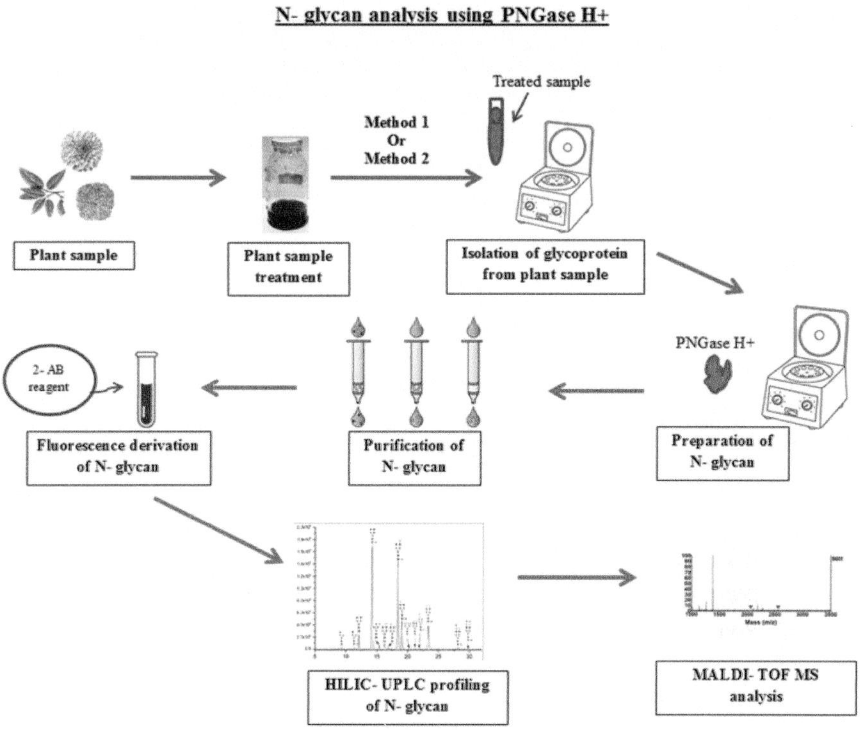

Fig. 6: N-Glycan analysis using PNGase-H+.

glycomic studies are very few in plants. This analysis method (Fig. 6) will help to solve the mystery of glycan and cell biology and also discover the interaction between antibodies and antigens of pathogen and host.

References

1. Yadav, S., D.K. Yadav, N. Yadav and S.M. Paul Khurana. (2015). Plant glycomics: Advances and applications. In: *PlantOmics: The Omics of Plant Science*. Springer India.
2. Bielik, A.M. and J. Zaia. (2010). Historical overview of glycoanalysis. *In*: J. Li. (ed.). *Functional Glycomics, Methods in Molecular Biology* (600). Humana Press.
3. Srisangsung, T., T. Phetphoung, S. Manopwisedjaroen, K. Rattanapisit, C.J.I. Bulaon, A. Thitithanyanont, V. Limprasutr, R. Strasser and W. Phoolcharoen. (2024). The impact of N-glycans on the immune response of plant-produced SARS-CoV-2 RBD-FC proteins. *Biotechnology Reports*, 43: e00847.
4. Bardor, M., G. Cabrera, P.M. Rudd, R.A. Dwek, J.A. Cremata and P. Lerouge. (2006). Analytical strategies to investigate plant N-glycan profiles in the context of plant-made pharmaceuticals. *Current Opinion in Structural Biology*, 16(5): 576–583.
5. Triguero, A., G. Cabrera, J.A. Cremata, T.C. Yuen, J. Wheeler and N.I. Ramírez. (2005). Plant-derived mouse IgG monoclonal antibody fused to KDEL endoplasmic reticulum-

retention signal is N-glycosylated homogeneously throughout the plant with mostly high-mannose-type N-glycans. *Plant Biotechnology Journal*, 3(4): 449–457.
6. Lerouge, P., M. Cabanes-Macheteau, C. Rayon, A.C. Fischette-Lainé, V. Gomord and L. Faye. (1998). N-Glycoprotein biosynthesis in plants: Recent developments and future trends. *Protein Trafficking in Plant Cells*, 38: 31–48.
7. Kornfeld, R. and S. Kornfeld. (1985). Assembly of asparagine-linked oligosaccharides. *Annual Review of Biochemistry*, 54: 631–664 .
8. Strasser, R. (2022). Recent developments in deciphering the biological role of plant complex N-glycans. *Frontiers in Plant Science*, 13: 897549.
9. Kaulfürst-Soboll, H., M. Mertens-Beer, R. Brehler, M. Albert and A. Von Schaewen. (2021). Complex N-glycans are important for normal fruit ripening and seed development in tomato. *Frontiers in Plant Science*, 12: 635962.
10. Higel, F., A. Seidl, F. Sörgel and W. Friess. (2016). N-glycosylation heterogeneity and the influence on structure, function, and pharmacokinetics of monoclonal antibodies and Fc fusion proteins. *European Journal of Pharmaceutics and Biopharmaceutics*, 100: 94–100.
11. Morelle, W., V. Faid, F.Chirat and Jean-Claude Michalski. (2009). Analysis of N- and O-linked glycans from glycoproteins using MALDI-TOF mass spectrometry. Glycomics, 79–92.
12. Fan, J.Q. and Y.C. Lee. (1997). Detailed studies on substrate structure requirements of glycoamidases A and F. *Journal of Biological Chemistry*, 272: 27058–27064.
13. Erika, L., B. Adela, B. Veronika, P. David, B. Jan and Z. Zbynek. (2015). N-Glycome profiling of patatins from different potato species of solanum genus. *Journal of Agricultural and Food Chemistry*, 63(12): 3243–3250.
14. Robbins, P.W., R.B. Trimble, D.F. Wirth, C. Hering, F. Maley, G.F. Maley, R. Das, B.W. Gibson, N. Royal and K. Biemann. (1984). Primary structure of the Streptomyces enzyme endo-beta-N-acetylglucosaminidase H. *The Journal of Biological Chemistry*, 259(12): 7577–7583.
15. Stanley, P., K.W. Moremen, N.E. Lewis, N. Taniguchi and M. Aebi. (2022). N-Glycans. *In*: A. Varki, R.D. Cummings, J.D. Esko et al. (eds.). *Essentials of Glycobiology* [Internet]. (4th Edn.). Cold Spring Harbor (NY): Cold Spring Harbor Laboratory Press; https://www.ncbi.nlm.nih.gov/books/NBK579964/ doi: 10.1101/glycobiology.4e.9.
16. O'Neill, R.A. (1996). Enzymatic release of oligosaccharides from glycoproteins for chromatographic and electrophoretic analysis. *Journal of Chromatography A*, 720(1-2): 201–215.
17. Du, Y.M., T. Xia, X.Q. Gu, T. Wang, H.Y. Ma, J. Voglmeir and L. Liu. (2015). Rapid sample preparation methodology for plant N-Glycan analysis using acid-stable PNGase H+. *Journal of Agricultural and Food Chemistry*, 63: 10550–10555.
18. Bigge, J., T. Patel, J. Bruce, P. Goulding, S. Charles and R. Parekh. (1995). Nonselective and efficient fluorescent labeling of glycans using 2- amino benzamide and anthranilic acid. *Analytical Biochemistry*, 230: 229–238.
19. Zhang, X., C.E. Reed, R.E. Birdsall, Y.Q. Yu and W. Chen. (2020). High-throughput analysis of fluorescently labeled N-glycans derived from biotherapeutics using an automated LC-MS-based solution. *Slas Technology: Translating Life Sciences Innovation*, 25(4): 380–387.
20. Ceroni, A., K. Maass, H. Geyer, R. Geyer, A. Dell and S.M. Haslam. (2008). GlycoWorkbench: A tool for the computer-assisted annotation of mass spectra of glycans. *Journal of Proteome Research*, 7: 1650–1659.
21. Kolarich, D. and F. Altmann. (2000). N-Glycan analysis by matrix-assisted laser desorption/ionization mass spectrometry of electrophoretically separated non mammalian proteins: Application to peanut allergen Ara h 1 and olive pollen allergen Ole e 1. *Analytical Biochemistry*, 285(1): 64–75.

22. Du, Y.M., S.L. Zheng, L. Liu, J. Voglmeir and G. Yedid. (2018). Analysis of N-glycans from *Raphanus sativus* cultivars using PNGase H+. *Journal of Visualized Experiments: JoVE*, 136: e57979.
23. Martial, S., C. Gleysin, T. Ada, B. Carole, L. Jérome, R. Christophe, P.V. Louis, M.A. Aoust, P.M. Rudd, L. Royle, R.A. Dwek, D.J. Harvey, P. Lerouge, J.A. Cremata and M. Bardor. (2008). Plant N-glycan profiling of minute amounts of material. *Analytical Biochemistry*, 379: 66–72.
24. Karg, S.R., A.D. Frey, C. Ferrara, D.K. Streich, P. Uman and P.T. Kallio. (2009). A small-scale method for the preparation of plant N-linked glycans from soluble proteins for analysis by MALDI-TOF mass spectrometry. *Plant Physiology and Biochemistry*, 47: 160–166.
25. Ruiz-May, E., S. Hucko, K.J. Howe, S. Zhang, R.W. Sherwood, T.W. Thannhauser and J.K. Rose. (2014). A comparative study of lectin affinity based plant N-glycoproteome profiling using tomato fruit as a model. *Molecular and Cellular Proteomics*, 13(2): 566–579.

Chapter 4

Glycan Signalling in Environmental Stress Perception and Response

Lekhavani R.,[1] *Suguna K.,*[1] *Ramalakshmi S.,*[1] *Helan Chandra J.,*[2,*] *Arvind Bharani R.S.,*[3] *Sivasubramanian K.*[1] and *Anand Raj L.F.A.*[4]

Glycans are structurally diverse carbohydrates that exist as characteristic components of the cell surface and extracellular matrices and thus are now emerging as competent modulators in the perception and response to environmental stresses. Glycans are acknowledged for their more active role in signalling pathways regulating cellular responses to biotic and abiotic stress. In both plants and animals, glycan patterns are being remodelled in response to diverse stimuli such as invasion by pathogens, drought, and salinity. These modulations are recognised via specific glycan-binding proteins like lectins and pattern-recognition receptors, which trigger complex intracellular signalling cascades to regulate gene expression, immune

[1] Centre for Materials Engineering and Regenerative Medicine, Bharath Institute of Higher Education and Research, Selaiyur, Chennai, Tamil Nadu-600 073, India.
[2] Professor, Ampersand Academy, Ashok Nagar, Chennai, Tamil Nadu-600 083, India.
[3] Research Scientist, Institute of Obstetrics and Gynaecology, Madras Medical College, Chennai, Tamil Nadu-600008, India.
[4] Associate Professor, St. Joseph's College of Engineering, OMR, Chennai, Tamil Nadu-600 119, India.
* Corresponding author: helananand@gmail.com

DOI: 10.1201/9781003635901-4

responses, and stress adaptation mechanisms. This elaborate glycan-mediated communication network connects the external environment with the internal cellular apparatus, ensuring an instantaneous stress response alongside a well-suited contextual environment. The latest developments in glycomics and systems biology have added substantial depth in the area of glycan signalling networks, leading to profound discoveries of new therapeutic targets. The changes in glycans are closely associated with inflammatory diseases, cancer, and other infectious diseases in the field of medical science. New types of therapies, such as glycan-based vaccines, anti-inflammatory agents, and cancer immunotherapies, are being developed by targeting glycan-protein interactions or modifying glycosylation patterns. Glycan signalling pathways can also be used in agriculture to enhance crop environmental resilience. This explains the signalling mechanisms underlying these processes and the emerging therapeutic applications of this knowledge. Understanding glycan signalling significantly expands our knowledge of cellular communication and holds potential for advancing biomedical and environmental resilience strategies.

1. Introduction

Glycans are chains of carbohydrates that are connected to lipids and proteins. Formerly, it was thought that glycans were mainly structural scaffolds that helped to give rigidity to cell walls, fluidity to membranes, and protective layers like animal mucins and simple microbial capsules. This view has been reversed by current glycoscience, which now considers glycans as dynamic elements that are involved in stress recognition, cell signalling and adaptation [17, 18]. Glycan biosynthesis does not have the templated constraints of DNA or proteins; rather, it is highly context-sensitive, reacting to disease and pathogens, metabolic states, environmental stresses, and cellular conditions. Glycans are dynamic molecular barcodes that are constantly modified to indicate the internal and external environment of the cell [6].

The class and structure of monosaccharide residues, the structure and localisation of glycosidic bonds, and the nature of the non-glycan part (the aglycone) to which it is connected all define the fundamental structure of a glycan [23]. O-linked glycans connect an amino acid (serine or threonine) to a monosaccharide, whereas N-linked glycans connect an amino acid (asparagine) to a monosaccharide [77]. Recent advances in the study of glycans have disclosed a wealth of structural diversity and functional significance of glycan molecules. Many practical machine learning designs, such as CandyCrunch can now define glycan structures from raw Liquid chromatography/mass spectrometry reports with almost complete accuracy. Besides this, cryo-electron microscopy (cryo-EM) techniques have been recently released, allowing us to view glycan conformations at near atomistic resolution [76]. Further progress has been made in codifying and interpreting glycomic data with tools like UniCarb-DB. Glycans are not only important structural components, but they also serve as important signalling molecules for immune response modulation, cell recognition, and environmental adaptation [81].

The glycan structural complexity allows us to look for coordinated explanations to explain how fungi and bacteria control the relationships between microbes and plants in terms of both influencing positive symbiotic relationships and pathogenic interactions [70, 27]. Glycosylation also impacts how animal immune systems respond to stress [12]. Glycans serve as sensors, messengers, and modulators during stress adaptation by performing chemical recognition, signal transduction, and regulatory feedback (Fig. 1). Thus, this chapter discusses the types of glycans, the effect of structural change of glycans to cellular signalling; environmental stress; glycan-mediated signal transduction mechanisms; and the nearly linked developments in functional glycomics by presenting case studies across various model systems. Ultimately, the multifarious potential for glycans' application and future research opportunities is explored.

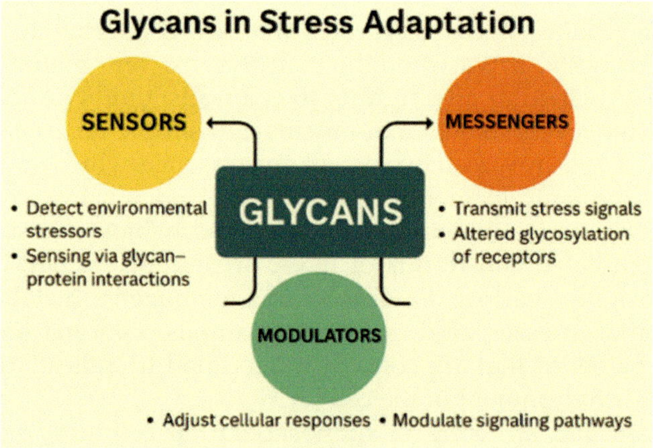

Fig. 1: Roles of Glycans in cellular stress adaptation.

2. Structural Units of Glycans

In all the stages of life, glycans, the complex carbohydrates, demonstrate an incredible diversity of structures and biological significance. Based on their attachment to proteins or lipids, they may be either N-linked or O-linked glycans, or Proteoglycans or glycolipids. N-linked-glycans generally attach to the nitrogen atom of the side chain of asparagine residues, while O-linked glycans are said to attach to the hydroxyl group of threonine or serine residues [69]. The section that follows gives an overview of the different types of glycans as indicated in Fig. 2.

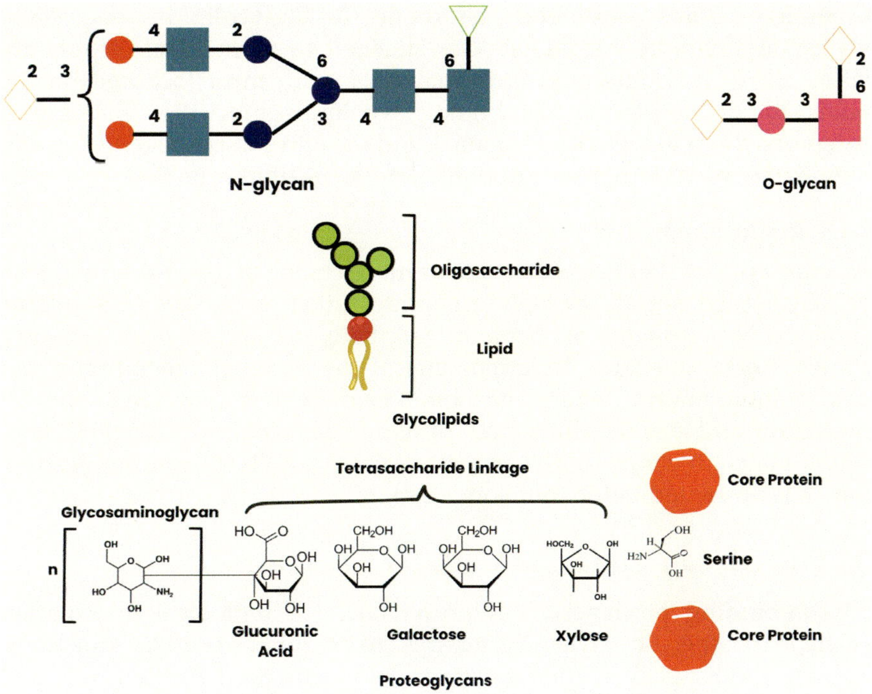

Fig. 2: Structure and major classes of Glycoconjugates.

2.1 *Types of Glycans*

2.1.1 N-linked glycans

The assembly of N-linked glycans begins in the endoplasmic reticulum (ER) when a preassembled oligosaccharide binds to the consensus sequence (Asn-X-Ser/Thr) amide nitrogen atom of asparagine residues. These modifications by the Golgi apparatus lead to the generation of complex, hybrid, or high-mannose N-glycan structures. These modifications are important in protein folding, quality control, and trafficking [48].

2.1.2 O-linked Glycans

The O-linked glycans are made solely within the Golgi apparatus and generally attach to the hydroxyl group of either serine or threonine residues. O-glycosylation is still more different than N-glycans as it does not require any pattern sequence. O-glycans also function in immune recognition, cell adhesion, and mucin secretion [30]. They may be as straightforward as pure monosaccharide add-ons or even as intricate as branching molecules, which often terminate with sialic acid or fucose [51].

2.1.3 Glycolipids

Ceramide lipids are the backbone of glycolipids, especially glycosphingolipids, and sugars are attached to them. These lipids are abundant in neuroglia and epithelial cells, with most localised on the outer side of the plasma membrane. Glycolipids affect the clustering of receptors, the transduction of signals, and the entry of pathogens through membrane microdomains, commonly known as lipid rafts [34].

2.1.4 Proteoglycans

A central protein is attached to one or more chains of glycosaminoglycan (GAG), which are linear polysaccharides such as dermatan sulphate, chondroitin sulphate, or heparan sulphate, to create proteoglycans. Proteoglycans are crucial components of the basement membrane and extracellular matrix, where they bind growth hormones, regulate cell migration and proliferation, and provide structural support [58]. One of the main functions of glycans is to facilitate cell-cell communication, which is presented in the following section.

3. The Glycan Code and Cellular Communication

Glycan-binding proteins can interpret encoded biological information carried by specific glycan structures to regulate cellular functions. This is a theory called the "glycan code". Compared to the linear and deterministic genetic code, the glycan code is fluid, branching and context-dependent; it reacts to cellular metabolome, substrate status, and expression levels of glycosylation enzymes. This non-genetic, biochemical language is also a mode of cell communication [16]. During infection, pathogens such as the influenza virus or *Helicobacter pylori* would bind to specific glycans of the host for adhesion and entry. In some cancers, abnormal glycosylation (such as the enhancement of sialylation and shorter O-glycans) affords an escape from immune recognition via interfering with how siglecs and galectins recognise their ligands, and thereby promote metastasis [80]. While we describe environmental stress in affiliation to this point further on, it is noted here that environmental stresses can have the potential to affect cellular signalling, protein folding, and immune signalling.

3.1 Environmental Stressors and Glycan Response Modulation

Environmental stressors are implicated in inducing changes in multiple aspects of cellular homeostasis and indecisively altering a number of biochemical and molecular signatures. While there has been a long history of research into stress-driven changes in DNA and proteins, there has been very little work conducted with respect to cellular glycomes. Instead of being structural components that are static or relaxed, glycans are functional biomolecules that participate in signalling transduction, organismal adaptability, and cellular defence owing to their complex and dynamic architecture. There is newly gained knowledge that programmed glycan architecture is readily modifiable by environmental perturbation and that this can have real consequences for pathophysiology, immunomodulation, and cellular viability. Because the glycome and glycosylation pathways are sensitive to environmental perturbations, the glycome also represents a prominent area in which to study environmental stress response, particularly from a systems biology perspective [26]. The following are the categories for the types of environmental stress.

3.1.1 Abiotic Stress

Biotic stress, including extreme biotic agents (fungi, bacteria, viruses, animals), results in drastic changes in glycosylation processes in both plant and animal systems, while abiotic stressors like UV radiation, salts, droughts, and extreme temperature also modify the glycosylation processes. Glycans influence stress-resistance fitness and the experience of stress because glycan shapes can alter protein conformation, cell wall structure, membrane stability, and even signalling. Since they have protective and supportive roles as polypeptides, glycans are also unique mediators of stress signalling cascades. Glycan remodelling, perhaps, modifies an organism's response to external stimuli, the stress response of immune systems, and subsequent developmental patterns.

3.1.2 Oxidative Stress

Oxidative stress happens when there is an imbalance between reactive oxygen species (ROS) and antioxidants. This imbalance can change how glycans are structured and how they work [33]. Both glycoproteins and the enzymes that help with glycosylation can be affected by ROS. When ROS levels go up, it throws off the balance in the ER, making it harder to form disulphide bonds and fold glycoproteins. When proteins do not fold right and do not have enough glycosylation, they pile up. This triggers a response to unfolded proteins, which can either lead to cell death or help restore balance. In cells of mammals, oxidative stress changes core fucosylation patterns and reduces the activity of sialyltransferases. These changes affect how cells talk to each other, how tissues heal, and how the immune system recognises things. This could play a role in how cancer and long-lasting inflammatory diseases develop [29, 60, 28].

3.1.3 Biotic Stress

Biotic stress from germs, viruses, fungi, and plant-eaters leads to big shifts in the host's glycome. Glycans play many roles during biotic stress, such as tweaking the immune system, forming physical barriers, and sending chemical signals. When under attack or infected, hosts change their glycosylation patterns as part of their built-in and learned defences.

3.1.4 Pathogen Recognition

Glycans help with the first line of defence: spotting invaders [25]. Host Pattern Recognition Receptors (HPRRs), which are often glycoproteins themselves, spot Pathogen Associated Molecular Patterns (PAMPs) like bacterial lipopolysaccharides or fungal chitin [56]. How well these receptors are glycosylated affects their ability to bind to targets and send signals. For instance, plants kick off PAMP Triggered Immunity (PTI) when receptor-like kinases have the right N-glycosylation. To weaken immunity, invaders often target the host's glycosylation machinery [5]. Some bacteria release enzymes that break down protective glycans on mucus layers or cell walls. At the same time, viruses hijack host glycosylation pathways to make glycoproteins that look like 'self' and dodge immune detection. In cases of biotic stress, the result of an arms race causes the host to change its glycosylation patterns quite rapidly [78, 24]. The structural changes of glycans due to the influence of environmental stress, as well as a description of the mechanisms of action, are summarised in the table below (Table 1). In addition to the structural changes of glycans due to environmental stress and several other factors to use in some kind of defence response, signalling is a major function of the glycan.

3.2 Mechanisms of Glycan-mediated Signal Transduction

Glycans have many functions, but perhaps most importantly, they can also work as biochemical signals in the process of signal transduction. As signal transducing biochemical signals, glycans form a network of signals that can regulate a variety of signalling cascades, especially in the context of stress responses and immune surveillance [32]. Glycans, as signals, may be added to proteins or lipids to form glycoproteins and glycolipids, respectively. The processes of glycan signal transduction are complex and involve, but are not limited to, intracellular trafficking, enzymatic alterations of glycosylation-specific patterns, and specific interactions with glycan-binding proteins [35]. The Glycan signal transduction mechanism is depicted in Fig. 3.

3.2.1 Glycan-Lectin Interactions during Stress Signalling and Adaptation

Lectins are carbohydrate-recognising and -binding proteins with an important role in glycan-mediated signalling. They can initiate cell

Table 1: Mechanistic action of glycan to environmental changes.

S. No.	Environmental Stressor	Mechanism of Action	Glycan Alterations	Associated Pathological Outcomes	References
1.	Oxidative Stress	ROS-mediated damage to glycosylation enzymes	↓ Sialylation, altered fucosylation, increased neo-epitopes	Chronic inflammation, autoimmunity, and cardiovascular disease	[29]
2.	Temperature Extremes	Modulation of glycosyltransferase activity	Hypothermia: ↓ glycan complexity; Hyperthermia: ↑ glycan branching	Membrane instability, impaired receptor signalling, stress-induced apoptosis	[60]
3.	Pathogen Exposure	Pathogen-driven hijacking of host glycome	Host glycan mimicry, overexpression of pathogen-binding glycans (e.g., sialic acid, heparan sulfate)	Immune evasion, enhanced infectivity, persistent infections	[78]
4.	Chemical Pollutants	ER stress, inhibition of glycan biosynthesis pathways	Misfolded glycoproteins, truncated glycans, aberrant glycan structures	Fibrosis, metabolic disorders, carcinogenesis	[1]
5.	UV Radiation	DNA and protein damage inducing stress responses	Dysregulation of glycosyltransferase gene expression	Premature skin ageing, mutagenesis, cancer development	(Jaillet et al., 2017)
6.	Hypoxia (Low Oxygen)	Disruption of oxygen-dependent glycosylation steps	↓ O-glycan elongation, accumulation of immature glycans	Tumour progression, angiogenesis, metastasis	[74]

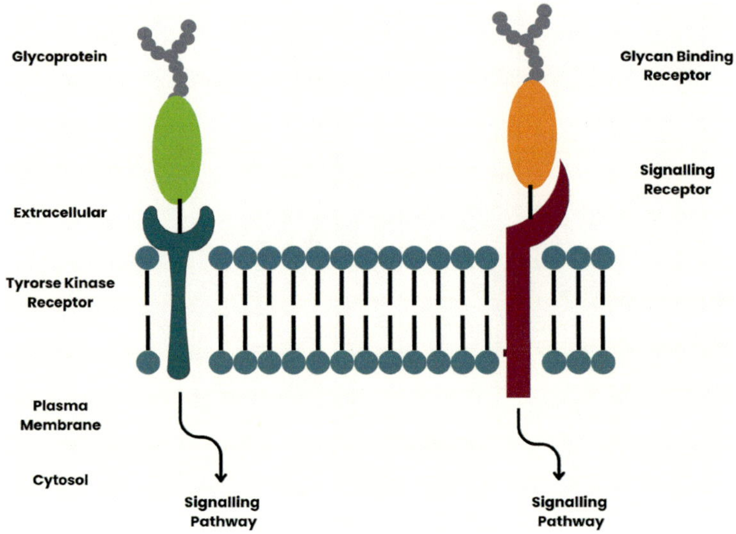

Fig. 3: Glycan-mediated receptor signalling mechanisms.

signalling pathways, i.e., Mitogen-Activated Protein Kinase (MAPK) and Nuclear Factor kappa B (NF-κB), involved in regulating inflammation, apoptosis, and responses to stress [83]. The binding of glycan structures on the cell surface to lectins will determine how cells will respond to external stimuli, including infection or environmental changes, by potentially altering gene expression [41]. In addition to initiating signalling pathways, glycans allow the cell to respond to constant stress by modifying the glycosylation pattern of glycoproteins and glycolipids. Glycan-lectin binding is critical to regulate many cellular processes, and some lectins initiate the aggregation of glycoproteins and glycolipids at the cell surface, which results in cellular changes that activate signalling cascades downstream [83]. The binding can be specific to major pathways such as NF-κB and MAPK that regulate numerous responses, including differentiation, apoptosis, and inflammation. Therefore, the signalling outcomes are determined by the type of cell, lectin, and glycan structures involved. In other words, this is a context dependent phenomenon.

3.2.2 Glycosylation and Intracellular Trafficking and Disease Implications

Our understanding of the pathway-modifying influence of glycan-lectin interactions creates a foundation upon which to functionally exploit these entities as potential therapeutic targets as well as providing us with valuable insight into host-pathogen relationships, cancer development and innate and adaptive immune responses [38]. Glycosylation, an enzyme-mediated modification of proteins and lipids with carbohydrate groups, is not only a normal physiological response in animals but an essential feedback response in the sensation and adaptation to stress. Glycosylation is dynamically regulated after cellular stress [11]. Cells will undergo alterations in glycosylation patterns, including glycan branching, sialylation and fucosylation, in direct response to environmental conditions such as oxidative stress, infection, or inflammation. Changes in the structure, location, and function of glycoproteins can occur, particularly when the glycoproteins are involved in important signalling pathways and microbial immune recognition [57]. Changes in glycosylation will ultimately alter the intracellular signalling circuitry and transcriptional responses of a cell due to some functional modification of the activity of stress sensors and receptors as the feedback signal [2].

3.3 Receptor Function and Glycan Remodelling

N-glycan changes accompanying their membrane receptors are liable to interrupt possible engagement with ligands and lectins, which in turn can participate in many critical processes, e.g., inflammation and the Unfolded Protein Response (UPR) [1]. This pathway between stress signalling and glycosylation shows how crucial glycan remodelling is when it comes to cell behaviour in connection with stress and affects diseases and consequent therapies [50]. The glycosidases and glycosyltransferases are the two major classes of enzymes that mediate glycan remodelling. Glycosyltransferases add specific monosaccharides to form different glycan chains with differences in structure, whereas glycosidases help in glycan maturation by removing sugar residues. Altogether, these enzymes allow the direct response of the glycome to stress and other signals, being vital for immune responses, cellular trafficking, and communication. Their pathophysiology relates to various conditions, including cancer and autoimmune disorders, making these enzymes an attractive therapeutic target [68]. Under stressful conditions, the glycan structure on the surface of the cell is a major regulator that controls many processes in cells, such as transport and endocytosis.

4. Model Systems and Case Studies

Glycosylation plays dynamic roles in cellular pathology and function [39]. Numerous case studies and model systems have encouraged the study of glycan-mediated processes under stress [3]. *Caenorhabditis elegans* is one of the extensively studied models in which glycosylation enzymes such as glycosyltransferases have been genetically altered to examine their effect on longevity, stress resistance, and development [53]. Correspondingly, mice in which certain glycosylation enzymes are overexpressed or knocked out. Glycan remodelling has been shown to play important roles in inflammation, tumour development, and immune defence (ST6GAL1 or MGAT5). Endoplasmic reticulum stress or hypoxia exposure has been shown to cause certain glycosylation changes in human cell lines, influencing signal transduction and protein trafficking. Specifically of note is how glycan modification resulting from stress can affect host-pathogen relationships, for instance, by employing the infection models for the influenza virus, which make use of the host's glycan receptors as a platform for virus entry [45]. Moreover, tumour-derived cell lines are also choice reagents with which one may investigate aberrant glycosylations' contribution towards immune escape and metastasis. These model systems, ranging from simple organisms to sophisticated mammalian models, inform fundamental biological knowledge and the development of novel therapies. They offer invaluable systems for understanding the mechanisms and consequences of glycan remodelling under stress [85].

4.1 Plants: Extensins and Arabinogalactan Proteins (AgPs) in Abiotic Stress Tolerance

Arabinogalactan Proteins (AgPs) and extensions are two examples of glycoproteins attached to the cell wall and are essential for the mediation of abiotic stress response in plants [6]. These hydroxyproline-rich, highly glycosylated glycoproteins play a crucial role in maintaining plant cell wall structural integrity as well as their signalling capability [37]. Apart from their established functions in signal transduction, cell differentiation, and proliferation, AgPs glycan moieties could also interact with cell surface receptors or engage in complex formations that regulate calcium signalling under stress conditions [61]. The induction of AgPs and their glycosylation patterns is frequently triggered by drought, salinity, and cold, underscoring their role in sensing and responding to stress. Extensions, however, are distinguished by their extensive O-glycosylation and the presence of repeating Ser-Pro4 motifs, which enable them to form cross-linked networks in the cell wall. This cross-linking provides cells with protection against mechanical stress and desiccation damage, as well as stiffening the cell wall.

AgPs (Arabinogalactan Proteins) and extensions are believed to play significant roles in regulating cell wall plasticity, maintaining ion balance, and influencing stress-related gene expression [59]. Their sensitivity to environmental conditions and adaptability indicate that they are crucial for helping plants endure abiotic stress. Hence, these proteins are potential targets for crop resistance enhancement strategies.

4.2 Bacteria: Antibiotic or Nutrient Deprivation-Induced LPS Modifications

Gram-negative bacteria possessing an outer membrane of Lipo-Poly-Saccharides (LPS) also acts as a determinant of immune evasion, environmental adaptation, and membrane stability. Bacteria transiently modify their LPS structures for survival improvement when exposed to antibiotic stress or starvation. These alterations tend to be directed against the O-antigen, the core oligosaccharide or lipid A moiety. For instance, the negative charge of lipids is decreased when 4-Amino-4-deoxy-L-arabinose (L-Ara4N) or phospho-Ethanolamine (pEtN) is added, lowering the affinity for cationic antimicrobial peptides and polymyxin antibiotics. Thus, these proteins are promising candidates for approaches to enhance crop resilience.

4.3 *Bacteria: LPS Modifications under Antibiotics or Nutrient Deprivation*

Adaptive LPS changes are usually modulated by two-component sensors like PhoP/PhoQ or PmrA/PmrB, which are involved in perceiving environmental signals and controlling the correct transcriptional response. In addition to providing insights into bacterial stress response mechanisms, knowledge of LPS remodelling during stress is relevant to the development of new antimicrobial strategies targeting the biosynthesis and regulation of LPS [36].

4.4 *Fungi: Glycoprotein Changes in Thermal and Osmotic Adaptation*

Fungi are subjected to fluctuating environmental conditions, including osmotic and heat stress, and need robust adaptive mechanisms to survive and operate. Modification of the glycoproteins—protein covalently bonded to carbohydrate moieties—and predominantly located in the cell wall and plasma membrane is one such crucial element engaged in these adaptations [75]. These glycoproteins are implicated in cell wall strength maintenance, signal transduction, and control of the interaction with the outside world. During osmotic stress, for example, desiccation or high salt, fungi turn on signalling pathways such as the High Osmolarity Glycerol (HOG) pathway, wherein glycoproteins act as osmosensors that trigger protective mechanisms, including cell wall remodelling and glycerol accumulation [8]. During heat stress, glycoproteins are structurally and functionally modified, including modifications in glycosylation patterns. These modifications stabilise membrane proteins and allow for correct protein folding in the endoplasmic reticulum [54]. Consequently, they increase the stress resistance of the cell by maintaining membrane integrity and inhibiting protein aggregation. Furthermore, the glycosylation machinery, such as mannosyl transferases and glucosidases, is frequently controlled under stress conditions. This points to the dynamic nature of glycoprotein biosynthesis in stress adaptation [49]. Glycoprotein modification offers useful information about fungal physiology and could prove valuable in the identification of candidate antifungal compounds and targets for industrial strain improvement.

4.5 Mammals: The Role of Glycan Shifts in Immune Cell Stress Responses

Glycans, structurally heterogeneous carbohydrates attached to proteins and lipids, play a crucial role in regulating immune responses. Oxidative stress, inflammation, and infection repeatedly alter the glycan composition of mammalian immune cell surfaces. These alterations in glycan **Regulate Cell–Cell Interaction:** Alterations in the glycosylation pattern affect immune cells' mode of interaction with pathogens and between cells.The immune signal transduction pathways are activated or inhibited by specific glycan forms through lectin-glycan interactions [36]. Stress-induced modifications may change the dynamic that allows pathogens to take advantage of or be recognised by host glycan forms. **Modulate Immune Cell Trafficking:** Glycosylation controls immune cell migration into sites of injury or inflammation. Getting to know such glycan alterations is needed for controlling the immune system in stress conditions and for aiding cancer infection therapy, as well as therapy of autoimmune disorders [42].

As stated earlier, Glycans are multi-functional carbohydrate molecules with various functions and, as such, have multiple applications in medicine and biotechnology, including vaccine preparation, cancer therapy, and drug targeting [40]. They are also used for disease diagnosis through biomarkers and glycan-based biosensors.

5. Applications of Glycans

5.1 Glycans in Vaccines

5.1.1 Microbial Vaccines

In general, polysaccharide vaccines that consist solely of glycan constituents induce poor immunity, especially in infants [19]. Glycans are not effective at inducing T-helper-dependent activation and class switching of B-cell-mediated immunity since they are T-cell-independent antigens [13]. Conjugate vaccines, which incorporate glycans along with carrier proteins, have proven to be effective. There are presently three conjugate vaccines available in the market as of now: Type B (*Haemophilus influenzae*) induces acute lower respiratory infection in children who are a high-risk group for H. infection due to influenza Type B. As a result, Capsular-Polysaccharides (CPs) protein conjugate vaccine preparations replaced the Hib polysaccharide vaccine that entered the market first in 1985, when in 1988 it was removed from the market [21, 66]. Routine immunisation regimens involve a conjugated derivative of a Hib-derived oligosaccharide attached to a protein carrier. The method has been so successful that infectious diseases from this bacterium have been virtually eradicated from vaccinated subjects [63]. QuimiHib, the first semisynthetic glycoconjugate vaccine marketed in Cuba, comprises glycan chains with an average length of 16 monosaccharide residues [62]. Today, two products of pneumococcal conjugate vaccines have been developed, the 10-valent product Synflorix by GlaxoSmithKline, and the 13-valent product Prevnar13 from Pfizer. Such vaccines have come in to guard against a larger number of serotypes. The highest-selling product is Prevnar 13, which captured six billion dollars in sales revenue in 2019, guarding against serotypes causing over 70% of invasive pneumococcal diseases globally [4]. The market is also witnessing success with conjugate vaccines against *Neisseria meningitidis* [55]. Vaccination is needed to avoid meningococcal infections because of their fast progression and onset. Different conjugated capsular polysaccharide (CPS) vaccines are available in various nations. This comprises some monovalent vaccines that are prepared from serogroup-C CPS (e.g., Meningitec, Menjugate, and NeisVac-C) as well as tetravalent vaccines for serogroups A, C, W, and Y (e.g., Menactra, Menveo, and Nimenrix) [14]. There are also combination vaccines for meningococcal serogroups C and Y (MenHibrix) and serogroup C (Menitorix) and *Haemophilus influenzae* Type b (Hib). In the African sub-Saharan meningitis belt, a highly used monovalent serogroup A vaccine, MenAfriVac, is used [10]. To avoid bacterial infections in the elderly and children, some new vaccines based on synthetic oligosaccharide antigens are under development. Vaccines for nosocomial infections against antibiotic-resistant *Clostridium difficile* and *Klebsiella pneumoniae* are under preclinical investigation [15].

5.2 Cancer Vaccines

Numerous cancer vaccines based on carbohydrates are in various stages of development. Researchers are investigating ganglioside immunogens found on specific cancer cell types, such as Globo H in breast cancer and gangliosides GM2 and GD2 in melanomas [67]. In the past two decades, there has been minimal advancement in treatment approaches targeting the shorter glycan sequences on cancer mucins, such as sialyl-Tn (sialylα2-6GalNAcα-). Despite reaching Phase 3 clinical trials, the synthetic Globo H hexasaccharide, which resembles the antigen for breast and prostate cancer, was not approved for sale [9]. Clinical trials on humans are still being conducted. Preclinical studies are currently investigating the use of humanised anti-glycan antibodies (passive immunisation) in place of active immunisation of immune-compromised patients [31]. Unlike glycomimetic, bio mimics (Natural carbohydrates) from pathogens are engineered to improve qualities like stability and conformation, enhancing antigens' property to elicit an immune response [65]. Vaccines based on glycans are designed to create immunity against particular infections. In contrast, glycomimetics can be applied to several tasks, such as preventing the attachment of pathogens or creating novel treatments [73].

6. Therapeutic Applications of Glycomimetics

Glycomimetics are synthetic carbohydrate structures that mimic the action of glycans, possibly for therapeutic purposes. Glycans are used in vaccines to stimulate an immune response against particular bacterial or viral antigens. To increase immunogenicity, glycans are joined to a protein carrier in glycoconjugate structures, which are used in current vaccines. Conversely, Glycomimetics can prevent bacteria and toxins from adhering to cell surfaces in the first place by imitating the glycan ligands they are familiar with [20]. Glycomics provides avenues for biology, chemistry, and medicine. Using iminosugars, González-Cuesta et al. [44] constructed human glucocerebrosidase inhibitors acting as picomolar chaperones for Gaucher disease, which is one of the most prominent lysosomal storage disorders. The hallmark of this class of uncommon genetic disorders is a lack of glycosidases, which are necessary for the lysosome's breakdown of glycosphingolipids. The foundation of pharmacological chaperone therapy is reversible inhibitors that can alter the three-dimensional structure of the catalytically active but misfolded glycosidases implicated in lysosomal storage disorders. Because of this interaction, they are not broken down by the endoplasmic reticulum-associated degradation (ERAD) pathway prior to being transported to lysosomes.

6.1 Glycomimetics in Blocking Infections

As was previously mentioned, various bacteria and poisons attach themselves to the tissues of mammals by identifying particular glycan ligands [74]. Therefore, it is possible to prevent or suppress infection by using small soluble glycans or glycan mimetics to stop microorganisms and toxins from initially adhering to cell surfaces (or from releasing them) [79]. Glycan-based medications can be administered directly rather than systemically because many of these organisms naturally enter through the gut or airways. For example, infants' intestinal infections are naturally repelled by milk oligosaccharides [46]. Glycosylated polymers prevent influenza and other viruses from binding. Despite having a solid scientific foundation and solid *in vitro* research, these 'antiadhesive' or 'mimicry' treatments have not yet seen widespread use [43].

7. Transfusion and Transplantation Rejection by Antiglycan Antibodies

Many glycans as a determinant can act as barriers to organ transplantation and blood transfusion. *In vitro,* bacterial enzymes can be utilised to transform A and B red blood cells to universal donor O RBC by removing their blood group determinants [7]. A similar problem arises with xenotransplantation. An active solution to the patient shortage of human organs is organ transplantation between species. Pigs are the preferred animal donors because many of their organs are similar to human organs in size, physiology, and anatomy [20]. The terminal 'α-Gal' epitope on glycoproteins and glycolipids is produced by pigs and the majority of other mammals, in contrast to humans and some other primates. Due to a reaction between the antibodies and blood vessel endothelial cells, human blood contains naturally occurring high-titer antibodies that are directed against this epitope, which causes hyperacute rejection of pig organ transplants [52]. Attempts to stop this reaction include blocking the interaction by injecting soluble competing oligosaccharides and filtering blood over glycan affinity columns to eliminate xenoreactive antibodies [71].

Additionally, transgenic pigs with an excess of complement-controlling proteins on their cell surfaces and pigs lacking the reactive epitope have been created. High concentrations of nonhuman sialic acid (Neu5Gc), which most people have antibodies against, are also found in pig organs. Even if this issue is resolved, additional structural variations in glycans and proteins between humans and pigs lead to later stages of graft rejection, which calls for immunosuppression [72]. Table 2 provides information on the drugs which are based on glycans, targeted diseases, and their mode of action.

8. Challenges and Opportunities of Future Research

Glycan research, or glycomics, is the most challenging field in biology due to structural complexity and variation. The factors that contribute to this complexity are the absence of standard methods of analysis and large-scale databases, which considerably restrict the understanding of glycan functions and their interactions. In particular, the lack of template meaning because the process is not directly template-driven leads to a dynamic and contextually dependent glycome that changes with cell type and environmental conditions. These issues are essentially hurdles for glycan-based diagnostic and therapeutic works [64].

However, despite these challenges, glycomics possess much promise for future scientific and medical breakthroughs. Techniques in analysing, such as mass spectrometry and glycan microarrays, are making great strides in understanding the structures and functions of glycans. The combination of glycomics into all other omics endeavours, such as genomics and proteomics, takes biological systems into an integrative whole [84]. In addition, developing glycan-based biomarkers and therapeutics could be a promising way toward personalised medicine, especially in cancer and infectious diseases. Glycoproteins constitute the basis of many FDA-approved protein-based cancer biomarkers, such as the CA 15-3 antigen for breast cancer, CA 19-9 for pancreatic cancer, and the CA 125 antigen for ovarian cancer [47]. These antigens' ability to detect cancer can be enhanced by researching their glycosylation pattern. Furthermore, research on glycans, glycomics, and glyconanoparticles will soon form the core of biomedical innovations [32]. As research continues to identify and describe more of the glycome's workings, glycomics will soon form the core of biomedical innovation.

9. Conclusion

Glycans have a number of significant functions in cellular development and differentiation, along with how cells react to their surroundings. Certain glycan alterations are also directly associated with a number of disorders. They play a role both in biotic and abiotic stress pathways as modulators and sensitive molecular sensors. They function at the interface between cell and environment, and involvement includes adaptive response, pathogen identification, immunological surveillance, and signal transduction. Furthermore, glycomimetics display better stability and interesting therapeutic uses compared to their natural counterparts. Typical examples of drugs that exclusively inhibit neuraminidase are Oseltamivir and Zanamivir, which are used to treat and prevent influenza virus infections. Glyco-nanotechnology is improving in specificity and effectiveness in biomedical applications through the use of nano systems by simulating natural glycan interactions. They are involved in diagnostics, targeted drug delivery, and vaccine development. Glycan-functionalised nanoparticles have been applied to vaccine development, targeted medication delivery, and diagnostics. In addition, there are strong disease markers in each of the three types of glycoproteins, proteoglycans, and glycolipids. Integration of glycans with multi-omics study, Artificial Intelligence, and CRISPR-based gene editing platforms accelerates the improvement of biomarkers and novel therapeutic agents. Upcoming innovations on glycans, glycomics, and glyco-nanotechnology provide the cutting edge in biotechnology and medicine, which may help

Table 2: Glycan-based drugs, their target diseases, and modes of action.

Type of Drug	Source of Drug	Disease	Mechanism of Action	References
Sialic acid-based targeting				
Zanamivir (Relenza)	Biota/GlaxoSmithKline	Influenza Type A and B	Inhibition of neuraminidase	(Gao et al. 2024)
Oseltamivir (Tamiflu)	Roche/Gilead	chemoprophylaxis	Inhibition of neuraminidase	(Gao et al. 2024)
Targeting glycosaminoglycans				
Heparin	Several brands	Anticoagulant; potential benefit in preventing cancer metastases	It inhibits heparanase and selectins, activates antithrombin, and prevents growth factors and heparan sulfate from interacting	[9]
Hyaluronan	Several brands	Plastic surgery; osteoarthritis; ocular surgery	Anti-inflammatory agent; tissue space filler	[77]
Laronidase	Genzyme	α-idu-ronidase deficiency; mucopolysaccharidosis type I (MPSI)	Enzyme replacement therapy	(Perez-Lopez 2017)
Galsulfase	Biomarin	Arylsulfatase B deficiency; mucopolysaccharidosis	Enzyme replacement therapy	[19]
hyaluronidase	Halozyme	Being developed as a supplement to chemotherapy for cancer; *in vitro* fertilisation	Breaks down hyaluronidase surrounding oocytes to enhance fertilisation; breaks down hyaluronidase in tumours to lower intratumor pressure	[65]
ALZ-801	In phase III trials by Alzheon, Inc.	Alzheimer's disease, amyloid disorders, and maybe other amyloidoses	Binds to the amyloid plaque and prevents its production	[5]

Table 2 contd. ...

...Table 2 contd.

Type of Drug	Source of Drug	Disease	Mechanism of Action	References
Targeting glycosphingolipids				
N-butyl-deownxynojirimycin	Acetelion	Niemann-Pick's disease Type C, late-onset Tay-Sach's disease, Type 1 Gaucher's disease, Type 3 Gaucher's disease	Substrate reduction therapy that inhibits glucosylceramide synthase	(Mueed et al. 2021)
imiglucerase (Cerezyme)	Genzyme	Type 1 Gaucher's disease	Enzyme replacement therapy	(Andersson et al. 2005)
β-agalsidase	Genzyme	α-galactosidase A deficiency; Fabry disease	Enzyme replacement therapy	(Wraith et al. 2008)
Others				
acarbose	Bayer	Diabetes (Type 2)	Inhibits intestinal α-glucosidases, which aids in the digestion of dietary glycans	[22]
Alglucosidase alfa	Genzyme	α-glucosidase A deficiency; Pompe's disease	Enzyme replacement therapy	(Diaz-Manera et al. 2021)
allosamidin	In large-scale research	Insecticide	Inhibitor of chitinase	(Suzuki et al. 2008)

to control stress reactions in order to maintain a sustainable and healthy life.

Acknowledgement: The authors express sincere gratitude to the editors for their kind invitation and guidance, and to all colleagues and collaborators whose insights and support contributed to the completion of this chapter.

Ethical Statement: The authors declare no ethical concerns related to the content presented in this manuscript.

References

1. Adeniyi, M., C.D. Gutierrez Reyes, J. Chávez-Reyes, B.A. Marichal-Cancino, J. Solomon, M. Fowowe and Y. Mechref. (2024). Serum N-Glycan changes in rats chronically exposed to glyphosate-based herbicides. *Biomolecules*, 14(9): 1077.
2. Agard, N.J., C.R. Bertozzi and J.A. Prescher. (2020). Glycans in cancer and inflammation: Potential for therapeutics and diagnostics. *Nature Reviews Immunology*, 20(7): 402–416. https://doi.org/10.1038/s41577-020-0308-7.
3. Akune-Taylor, Y., A. Kon and K. Aoki-Kinoshita. (2024). *In silico* simulation of glycosylation and related pathways. *Analytical and Bioanalytical Chemistry*, 416(16): 3687–3696. https://doi.org/10.1007/s00216-024-05331-8.
4. Alderson, M.R., V. Sethna, L.C. Newhouse, S. Lamola and R. Dhere. (2021). Development strategy and lessons learned for a 10-valent pneumococcal conjugate vaccine. *Human Vaccines &Immunotherapeutics*, 17(8): 2670–2677.
5. Alzheon, Inc. (2023). Alzheon announces early completion of enrollment of APOLLOE4 Phase 3 clinical trial evaluating oral ALZ-801 (valiltramiprosate) in APOE4/4 homozygotes with early Alzheimer's Disease.
6. Ardá, A. and J. Jiménez-Barbero. (2018). The recognition of glycans by protein receptors. Insights from NMR spectroscopy. *Chemical Communications*, 54(38): 4761–4769.
7. Arthur, C.M., M. Hollenhorst, S.C. Wu, R. Jajosky, H. Nakahara, H.M. Jan, L. Zheng, M. Covington, S. Rakoff-Nahoum, M. Yeung, W. Lane, C. Josephson, R.D. Cummings and S.R. Stowell. (2024, July). ABO blood groups and galectins: Implications in transfusion medicine and innate immunity. In: *Seminars in Immunology*, 74: 101892. Academic Press.
8. Ball, B., A. Sukumaran, S. Pladwig, S. Kazi, N. Chan, E. Honeywell, M. Modrakova and J. Geddes-McAlister. (2024). Proteome signatures reveal homeostatic and adaptive oxidative responses by a putative co-chaperone, Wos2, to influence fungal virulence determinants in cryptococcosis. *Microbiology Spectrum*, 12(8). https://doi.org/10.1128/spectrum.00152-24.
9. Bendas, G. and L. Borsig. (2020). Heparanase in cancer metastasis–heparin as a potential inhibitor of cell adhesion molecules. *Heparanase: From Basic Research to Clinical Applications*, pp. 309–329. Elsevier.
10. Berti, F., M.R. Romano, F. Micoli and R. Adamo. (2021). Carbohydrate-based meningococcal vaccines: Past and present overview. *Glycoconjugate Journal*, 38(4): 401–409.
11. Bindeman, W.E. and B. Fingleton. (2021). Glycosylation as a regulator of site-specific metastasis. *Cancer and Metastasis Reviews*, 41(1): 107–129. https://doi.org/10.1007/s10555-021-10015-1.
12. Collin, A. and A. Daszkowska-Golec. (2025). Navigating plant stress response: From nucleus to peroxisomes. *Trends in Plant Science*. https://doi.org/10.1016/j.tplants.2025.03.016.

13. Del Bino, L., K.E. Østerlid, D.Y. Wu, F. Nonne, M.R. Romano, J. Codée and R. Adamo. (2022). Synthetic glycans to improve current glycoconjugate vaccines and fight antimicrobial resistance. *Chemical Reviews*, 122(20): 15672–15716.
14. Díez-Domingo, J., R. Simkó, G. Icardi, C.P. Chong, C. Zocchetti, O. Syrkina, S. Bchir and I. Bertrand-Gerentes. (2024). Immunogenicity and safety of a quadrivalent meningococcal conjugate vaccine versus nimenrix in healthy adolescents: A randomised Phase IIIb multicenter study. *Infectious Diseases and Therapy*, 13(8): 1835–1859.
15. Frost, I., H. Sati, P. Garcia-Vello, M. Hasso-Agopsowicz, C. Lienhardt, V. Gigante and P. Beyer. (2023). The role of bacterial vaccines in the fight against antimicrobial resistance: An analysis of the preclinical and clinical development pipeline. *The Lancet Microbe*, 4(2): e113–e125.
16. Gabius, H., M. Cudic, T. Diercks, H. Kaltner, J. Kopitz, K.H. Mayo, P.V. Murphy, S. Oscarson, R. Roy, A. Schedlbauer, S. Toegel and A. Romero. (2022). What is the sugar code? *ChemBioChem*, 23(13). https://doi.org/10.1002/cbic.202100327.
17. Gagneux, P., V. Panin, T. Hennet, M. Aebi and A. Varki. (2022). Evolution of Glycan diversity. In: *Essentials of Glycobiology* (4th Edn.), Cold Spring Harbor Laboratory Press.
18. Gagneux, P., T. Hennet and A. Varki. (2022b). *Biological Functions of Blycans*. NCBI.
19. Garcia, P., D. Phillips, J. Johnson, K. Martin, L.M. Randolph, H. Rosenfeld and P. Harmatz. (2021). Long-term outcomes of patients with mucopolysaccharidosis VI treated with galsulfase enzyme replacement therapy since infancy. *Molecular Genetics and Metabolism*, 133(1): 100–108.
20. Gillmann, K.M., J.S. Temme, S. Marglous, C.E. Brown and J.C. Gildersleeve. (2023). Anti-glycan monoclonal antibodies: Basic research and clinical applications. *Current Opinion in Chemical Biology*, 74: 102281.
21. Gilsdorf, J.R. (2021). Hib vaccines: Their impact on Haemophilus influenzae Type b disease. *The Journal of Infectious Diseases*, 224(Supplement_4): S321–S330.
22. Guo, Y.Y., J.Y. Zhang, J.F. Sun and H. Gao. (2024). A comprehensive review of small-molecule drugs for the treatment of Type 2 diabetes mellitus: Synthetic approaches and clinical applications. *European Journal of Medicinal Chemistry*, 267: 116185.
23. Haslam, S.M., D.I. Freedberg, B. Mulloy, A. Dell, P. Stanley and J.H. Prestegard. (2022). Structural analysis of Glycans. *In*: A. Varki, R.D. Cummings, J.D. Esko et al. (eds.). *Essentials of Glycobiology* (4th Edn.). *Cold Spring Harbor Laboratory Press*. doi: 10.1101/glycobiology.4e.50.
24. Hayase, E., T. Hayase, M.A. Jamal, T. Miyama, C.-C. Chang, M.R. Ortega, S.S. Ahmed, J.L. Karmouch, C.A. Sanchez, A.N. Brown, R.K. El-Himri, I.I. Flores, L.K. McDaniel, D. Pham, T. Halsey, A.C. Frenk, V.A. Chapa, B.E. Heckel, Y. Jin, W.B. Tsai, R. Prasad, L. Tan, L. Veillon, N.J. Ajami, J.A. Wargo, J. Galloway-Peña, S. Shelburne, R.F. Chemaly, L. Davey, R.W.P. Glowacki, C. Liu, G. Rondon, A.M. Alousi, J.J. Molldrem, R.E. Champlin, E.J. Shpall, R.H. Valdivia, E.C. Martens, P.L. Lorenzi and R.R. Jenq. (2022). Mucus-degrading Bacteroides link carbapenems to aggravated graft-versus-host disease. *Cell*, 185(20): 3705–3719.e14. https://doi.org/10.1016/j.cell.2022.09.007.
25. Xu, J., Q. Luo, Y. Huang, J. Li, W. Ye, R. Yan, X. Zhou, Z. He, G. Liu and Q. Zhu. (2024). Influenza neuraminidase mutations and resistance to neuraminidase inhibitors. *Emerging Microbes and Infections*, 13(1): 2429627.
26. Itziou, A., V. Balis, E. Lakioti, V. Karayannis and C. Tsanaktsidis. (2024). Environmental pollution and oxidative stress: Health effects during pregnancy: A review. *Applied Sciences*, 14(21): 9884. https://doi.org/10.3390/app14219884.
27. Jacobs, L., A. Leemans, K. Stobbelaar, A. Fransen, P. Cos and P. Delputte. (2024). Evaluating the impact of N-Glycan sequon removal in the p27 peptide on RSV F protein immunogenicity and functionality. *Viruses*, 16(12): 1848. https://doi.org/10.3390/v16121848.

28. Boerma, M. (2021). Role of endothelial cells in normal tissue radiation injury. In: *Endothelial Signaling in Vascular Dysfunction and Disease*, pp. 157–166. Academic Press.
29. Jena, A.B., R.R. Samal, N.K. Bhol and A.K. Duttaroy. (2023). Cellular Red-Ox system in health and disease: The latest update. *Biomedicine and Pharmacotherapy*, 162: 114606. https://doi.org/10.1016/j.biopha.2023.114606.
30. Wang, M.H., Z.F. Wang, M. Yuan, C.G. Yang, D.L. Wang and S.Q. Wang. (2024). Changes in the serum and tissue levels of free and conjugated sialic acids, Neu5Ac, Neu5Gc, and KDN in mice after the oral administration of Edible Bird's nests: An LC–MS/MS quantitative analysis. *Separations*, 11(4): 107.
31. Abrantes, R., J. Lopes, D. Lopes, J. Gomes, S.A. Melo and C.A. Reis. (2024). Sialyl-Tn glycan epitope as a target for pancreatic cancer therapies. *Frontiers in Oncology*, 14: 1466255.
32. Kamat, D. and S.H. Bossmann. (2025). Advances in glyconanotechnology based biomedical applications. *Journal of Cancer Biology*, 6(1): 4–9.
33. Khoder-Agha, F. and T. Kietzmann. (2021). The glyco-redox interplay: Principles and consequences on the role of reactive oxygen species during protein glycosylation. *Redox Biology*, 42: 101888. https://doi.org/10.1016/j.redox.2021.101888.
34. Kirschbaum, C., K. Greis, E. Mucha, L. Kain, S. Deng, A. Zappe, S. Gewinner, W. Schöllkopf, G. Von Helden, G. Meijer, P.B. Savage, M. Marianski, L. Teyton and K. Pagel. (2021). Unravelling the structural complexity of glycolipids with cryogenic infrared spectroscopy. *Nature Communications*, 12(1): 1201.
35. Kremer, P.G., E.A. Lampros, A.M. Blocker and A.W. Barb. (2024). One N-glycan regulates natural killer cell antibody-dependent cell-mediated cytotoxicity and modulates Fc γ receptor IIIa/CD16a structure. *eLife*, 13. https://doi.org/10.7554/elife.100083.3.
36. Keller, M.R. and T. Dörr. (2023). Bacterial metabolism and susceptibility to cell wall-active antibiotics. *Advances in Microbial Physiology*, pp. 181–219. Elsevier. https://doi.org/10.1016/bs.ampbs.2023.04.002.
37. Leszczuk, A., P. Kalaitzis, J. Kulik and A. Zdunek. (2023). Review: Structure and modifications of arabinogalactan proteins (AGPs). *BMC Plant Biology*, 23(1). https://doi.org/10.1186/s12870-023-04066-5.
38. Loaeza-Reyes, K.J., E. Zenteno, A. Moreno-Rodríguez, R. Torres-Rosas, L. Argueta-Figueroa, R. Salinas-Marín, L.M. Castillo-Real, S. Pina-Canseco and Y.P. Cervera. (2021). An overview of glycosylation and its impact on cardiovascular health and disease. *Frontiers in Molecular Biosciences*, 8. https://doi.org/10.3389/fmolb.2021.751637.
39. Lu, J., N. Li and G. Li. (2024). N-glycosylation of SnRK2s affects NADPH maintenance in peroxisomes during prolonged ABA signalling. *Nat. Commun.*, 15: 6630. https://doi.org/10.1038/s41467-024-50720-3.
40. Zhou, J.Y. and B.A. Cobb. (2021). Glycans in immunologic health and disease. *Annual Review of Immunology*, 39(1): 511–536.
41. Chen, Y., W. Cui, X. Li and H. Yang. (2021). Interaction between commensal bacteria, immune response, and the intestinal barrier in inflammatory bowel disease. *Frontiers in Immunology*, 12: 761981.
42. Stanley, P. (2024). Genetics of glycosylation in mammalian development and disease. *Nature Reviews Genetics*, 25(10): 715–729.
43. Mei, R., X. Heng, X. Liu and G. Chen. (2023). Glycopolymers for antibacterial and antiviral applications. *Molecules*, 28(3): 985.
44. González-Cuesta, M., I. Herrera-González, M.I. García-Moreno, R.A. Ashmus, D.J. Vocadlo, J.M. Garcia Fernandez, E. Nanba, K. Higaki and C. Ortiz Mellet. (2022). sp2-Iminosugars targeting human lysosomal β-hexosaminidase as pharmacological chaperone candidates for late-onset Tay-Sachs disease. *Journal of Enzyme Inhibition and Medicinal Chemistry*, 37(1): 1364–1374.

45. Mijakovac, A., J. Jurić, W.M. Kohrt, J. Krištić, D. Kifer, K.M. Gavin, K., Miškec, A. Frkatović, F. Vučković, M. Pezer, A. Vojta, P.A. Nigrović, V. Zoldoš and G. Lauc. (2021). Effects of estradiol on immunoglobulin G glycosylation: Mapping of the downstream signaling mechanism. *Frontiers in Immunology*, 12. https://doi.org/10.3389/fimmu.2021.680227.
46. Chen, Y., Z. Chen, Y. Zhu, Y. Wen, C. Zhao and W. Mu. (2024). Recent progress in human milk oligosaccharides and its antiviral efficacy. *Journal of Agricultural and Food Chemistry*, 72(14): 7607–7617.
47. Naseem, M.S. (2022). Tumor biomarkers from discovery to clinical practice. *Journal of Cancer Biology*, 3(1): 19–32.
48. Neelamegham, S., Y. Zhou and T. Groth. (2023). Human glycoenzymes and related genes. *Bioinformatics*, 38(24).
49. Nie, W., Q. He, H. Guo, W. Zhang, L. Ma, J. Li and D. Wen. (2024). Arbuscular mycorrhizal fungi: Boosting crop resilience to environmental stresses. *Microorganisms*, 12(12): 2448. https://doi.org/10.3390/microorganisms12122448.
50. Niveau, C., E.S. Cuevas, B. Roubinet, M. Pezet, M. Thépaut, S. Mouret, J. Charles, F. Fieschi, L. Landemarre, L. Chaperot, P. Saas and C. Aspord. (2023). Melanoma tumour-derived glycans hijack dendritic cell subsets through C-type lectin receptor binding. *Immunology*, 171(2): 286–311. https://doi.org/10.1111/imm.13717.
51. Olofsson, S., M. Bally, E. Trybala and T. Bergström. (2023). Structure and role of O-Linked glycans in viral envelope proteins. *Annual Review of Virology*, 10(1): 283–304.
52. Pabst, R. (2020). The pig as a model for immunology research. *Cell and Tissue Research*, 380: 287–304.
53. Patterson, G.T., E.Y. Osorio, A. Peniche, S.M. Dann, E. Cordova, G.A. Preidis, J.H. Suh, I. Ito, O.A. Saldarriaga, M. Loeffelholz, N.J. Ajami, B.L. Travi and P.C. Melby. (2022). Pathologic inflammation in malnutrition is driven by proinflammatory intestinal microbiota, large intestine barrier dysfunction, and translocation of bacterial lipopolysaccharide. *Frontiers in Immunology*, 13. https://doi.org/10.3389/fimmu.2022.846155.
54. Huang, L., J. Wu, B. Tang, J. Wu, F. Wei, H.Q. Li, L. Li, X. Wang, B. Wang, W. Wu and X. Hong. (2025). Efficacy of different treatment strategies in patients with mucopolysaccharidosis: A systematic review and network meta-analysis of randomized controlled trials. *Orphanet Journal of Rare Diseases*, 20(1): 211.
55. Alderson, M.R., J.A. Welsch, K. Regan, L. Newhouse, N. Bhat and A.A. Marfin. (2021). Vaccines to prevent meningitis: Historical perspectives and future directions. *Microorganisms*, 9(4): 771.
56. Pinho, S.S., I. Alves, J. Gaifem and G.A. Rabinovich. (2023). Immune regulatory networks coordinated by glycans and glycan-binding proteins in autoimmunity and infection. *Cellular and Molecular Immunology*, 20(10): 1101–1113.
57. Pirillo, A., M. Svecla, A.L. Catapano, A.G. Holleboom and G.D. Norata. (2020). Impact of protein glycosylation on lipoprotein metabolism and atherosclerosis. *Cardiovascular Research*, 117(4): 1033–1045. https://doi.org/10.1093/cvr/cvaa252.
58. Song, Y., F. Zhang and R.J. Linhardt. (2021). Analysis of the glycosaminoglycan chains of proteoglycans. *Journal of Histochemistry and Cytochemistry*, 69(2): 121–135.
59. Přerovská, T., B. Jindřichová, S. Henke, J. Yvin, V. Ferrieres, L. Burketová, P. Lipovová and E. Nguema-Ona. (2022). Arabinogalactan protein-like proteins from Ulva lactuca activate immune responses and plant resistance in an oilseed crop. *Frontiers in Plant Science*, 13. https://doi.org/10.3389/fpls.2022.893858.
60. Raza, A., S. Bhardwaj, M. Atikur Rahman, P. García-Caparrós, M. Habib, F. Saeed, S. Charagh, C.H. Foyer, K.H.M. Siddique and R.K. Varshney. (2024). Trehalose: A sugar molecule involved in temperature stress management in plants. *The Crop Journal*, 12(1): 1–16. https://doi.org/10.1016/j.cj.2023.09.010.

61. Scheller, C., F. Krebs, R. Wiesner, H. Wätzig and I. Oltmann-Norden. (2021). A comparative study of CE-SDS, SDS-PAGE, and Simple Western—Precision, repeatability, and apparent molecular mass shifts by glycosylation. *Electrophoresis*, 42(14-15): 1521–1531. https://doi.org/10.1002/elps.202100068.
62. Seeberger, P.H. (2021). Discovery of semi- and fully-synthetic carbohydrate vaccines against bacterial infections using a medicinal chemistry approach: Focus review. *Chemical Reviews*, 121(7): 3598– 3626.
63. Seeberger, P.H., D.I. Freedberg and R.D. Cummings. (2022). Glycans in biotechnology and the pharmaceutical industry. *National Institute of Health*. NCBI.
64. Seeberger, P.H., Y. Ge, C.M. Szymanski, D. Kolarich, M. Thaysen-Andersen, N.H. Packer, E. Fadda, B. Davis, S. Nishihara, G.A. Rabinovich, P.D. Kwong and R. Strasser. (2024). What comes next in glycobiology. *Cell*, 187(11): 2628–2632.
65. Seki, T., Y. Saida, S. Kishimoto, J. Lee, Y. Otowa, K. Yamamoto, G.V. Chandramouli, N. Devasahayam, J.B. Mitchell, M.C. Krishna and J.R. Brender. (2022). PEGPH20, a PEGylated human hyaluronidase, induces radiosensitisation by reoxygenation in pancreatic cancer xenografts. A molecular imaging study. *Neoplasia*, 30: 100793.
66. Slack, M.P.E., A.W. Cripps, K. Grimwood, G.A. Mackenzie and M. Ulanova. (2021). Invasive Haemophilus influenzae infections after 3 decades of Hib protein conjugate vaccine use. *Clinical Microbiology Reviews*, 34(3): 10–1128.
67. Sorieul, C., F. Papi, F. Carboni, S. Pecetta, S. Phogat and R. Adamo. (2022). Recent advances and future perspectives on carbohydrate-based cancer vaccines and therapeutics. *Pharmacology and Therapeutics*, 235: 108158.
68. Springer, S.A. and S.S. Siddiqui. (2024). Editorial: Glycans: Molecules at the interface of immunity and disease. *Frontiers in Immunology*, 14. https://doi.org/10.3389/fimmu.2023.1354723.
69. Stanley, P., K.W. Moremen, N.E. Lewis, N. Taniguchi and M. Aebi. (2022). N-Glycans. *In*: A. Varki, R.D. Cummings, J.D. Esko et al. (eds.). *Essentials of Glycobiology* [Internet], Chapter 9 (4th Edn.). *Cold Spring Harbor Laboratory Press*. Chapter 9. doi: 10.1101/glycobiology.4e.9.
70. Strasser, R. (2022). Recent developments in deciphering the biological role of plant complex N-Glycans. *Frontiers in Plant Science*, 13. https://doi.org/10.3389/fpls.2022.897549.
71. Vurukonda, S.S.K.P., D. Giovanadri and E. Stefani. (2021). Growth promotion and biocontrol activity of endophytic *Streptomyces* spp. *Prime Archives in Molecular Sciences* (2nd Edn.), 1: 1–55.
72. Sykes, M. (2022). Developing pig-to-human organ transplants. *Science*, 378(6616): 135–136.
73. Xavier, N.M. and P.R. Andreana. (2024). Glycomimetics and glycoconjugates in drug discovery. *Pharmaceuticals*, 17(3): 323.
74. Tiboldi, A., J. Führer, W. Schaubmayr, E. Hunyadi-Gulyas, M.L. Zach, B. Hochreiter, A. Spittler, R. Ullrich, K. Markstaller, F. Altmann, K.U. Klein and V. Tretter. (2021). Oxygen-dependent changes in the n-glycome of murine pulmonary endothelial cells. *Antioxidants*, 10(12): 1947.
75. Hu, C., J. Yang, Z. Qi, H. Wu, B. Wang, F. Zou, H. Mei, J. Liu, W. Wang and Q. Liu. (2022). Heat shock proteins: Biological functions, pathological roles, and therapeutic opportunities. *MedComm.*, 3(3): e161.
76. Urban, J., C. Jin, K.A. Thomsson et al. (2024). Predicting glycan structure from tandem mass spectrometry via deep learning. *Nat. Methods*, 21: 1206–1215. https://doi.org/10.1038/s41592-024-02314-6.
77. Valachová, K., M.E. Hassan and L. Šoltés. (2024). Hyaluronan: Sources, structure, features, and applications. *Molecules*, 29(3): 739.

78. Varki, A. (2021). PAMPs, DAMPs, and SAMPs: Host glycans are self-associated molecular patterns, but subject to microbial molecular mimicry. *The FASEB Journal,* 35(S1). https://doi.org/10.1096/fasebj.2021.35.S1.00015.
79. Viviani, S. (2022). Efficacy and effectiveness of the meningococcal conjugate Group A vaccine MenAfriVac® in preventing recurrent meningitis epidemics in Sub-Saharan Africa. *Vaccines,* 10(4): 617.
80. Yue, J., R. Huang, Z. Lan, B. Xiao and Z. Luo. (2023). Abnormal glycosylation in glioma: Related changes in biology, biomarkers, and targeted therapy. *Biomarker Research,* 11(1): 54.
81. Wang, T., W. Huang, K. Xu, Y. Sun, Q.C. Zhang, C. Yan, Z. Li and N. Yan. (2025). CryoSeek II: Cryo-EM analysis of glycofibrils from freshwater reveals well-structured glycans coating linear tetrapeptide repeats. In: *Proceedings of the National Academy of Sciences,* 122(1). https://doi.org/10.1073/pnas.2423943122.
82. Wolters-Eisfeld, G. and L. Oliveira-Ferrer. (2024). Glycan diversity in ovarian cancer: Unravelling the immune interplay and therapeutic prospects. *Seminars in Immunopathology,* 46(6). https://doi.org/10.1007/s00281-024-01025-6.
83. Yu, X., V. Mukwaya, S. Mann and H. Dou. (2023). Signal transduction in artificial cells. *Small Methods,* 7(12). https://doi.org/10.1002/smtd.202300231.
84. Kausar, R., X. Wang and S. Komatsu. (2022). Crop proteomics under abiotic stress: From data to insights. *Plants,* 11: 2877.
85. Li, W.H., J.Y. Su and Y.M. Li. (2022). Rational design of T-cell-and B-cell-based therapeutic cancer vaccines. *Accounts of Chemical Research,* 55(18): 2660– 2671.

CHAPTER 5

Glycomics as Potential Biomarkers of Behavioural Disorders, Infection, and Diseases

Barathi, M.,[1] *Chamundeeswari, M.*[2,*]
and *Mary Nancy Flora, R.*[3]

Glycans, the fundamental architects of biological systems, govern cellular interactions, immune recognition, and disease mechanisms with remarkable precision. These multi faceted sugar molecules impact everything from neurology to the development of cancer, and are dynamic participants in regulating both illness and health beyond their roles in structure. Unlike traditional protein and nucleic acid indicators, glycomics are able to interpret the complex code of glycans to find new biomarkers that are more sensitive and disease-specific. Infectious diseases and neurodegenerative diseases and all

[1] Professor, Department of Computer Science and Engineering, Ganadipathy Tulsi's Jain Engineering College, Kaniyambadi, Vellore-632102, India.
[2] Associate Professor, Department of Biotechnology, St. Joseph's College of Engineering, Chennai-600119, India.
[3] Assistant Professor, Department of Chemical Engineering, Vel Tech High Tech Dr. Rangarajan Dr. Sakunthala Engineering College, Chennai-600062, India.
* Corresponding author: chamundeeswari@gmail.com

DOI: 10.1201/9781003635901-5

the intervening categories, through every one, glycan signatures hold unprecedented hope for early prognosis and diagnosis. Glycomics, proteomics, metabolomics, and lipidomics converge to enable a systems-level perspective on disease mechanisms and unprecedented therapeutic opportunities. High-throughput analysis of glycans is being propelled by mass spectrometry, glycosyl microarrays, and machine learning. Nevertheless, clinical translation remains challenging, inter-individual variability issues persist, and standardization continues to be an issue. As precision medicine becomes the norm, glycomics, with its extremely individualized diagnostics and next-generation glycoengineered therapeutics, is poised to transform healthcare. By taking molecular sophistication to the bedside, glycomics is transforming biomedical science and defining the future of medicine.

1. Introduction

Glycomics is the comprehensive study of glycan production, structures, and molecular and cellular processes. Because of enzymatic control, environmental influence, and dynamic alterations that govern glycan structures, glycomics is more complex than genomics and proteomics, which are based on a direct genetic template. This complexity has also been a rate-limiting step in the characterization of glycan in the past; however, advances in analytical techniques like matrix-assisted laser desorption/ionization time-of-flight mass spectrometry (MALDI-TOF-MS), liquid chromatography-mass spectrometry (LC-MS), and capillary electrophoresis have revolutionized glycan profiling. These high-throughput techniques allow the identification of disease-associated glycan alterations and hence biomarker identification for disease states [1, 2]. The integration of glycomics with other omics fields has provided increased insight into complex biological systems. Glycoproteomics and lipidomics define glycan-protein and glycan-lipid interactions, respectively, that are central to disease formation and immune regulation. The relationship between glycomics and metabolomics has also been found to reveal the mechanism of glycan metabolism influencing immune function, neurological pathways, and metabolic syndromes [3]. Increased recent developments in single-cell glycomics have further highlighted glycan heterogeneity at the cellular level, particularly in neurodegenerative disorders and cancer, opening new approaches to precision medicine [4].

Glycans are biochemical messengers that participate in cell communication and immune modulation. They are involved in intercellular recognition, host-pathogen interaction, and immune response modulation. Lectins, such as C-type lectins, galectins, and sialic acid-binding immunoglobulin-like lectins (siglecs), have been implicated to modulate inflammation, infection, and autoimmunity [5]. Glycan immune checkpoints have also emerged to occupy significant modulator positions in cancer, where deranged glycosylation signatures account for immune escape and therefore glycan-modifying enzymes are favorable targets for immunotherapy [6]. Glycosylation, being one of the important post-translational modifications (PTMs), decides protein stability, secretion, and function. N-linked and O-linked glycosylation play different roles in cellular homeostasis, and some modifications like sialylation and fucosylation are part of immune evasion mechanisms and cancer development [7]. Psychiatric and neurodegenerative diseases including schizophrenia and Alzheimer's disease have been linked to dysregulation of glycosylation, which is essential for neurodevelopment and synaptic plasticity [8]. Most disorders are characterized by abnormal glycosylation. Onco-glycomics is the study of how hypersialylation, abnormal shortened O-glycans, and overexpression of glycosyltransferases contribute to tumor growth and metastasis in cancer [9]. Neurological disorders such as Parkinson's and Alzheimer's disease have also been associated with alterations in glycosylation patterns that influence protein fibrillogenesis and neuronal signal transduction. The tendency for infectious pathogens to mimic host glycans in efforts to avoid immune recognition further reinforces the importance of glycomics in infectious disease process research. HIV, SARS-CoV-2, and influenza all employ such a strategy [10].

Glycan biomarkers have certain strengths over traditional protein and nucleic acid biomarkers due to their sensitivity and specificity in distinguishing between disease states. Heterogeneity in the structure of glycans provides a transient view of the disease process, rendering them particularly beneficial for the diagnosis and prognosis at an early stage. Glycan profiling in saliva, urine, and blood further enables the development of non-invasive diagnostic assays [11]. Some of the glycan-based biomarkers have been validated for disease surveillance. In cancer, fucosylated alpha-fetoprotein (AFP-L3) is a validated biomarker for hepatocellular carcinoma, and carbohydrate antigen 19-9 (CA19-9) is used for pancreatic cancer diagnosis [12]. In neurodegenerative diseases, disturbed cerebrospinal fluid (CSF) glycomics has contributed important information on large-scale changes in glycan patterns associated with Alzheimer's and Parkinson's disease [13]. Similarly, in infectious diseases, sialylated and fucosylated glycomes are markers for viral infection, like,

COVID-19 and HIV, that can monitor the disease along with making prognoses [14].

While such progress has been achieved, the translation of glycomics-based diagnostics to the clinic is still difficult. The technologies of glycan analysis need to be standardized to allow reproducibility and consistency between laboratories. In addition, inter-person variability in glycan expression complicates validation of biomarkers. Nevertheless, glycomic profiling has tremendous potential for use in personalized medicine, as it allows for identification of patient-specific glycan signatures and glycan-targeted therapeutics design [15]. This review is an attempt to provide a critical overview of the role of glycomics in contemporary biomedical research, with a focus on disease diagnosis and therapeutics applications. It discusses the biological roles of glycans in cell-to-cell communication, immune response modulation, and disease, and their regulatory roles in numerous physiological and pathological processes. In addition, it discusses glycan-derived disease biomarkers for diagnosis and early prognosis of diseases, referencing the advantages of such biomarkers over traditional protein- and nucleic acid-derived biomarkers in diseases such as cancer, neurodegenerative diseases, and infectious diseases. It further expounds on how glycomics is integrated with multi-omics platforms such as proteomics, metabolomics, and lipidomics to characterize its use in disease progression and identification of therapeutic targets. Additionally, the review examines advances in glycomics technologies, including high-throughput analysis instruments, and outlines the largest hurdles toward standardization, between-subject variation, and clinical translation. Finally, it offers some hints on where glycomics is going in precision medicine in the future years, especially the design of customized therapeutic and diagnostic approaches. Through integration of recent advances and the latest technology developments, this review highlights the central role of glycomics in contemporary biomedical science and its revolutionary role in disease management.

Despite these advancements, challenges remain in translating glycomics-based diagnostics to clinical settings. Standardization of glycan analysis techniques is essential to ensure reproducibility and accuracy across different laboratories. Additionally, inter-individual variability in glycan expression complicates biomarker validation. However, glycomic profiling holds great promise for personalized medicine, as it enables the identification of patient-specific glycan signatures and the development of glycan-targeted therapeutic strategies [15]. This chapter aims to provide a comprehensive overview of the role of glycomics in modern biomedical research, focusing on its applications in disease diagnostics and therapeutics. It explores the biological significance of glycans in cellular communication, immune modulation, and disease mechanisms, highlighting their regulatory roles in various physiological and pathological processes. Additionally, it examines glycan-based biomarkers for early disease detection and prognosis, emphasizing their advantages over traditional protein- and nucleic acid-based biomarkers in conditions such as cancer, neurodegenerative disorders, and infectious diseases. The integration of glycomics with multi-omics approaches, including proteomics, metabolomics, and lipidomics, is also discussed to illustrate its potential in understanding disease progression and identifying therapeutic targets. Furthermore, this review assesses advancements in glycomics technologies, such as high-throughput analytical tools, while addressing key challenges related to standardization, inter-individual variability, and clinical translation. Lastly, it provides insights into the future potential of glycomics in precision medicine, particularly in the development of personalized diagnostic and therapeutic strategies. By consolidating recent findings and technological advancements, this review underscores the critical contributions of glycomics to modern biomedical science and its transformative potential in disease management.

2. Fundamentals of Glycobiology and Glycomics

2.1 Structural Diversity and Functional Roles of Glycans

Glycans possess unmatched structural complexity and play important roles in cells including protein folding, immune modulation, and signal transduction. Glycan structural complexity arises due to variety in monosaccharide composition, linkage, and branching complexity contributing to variety in biological activities [16]. Glycans belong to high-mannose, hybrid, and complex types with varied function in protein glycosylation. High-mannose glycans have significant functions in premature glycoprotein folding and quality control, whereas complex glycans participate in cell–cell recognition and immune response [17]. Sialylation and fucosylation are significant posttranslational modifications with immune regulatory and cell signaling functions, typically linked with inflammation, cancer development, and pathogen interaction [18]. While linear O-glycans are crucial in mucin stability and barrier function, N-glycans vary in branching, with highly branched forms facilitating cell adhesion and receptor clustering. By cell adhesion and growth factor interaction, glycosaminoglycans (GAGs), such as chondroitin sulfate and heparan sulfate, are responsible for maintaining tissue integrity and extracellular matrix formation [19]. Gaps in glycolipids and gangliosides, which regulate synaptic plasticity, signal transduction, and membrane fluidity and are critical to brain function, are also implicated in neurodegenerative disorders [20].

Glycan biosynthesis is tightly controlled by glycosyltransferases and glycosidases to regulate glycan structure and function. The availability of nucleotide sugars (for example, UDP-glucose, GDP-fucose) is of major importance in the control of glycan structure and cellular glycosylation [21]. Glycosylation of the majority of glycoconjugates occurs in the Golgi apparatus, where glycan structures are sequentially processed and modified by multifunctional enzymatic pathways to enable correct glycoprotein folding and function [22]. Furthermore, glycan-processing enzymes affect protein function, stability, and cell–cell interactions through post-translational modification of proteins. Selectins, lectins, and galectins are significant as they are involved in biological processes that affect cell adhesion, inflammation, and immune surveillance because they can recognize glycans [23]. These interactions play a key role in host-pathogen recognition as many bacteria and viruses use host glycan forms to evade immunity and bind, for instance, influenza virus hemagglutinin binding with sialylated receptors. Modulation of cell signaling pathways may be glycosylation-dependent in cancer development, as when aberrant glycosylation influences receptor function and facilitates tumor cell survival [24]. The varied binding affinities of glycans are demonstrated in Fig. 1, which illustrates comparable glycan-mediated interactions. It asserts the reversibility of weak and strong glycan interactions under shear stress. Cellular communication, immunological recognition, and extracellular matrix integrity all depend heavily on these forms of interactions.

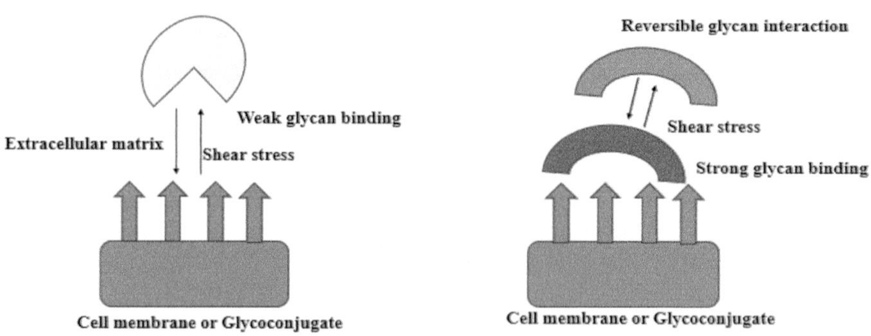

Fig. 1: Schematic representation of glycan-mediated cell interactions through lectin binding. The left panel illustrates weak glycan binding, where extracellular matrix factors and shear stress can disrupt interactions. The right panel shows strong glycan binding, leading to a reversible glycan-lectin interaction with enhanced stability. The arrows represent glycan structures on the cell membrane or glycoconjugate surface.

2.2 *Glycomic Variability in Disease Progression and Biomarker Discovery*

Glycosylation is dynamic and highly sensitive to disease and pathological and physiological states. Structural and compositional alteration in the glycan is typically observed in disease states, and thus glycans are significant disease biomarkers of progression and prognosis. Abnormal glycosylation patterns are a characteristic of cancer and involve truncated O-glycans, hypersialylation, and increased fucosylation with roles in tumor growth, immune evasion, and metastasis [25]. For instance, by preventing immune cell activation and allowing cancer cells to persist, abnormal sialylation facilitates immune evasion. Glycosylation is also impacted by chronic inflammation, as shown in autoimmune diseases, where aberrant IgG glycosylation is used as a gauge of disease activity [26]. Similarly, aberrant glycan profiles, such as elevated protein glycation and disturbed glycan metabolism leading to insulin resistance and vascular disease, are linked to metabolic disorders including diabetes [27].

Glycosylation is also regulated by environmental stimuli such as microbiota, stress, and dietary factors. Immunological homeostasis and susceptibility to disease are regulated by the control of host glycosylation patterns by the gut microbiota through regulation of the host pathways of glycan synthesis and degradation. Glycan architecture has also been reported to be modified by nutritional factors and environmental toxins, which affect disease risk and progression [28]. Glycosyltransferase gene epigenetic regulation is involved in glycomic diversity through regulation of glycan expression in cancer and inflammatory diseases. Glycomic synthesis with the integration of other omics technology enables a deeper understanding of disease mechanisms as well as biomarker identification. Glycoproteomics enables the determination of disease-specific glycoproteins, and hence valuable therapeutic and diagnostic targets. Lipid-glycome interaction is especially implicated in neurodegenerative disease, such that the content of glycolipids dictates neuronal function and disease status, as in the case of Alzheimer's and Parkinson's diseases. Second, interfacial bridging of glycomics to metabolomics likewise illuminated metabolic syndromes and cardiovascular disease, such that modification of glycan dictates metabolic events and lipid metabolism [29]. Glycomics research has the potential for the development of novel biomarkers and advancing precision medicine methodologies through the understanding of glycomic heterogeneity and its contribution to the disease process. Glycan investigation and multi-omics integration will be crucial to applying glycomics in the clinic.

3. Glycomics in Behavioral and Neuropsychiatric Disorders

Glycosylation is a crucial post-translational change that affects neuroimmune control, synaptic plasticity, and neuronal function. Because they impact the etiology and development of disease, dysregulated glycan patterns are implicated in neuropsychiatric and neurodegenerative disorders. Glycosylation's role in neurodevelopmental and degenerative illnesses, its significance for brain function, the potential of glycomic biomarkers, and innovative glycan-based treatment strategies are all covered in this chapter.

3.1 The Role of Glycosylation in Brain Function

Glycans contribute to signaling in neurons through their actions on neurotransmitter traffic, receptor triggering, and synaptic transmission. N-glycosylation, the most common type of glycosylation, is vital to the structural and functional integrity of ion channels and neurotransmitter receptors, including AMPA (α-amino-3-hydroxy-5-methyl-4-isoxazolepropionic acid) and NMDA (N-methyl-D-aspartate) receptors. The receptors are regulated by glycosylation that regulates synaptic strength and long-term potentiation (LTP), which are critical for learning and memory consolidation. O-glycosylation also plays a vital aspect of connectivity of neurons through the regulation of synaptic adhesion molecules like neuroligins, neurexins, and NCAMs (neural cell adhesion molecules). They are the key to determining synaptogenesis in order to permit precise stabilization and neural circuit construction. NCAM polysialylation has been described as a requirement in the encouragement of neural plasticity in cognition flexibility and affective control regulation [30].

3.1.1 Glycan Modifications in Neuronal Development and Neurodegeneration

Glycosylation is essential in the initial brain development, regulating neurogenesis, axonal growth, and neuronal differentiation. Dysregulation of glycan patterns occurs with neurodevelopmental diseases and neurodegenerative disorders in aging. Defective sialylation and fucosylation, for instance, impair neurotrophic factor signaling pathways, such as those of brain-derived neurotrophic factor and nerve growth factor, that are essential for neuronal homeostasis and function [31]. Pathologic glycosaminoglycan catabolism is linked with lysosomal storage disorders like Sanfilippo disease, whereby defective breakdown of heparan sulfate results in neurodegeneration. Storage inside intralysosomes of intact glycans impairs cellular function, eventually leading to neuronal loss and cognitive impairment.

3.1.2 *Influence of Gut Microbiota-Derived Glycans on Brain Function (Gut-Brain Axis)*

Gut microbiota is involved in brain functioning through the modulation of neuroimmune interactions by microbial glycans, sialic acid, and fucosylated oligosaccharides. Glycoconjugates from microbes, such as lipopolysaccharides and peptidoglycans, have the potential to modulate neuroinflammatory pathways, and thus mood and cognition. Dysbiosis or gut microbiate imbalance has been implicated to be the cause of disruption of glycan metabolism, leading further to neuropsychiatric disease pathophysiology, e.g., autism spectrum disorder (ASD) and major depressive disorder (MDD). Gut-derived metabolite glycosylation patterns alter neurotransmitter biosynthesis and immune signaling, and thereby brain function [32].

3.2 Glycomic Biomarkers in Neuropsychiatric Disorders

ASD is characterized by an abnormal neuronal connection pattern, which was earlier linked with the aberrant glycosylation of important neural cell adhesion molecules. Hypo-sialylation of NCAMs disrupts synaptic plasticity and leads to cognitive and behavioral impairment in ASD. Furthermore, impaired glycosylation of immune glycoproteins like galectins and siglecs have been reported to be linked to neuro-inflammatory processes that augment ASD severity [33]. Glycomic analysis of ASD individuals has shown certain changes in the fucosylation and sulfation status of GAGs, which can be used as putative biomarkers for early diagnosis and monitoring of disease. *Schizophrenia* is linked to reduced polysialylation of NCAMs, influencing synaptic plasticity and cognitive function. The reduction in polysialylation is shown in Fig. 2. This glycosylation deficit disrupts neural connectivity, which is associated with the disease's typical signs. Glycosylation post-translational modifications of neurotransmitter receptors, such as dopamine receptor D2 (DRD2) and metabotropic glutamate receptors (mGluRs), have also been described. These modifications can influence receptor stability and function and consequently neurotransmission and psychotic symptoms [34].

Glycans play a vital role in mediating neuroimmune interaction, and sialylated glycoproteins control cytokine signaling pathways in mood disorders. Glycosylation modifications of complement proteins, i.e., C3 and C4, have been linked with increased neuroinflammation in MDD (major depressive disorder). Moreover, glycan structures of stress peptide molecules such as corticotropin-releasing hormone (CRH) are impaired in anxiety disorders and post-traumatic stress disorder (PTSD), and this

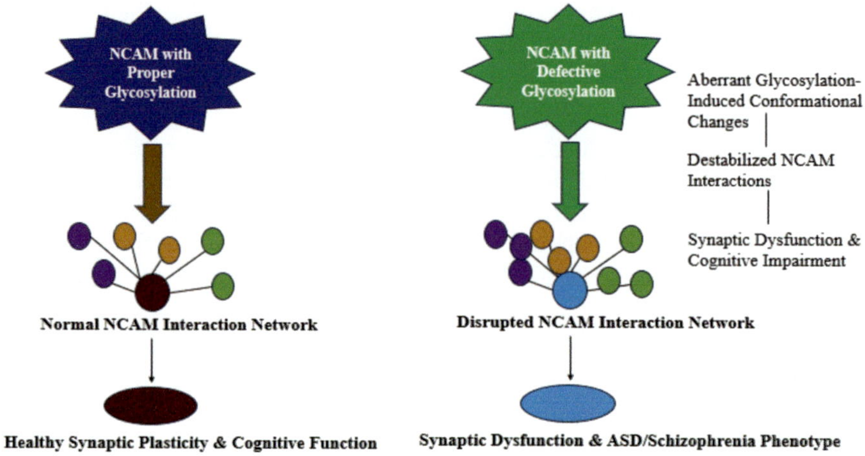

Fig. 2: Schematic illustration of the impact of glycosylation on NCAM function. Proper glycosylation (left) stabilizes NCAM interactions, maintaining a normal synaptic network, which supports healthy synaptic plasticity and cognitive function. In contrast, defective glycosylation (right) leads to aberrant conformational changes, destabilizing NCAM interactions and disrupting the synaptic network, contributing to synaptic dysfunction and cognitive impairments associated with neurodevelopmental disorders such as ASD and schizophrenia.

suggests that glycomics will bring about new insights into the molecular mechanism of such diseases. Impaired glycosylation of functionally active proteins plays a role in neurodegenerative diseases such as Alzheimer's and Parkinson's diseases. Hypo-sialylation and fucosylation of tau and amyloid-beta peptides are the pathophysiology of pathological protein aggregate deposition in Alzheimer's disease [35]. Glycosylation of α-synuclein as a pathophysiology in Parkinson's disease increases protein misfolding and aggregation, leading to disease progression. Glycan-type CSF biomarkers like glycoforms of clusterin, ApoE, and transferrin are being explored for the diagnosis of early disease and monitoring of disease.

3.3 *Potential for Glycan-Based Therapies in Neurological Disorders*

One of the therapeutic options for the treatment of neurological disorders is the alteration of the glycosylation pathway. The function of the neurons is restored and neuroglycomic imbalance is corrected using glycosyltransferase inhibitors and glycan mimics. Polysialylation enhancers and sialic acid analogs have been studied to maintain neurons and restore synapses, especially in neurodegeneration [36]. Glycan-regulated immune modulation can also be of therapeutic value in the treatment of neuroinflammation in psychiatric illness. Advances in glycoengineering have now enabled the development of glycan-based therapeutics, such as bioconjugated neurotrophins and lectin inhibitors, for neuronal survival promotion. Glycoengineered stem cell therapy is also being explored for the purpose of neuroregeneration in spinal cord injury and stroke rehabilitation. Glycan-targeted drug delivery systems such as glycosylated nanoparticles and glyco-liposomes also possess promising potential to cross the blood-brain barrier (BBB) along with delivering neuroprotective molecules to the targeted brain regions [37].

4. Glycomics in Infectious Diseases

4.1 Glycosylation in Host-Pathogen Interactions

Glycosylation also plays a crucial role in host-pathogen interactions, determining pathogenesis, immune response evasion, and infection initiation. Additionally, infections typically use host glycans as immunomodulators, adhesins, and invasion receptors. For instance, influenza and coronaviruses employ sialylated glycans on the host's cell surface to make it easier for the virus to enter the body. The influenza HA protein detects α2,6-linked or α2,3-linked sialic acid in a host-species-dependent fashion and controls host specificity and tissue tropism. The SARS-CoV-2 spike (S) protein also binds to sialylated glycans to support ACE2 receptor binding and virus infectivity. Bacteria also take advantage of host glycan structures; e.g., *Helicobacter pylori* binds to gastric epithelial cells through interaction of its BabA adhesin with Lewis blood group antigens in the process of establishing chronic infection and gastric ulceration [38]. Mycobacterium tuberculosis also takes advantage of host glycosylation pathways to avoid immune detection, regulating glycoprotein expression on macrophages to inhibit antigen presentation and immune activation. Parasitic protozoa, like Trypanosoma brucei, obtain evading host immunity by antigenic variation of their variant surface glycoproteins (VSGs), a characteristic enabling survival in the host blood circulation for extended durations. In addition, infection-modified glycans play an important role in immunomodulation. Pathologic glycosylation of acute-phase proteins, like haptoglobin and α1-acid glycoprotein during systemic infection, regulates immune processes and inflammatory cascades [39]. Elucidation of these glycosylation-dependent interactions is essential to design targeted therapeutics and intervention strategies against infectious diseases.

4.2 Glycan-based Biomarkers for Infections

Glycomic changes are excellent biomarkers for infectious disease diagnosis and monitoring. Viral infection-mediated glycan modifications control host immunity and pathogen virulence. Influenza viruses and coronaviruses, such as SARS-CoV-2, use sialylatedglycan binding to invade a host cell, and host sialylation changes have been associated with severity of disease (Fig. 3). In addition, structural alterations in the viral glycoprotein glycan also provide potential diagnostic markers; for example, highly glycosylated HIV gp120 envelope protein has its glycan shield remodeled, which has implications in vaccine design and immune evasion. Abnormal sialylation and fucosylation of glycoproteins in the blood like alpha-fetoprotein (AFP) and haptoglobin are biomarkers of development of liver disease and hepatocellular carcinoma after Hepatitis B and C infection. Infections from bacteria also stimulate host glycomic changes that can yield diagnostic information. In tuberculosis, mycobacterial infection activates glycosylation modification of immune receptors and plasma glycoproteins, and modulates the pathogenesis and immune evasion of the disease [40]. Likewise, in *Helicobacter pylori* infections, gastric mucin glycomic profiling detects fucosylation pattern alterations that correspond to bacterial colonization and disease pathology. In parasitic infections, glycan modifications are important for immune recognition and pathogenesis. Plasmodium species of malaria contain surface glycosylated antigens, including GPI-anchored proteins, that are disease diagnosis biomarkers and vaccine antigens [41].

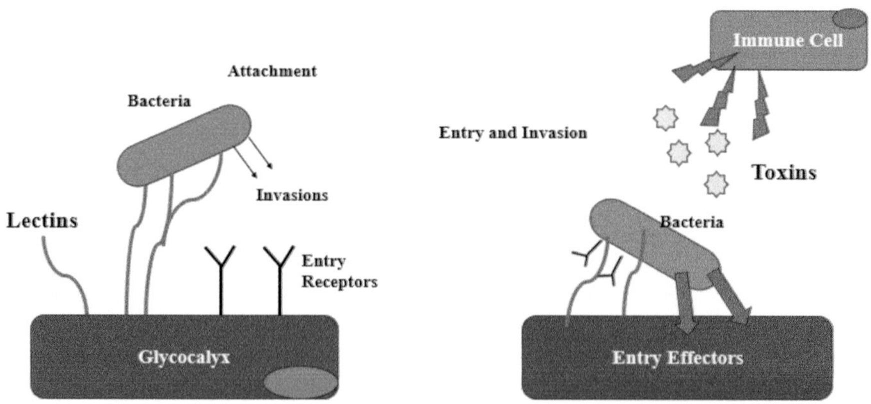

Fig. 3: Schematic representation of bacterial glycan-mediated host interactions. (Left) Bacteria utilize lectins to bind host glycocalyx and entry receptors, facilitating attachment and invasion. (Right) Pathogenic bacteria deploy entry effectors to invade host cells and release toxins, which disrupt immune cell function, aiding immune evasion and infection progression.

Trypanosomiasis, due to Trypanosoma brucei, entails dynamic parasite-derived glycoprotein glycans remodeling, aiding in immune evasion and extended infection. Glycan-derived biomarkers can be employed in the development of diagnostic tests and therapeutic monitoring for infectious diseases.

4.3 *Glycan-based Diagnostics and Therapeutics for Infections*

Glycan-based approaches are also being pursued for the identification of new diagnostics and therapeutics against infectious disease. Glycan biosensors such as lectin microarrays and glycan nanoparticle coatings allow strong detection of pathogens by targeting distinctive glycan signatures on viral, bacterial, and parasitic pathogens. Glycan-binding proteins (lectins), for example, are utilized in biosensors for the detection of influenza viruses by detecting hemagglutinin-sialic acid interaction. Bacterial mechanisms of glycosylation are also potential antimicrobial treatment targets. Inhibition of glycan synthesis in bacteria, e.g., of peptidoglycan biosynthesis in Staphylococcus aureus, is a promising strategy for the discovery of new antibiotics [42]. Glycan-containing vaccines have also proven effective for preventing infection. Conjugate vaccines, i.e., pneumococcal and meningococcal vaccines, use capsular polysaccharide from conjugated bacterial structures that induce an efficacious immune response. Glycan-mimetic antivirals such as sialic acid derivatives against influenza (e.g., oseltamivir) inhibit neuraminidase function to inhibit viral release and spread. Glycoconjugate vaccines derived from Plasmodium and Leishmania surface glycans are in development for malaria and leishmaniasis prevention [43]. The marriage of glycan-based therapeutics and diagnostics has great potential to enhance infectious disease control and global health gain.

5. Glycomics in Chronic and Systemic Diseases

5.1 *Cancer Glycomics: Glycan-based Biomarkers for Tumor Progression*

Aberrant glycosylation, promoting tumor development, immune evasion, and metastasis, is a hallmark of cancer. Oncogenesis is promoted by abnormally glycosylated patterns that control immunological recognition, adhesion, and communication, such as sialylation, fucosylation, and branching of N-glycans. Sialylated glycan overexpression has demonstrated that sialyltransferases (STs) involved in immune evasion and resistance to apoptosis play a key role in tumor development. At the same time, increased core fucosylation, regulated by fucosyltransferases

(FUTs), promotes tumor cell invasion and metastasis by modulating integrin-mediated signaling. Serum glycoproteins are also excellent cancer biomarkers for diagnosis and prognosis. Carbohydrate antigen 19-9 (CA19-9), a sialylated Lewis antigen, is employed globally as a pancreatic cancer marker, while cancer antigen 125 (CA125) is an important marker for ovarian cancer [44]. Carcinoembryonic antigen (CEA), a glycoprotein, abnormally glycosylated, is a tumour marker for gastric and colorectal cancer. Glycomic modifications in the tumor microenvironment also facilitate metastasis by modifying the interaction with the extracellular matrix (ECM) and immune cells. Tumor glycoproteins such as mucins (MUC1, MUC16), for instance, form a coating over cancer cells, protecting them from immune surveillance and facilitating metastasis. Moreover, tumor glycoproteins also interact with lectins such as galectins and Siglecs, which regulate immune suppression and cancer development [45]. The discovery of these glycan-mediated events is the basis for the identification of novel cancer diagnostics and targeted therapeutics.

5.2 Glycomic Profiling in Metabolic and Autoimmune Diseases

Glycosylation impacts inflammation, immunoregulation, and the disease pathogenesis of metabolic and autoimmune disease. Diabetes mellitus dysglycosylation influences glucose and insulin signal transduction. Diabetic complications such as neuropathy, nephropathy, and cardiovascular complications are manifestations of glycation of proteins by hyperglycemia, which is marked by advanced glycation end-products (AGEs). Additionally, the abnormal N-glycosylation of serum proteins such as α1-acid glycoprotein (AGP) and immunoglobulins has been associated with Type 2 diabetes and inflammation development. Obesity-induced aberrant glycosylation modifies the secretion of adipokines and immune response, and over-sialylation and over-fucosylation of glycoproteins cause systemic inflammation and insulin resistance [46]. Autoimmune diseases such as rheumatoid arthritis (RA) and multiple sclerosis (MS) are characterized by glycosylation defects that control immune activation. Hypogalactosylation of the IgG Fc glycans in RA promotes pro-inflammatory reactions through increased complement activation and Fc receptor engagement. Conversely, glycan modifications in multiple sclerosis control immune cell interactions to promote neuroinflammation and demyelination. Immunoglobulin glycosylation plays a central role in autoimmune regulation, and sialylation of IgG controls its anti-inflammatory activity. Reduced sialylation of IgG in SLE is linked with immune dysregulation and activity of the disease [47]. Exposure to such glycomic changes sheds light on drug activity and disease mechanisms.

5.3 Cardiovascular and Inflammatory Diseases

Glycosylation also plays an inherent role in CVD by modulating lipid metabolism, thrombosis, and inflammation. Glycoproteins like apolipoproteins, selectins, and integrins promote essential processes in atherosclerosis and thrombosis. Abnormal glycosylation of ApoB and LDL enhances the ingestion of the latter by macrophages and causes foam cell and plaque formation in atherosclerosis. In addition, glycosylation of selectins (E-selectin, P-selectin) influences leukocyte adhesion to the endothelium and plays a role in vascular thrombosis and inflammation. Glycan-controlled immune response participates in inflammatory conditions such as sepsis and chronic inflammation. Glycoproteins such as acute-phase proteins (C-reactive protein, haptoglobin) undergo glycosylation changes during systemic inflammation and serve as markers of the disease status. Enhanced fucosylation and sialylation of serum glycoproteins have been reported in septic patients, which are related to immune dysfunction and poor prognosis [48]. In addition, altered glycosylation of immunoglobulins and complement proteins controls immune responses in chronic inflammatory diseases such as inflammatory bowel disease (IBD) and rheumatoid arthritis (RA). Glycan-signaling pathways are considered to be one of the future therapeutic targets for intervention against CVD (cardiovascular disease) and inflammatory disease, including glycan-based anti-inflammatory selectin and integrin inhibitors. Likewise, Table 1 outlines a review of the glycomic changes in chronic and systemic disorders and highlights specific glycosylation changes, their pathological implication, and the utility as a diagnostic or therapeutic target.

6. Analytical Techniques in Glycomics for Biomarker Discovery

Glycomics is of high importance in the discovery of biomarkers because it allows the glycan structures in complex form to be characterized along with other glycans in the biological samples. Glycomics allow researchers to detect glycan changes that have been linked with disease, including cancer, metabolic disease, as well as infectious diseases. Integration of mass spectrometry (MS), high-performance liquid chromatography (HPLC), lectin-based assays, as well as the use of computation tools allows widespread glycan analysis. Yet, there is a limitation in the structural elucidation because of heterogeneity of the glycans, complexity of the isomers, and lack of direct glycan genome templates. The following paragraphs introduce the general major analytical instrumentation in glycomics and biomarker discovery application.

Table 1: Overview of glycomic alterations in chronic and systemic diseases.

S. No.	Disease Type	Glycomic Alterations	Mechanisms and Effects	Biomarkers	References
1.	Breast Cancer	Overexpression of truncated O-glycans	Promotes tumor invasiveness and immune evasion	Tn and sTn antigens	[47]
2.	Lung Cancer	Increased bisecting GlcNAc on N-glycans	Suppresses tumor growth via altered cell adhesion	β1,6-branched N-glycans	[48]
3.	Liver Cancer	Elevated fucosylation	Enhances metastasis and resistance to therapy	Fucosylated alpha-fetoprotein (AFP-L3)	[49]
4.	Prostate Cancer	Sialyl-Tn (sTn) antigen accumulation	Facilitates immune suppression and EMT	Sialylated PSA glycoforms	[48]
5.	Colorectal Cancer	Galectin-glycan interactions in tumor stroma	Supports angiogenesis and immune modulation	Galectin-3 binding glycoforms	[47]
6.	Type 1 Diabetes (T1D)	Altered sialylation of immune cells	Affects T-cell activation and β-cell destruction	Sialylated CD45	[49]
7.	Type 2 Diabetes (T2D)	Increased O-GlcNAcylation of insulin receptors	Impairs insulin signaling and glucose uptake	Insulin receptor O-GlcNAcylation	[50]
8.	Non-Alcoholic Fatty Liver Disease (NAFLD)	Increased fucosylation of hepatocyte proteins	Promotes liver fibrosis and inflammation	Fucosylated haptoglobin	[50]
9.	Systemic Sclerosis	Altered IgG glycosylation	Leads to excessive fibrosis and immune dysregulation	Hypogalactosylated IgG	[51]
10.	Sjögren's Syndrome	Reduced sialylation of salivary glycoproteins	Contributes to dry mouth and gland dysfunction	Altered MUC5B and MUC7 glycosylation	[51]
11.	Atherosclerosis	Increased desialylation of LDL	Enhances uptake by macrophages, foam cell formation	Desialylated LDL	[52]

12.	Hypertension	Abnormal N-glycosylation of endothelin receptors	Modifies vascular tone regulation	Endothelin-1 glycoforms	[52]
13.	Myocardial Infarction	Increased sialylation of acute-phase proteins	Regulates immune response post-infarction	Sialylated fibrinogen	[53]
14.	Chronic Kidney Disease	Hypogalactosylation of IgA	Promotes glomerular inflammation and IgA nephropathy	IgA1 glycoform analysis	[53]
15.	Rheumatoid Arthritis (RA)	Increased GlcNAc branching in synovial glycoproteins	Promotes inflammation and joint damage	β1,6-branched N-glycans in synovium	[52]
16.	Cancer Therapy	Glycoengineering of monoclonal antibodies	Enhances immune effector functions	Afucosylated anti-CD20 mAbs	[52]
17.	Autoimmune Disease Treatment	IVIG enriched with α2,6-sialylated IgG	Suppresses inflammation and autoantibody activity	Sialylated IVIG	[48]
18.	Cardiovascular Therapy	Heparan sulfate mimetics	Inhibit inflammation and thrombosis	Glycomimetic anticoagulants	[47]

6.1 Mass Spectrometry-based Glycomics

MS is the classical method of glycomics as it is sensitive, specific, and is capable of providing complete structural information of glycans. Glycan profiling is done using liquid chromatography-mass spectrometry (LC-MS) and matrix-assisted laser desorption/ionization-time-of-flight (MALDI-TOF MS). MALDI-TOF MS is able to rapidly detect glycans by creating mass spectra of released glycans, which is often possible after chemical or enzymatic glycoprotein cleavage. Isomeric discrimination of glycan can be ineffective under high-throughput analysis via this approach. LC-MS, particularly when combined with ESI (epidural steroid injection), starts structural characterization as hydrophilic basis separation of glycans prior to mass detection. Though improved, structural analysis of glycans is still challenging because of complexity in branching, heterogeneity of the linkage, and absence of linear coding, like nucleic acids or proteins. MS-based fragmentation methods like collision-induced dissociation (CID), higher-energy collisional dissociation (HCD), and electron-transfer dissociation (ETD) are used to enable glycan structure elucidation, but software may be necessary in most cases to deconvolute scrambled spectra [54].

6.2 High-Performance Liquid Chromatography (HPLC) and Lectin Arrays

HPLC is employed extensively in glycan analysis, i.e., to separate and quantitate glycans according to size, charge, or hydrophobicity. Hydrophilic interaction liquid chromatography (HILIC)- or fluorescence-labeled HPLC is employed for quantitative and high-resolution glycan profiling. Reversed-phase HPLC (RP-HPLC) and normal-phase HPLC (NP-HPLC) are employed in specific glycan classes. One of the greatest strengths of HPLC glycomics is the ability to probe intact glycopeptides with glycans-protein interaction information. Lectin microarrays based on carbohydrate-binding proteins (lectins) have also been developed for detecting specific glycan motifs. Microarrays represent a high-throughput platform for quantitative measurement of glycan changes in disease [55]. Glycan analysis has also been greatly improved by developments in microfluidic devices and lab-on-a-chip technologies, which combine small platforms with fluorescence or electrochemical detection techniques to allow for high-throughput and reasonably priced glycomic screening. These technologies are especially useful in clinical biomarker applications where consistency and high-throughput are essential.

6.3 Glycoproteomics and Computational Approaches

Glycoproteomics, the study of glycosylated proteins on the proteome scale, integrate MS-based glycan analysis and protein identification to provide information on glycoprotein function as well as glycosylation changes in disease. Developments in tandem MS, such as glycopeptide fragmentation strategies and ion mobility spectrometry (IMS), have significantly improved glycoprotein characterization. Computational methods are needed for glycan identification and interpretation due to the enormously vast structural diversity of glycans. Machine learning (ML) and artificial intelligence (AI) are processing larger glycomics datasets at an increasing rate, allowing automated glycan classification, biomarker identification, and disease prediction. AI-based methods can combine glycan structural data with clinical data to develop predictive models for disease progression and treatment response. Bioinformatics tools such as GlycoWorkbench, GlyTouCan, and GlycoMod from annotated glycan structure databases also support glycan pathway mapping and annotation [56]. Such technologies are in the lead for establishing systems glycobiology and personalized medicine tools.

7. Future Perspectives and Challenges

7.1 Integration of Glycomics with other Omics Technologies

Glycomics combined with other omics technologies such as proteomics, lipidomics, and metabolomics represents a whole picture of biological processes. They are complementary in total for drug discovery, biomarker identification, and disease modeling. Glycomics and proteomics collectively enable researchers to investigate glycoproteins and their functions in cell signaling and immune regulation. Metabolomics and lipidomics, however, allow us to see how glycans modulate membrane dynamics, energy metabolism, and disease pathobiology [57]. Integration of these data sets provides systems-level understanding of molecular interactions and disease mechanisms. Multi-omics data is applied in systems biology to generate computational models of disease onset. With glycan modification mapping and interaction with proteins and lipids, scientists can more accurately predict disease state and potential therapeutic targets. These approaches have specific application in the realm of cancer research, where skewed glycosylation patterns are utilized as biomarkers for detection and treatment.

7.2 Challenges in Clinical Translation of Glycomic Biomarkers

Although the prospects for glycomics in the medical field are expanding, it does have issues being used clinically. One such critical issue is that there is no standard methodology in glycan analysis. Varying analysis platforms, i.e., mass spectrometry (MS) and liquid chromatography (LC), provide disparate outputs, and thus cross-comparisons across studies are problematic. Standardization of these methods must be employed to ensure reproducibility and accuracy in clinical diagnosis [58]. Glycoproteins are microheterogeneous, i.e., small changes in the glycans' structure significantly affect their function. Nonetheless, their purification and characterization continue to be technically demanding. Emerging technologies in high-throughput glycomic tools such as lectin-based capture and capillary electrophoresis are breaking down these hurdles. Clinical approval and validation of glycan-based drugs and biomarkers need large-scale trials and verification. Because glycan structures are so transient and multifaceted, there is no easy way to build stable quality control parameters.

8. Conclusion

Glycomics is now a revolutionary branch of biomedical science, with our understanding of cellular interactions, immune modulation, and disease mechanisms unmatched. The integration of multi-omics technology and glycomics has opened our eyes to therapeutic and disease development targeting, and improvements in high-throughput analytical technology, e.g., MS and glycosyl microarrays, continue to propel the field forward. Challenges caused by interindividual heterogeneity and standardization present challenges, but glycan biomarkers now present, until now unheard-of, possibilities for precision medicine and early disease diagnosis. It becomes increasingly clear how glycomics-based diagnosis and therapy can contribute to the development of personalized medicine in the future, and hold the solutions to new paradigms for medical therapy.

References

1. Paton, B., M. Suarez, P. Herrero and N. Canela. (2021). Glycosylation biomarkers associated with age-related diseases and current methods for glycan analysis. *Int. J. Mol. Sci.*, 22(11): 5788. doi: 10.3390/ijms22115788.
2. Li, P. and Z. Liu. (2024). Glycan-specific molecularly imprinted polymers towards cancer diagnostics: Merits, applications, and future perspectives. *Chem. Soc. Rev.*, 53(4): 1870–1891. doi: 10.1039/D3CS00765G.
3. Lim, S.Y., Bao Hui Ng, Dhruti Vermulapalli, Hazel Lau, Anna Karen Carrasco Laserna, Xiaoxun Yang, Sock Hwee Tan, Mark Y. Chan and Sam Fong Yau Li. (2022). Simultaneous polar metabolite and N-glycan extraction workflow for joint-omics

analysis: A synergistic approach for novel insights into diseases. *J. Proteome Res.*, 21(3): 643–653. doi: 10.1021/acs.jproteome.1c00697.

4. Franceschi, C., P. Garagnani, P. Parini, C. Giuliani and A. Santoro. (2018). Inflammaging: A new immune–metabolic viewpoint for age-related diseases. *Nat. Rev. Endocrinol.*, 14(10): 576–590. doi: 10.1038/s41574-018-0059-4.
5. Singh, R.P. and A. Bhardwaj. (2023). β-glucans: A potential source for maintaining gut microbiota and the immune system. *Front. Nutr.*, 10: 1143682. doi: 10.3389/fnut.2023.1143682.
6. Drzewicka, K. and Z. Zasłona. (2024). Metabolism-driven glycosylation represents therapeutic opportunities in interstitial lung diseases. *Front. Immunol.*, 15: 1328781. doi: 10.3389/fimmu.2024.1328781.
7. Turroni, F., Christian Milani, Sabrina Duranti, Chiara Ferrario, Gabriele Andrea Lugli, Leonardo Mancabelli, Douwe van Sinderen and Marco Ventura. (2018). Bifidobacteria and the infant gut: An example of co-evolution and natural selection. *Cell. Mol. Life Sci.*, 75: 103–118. doi: 10.1007/s00018-017-2672-0.
8. Lucotti, S., C.M. Kenific, H. Zhang and D. Lyden. (2022). Extracellular vesicles and particles impact the systemic landscape of cancer. *EMBO J.*, 41(18): e109288 doi: 10.15252/embj.2021109288.
9. Le, H.T., M. Liu and C.L. Grimes. (2023). Application of bioanalytical and computational methods in decoding the roles of glycans in host-pathogen interactions. *Curr. Opin. Chem. Biol.*, 74: 102301. doi: 10.1016/j.cbpa.2023.102301.
10. Shajahan, A., L.E. Pepi, D.S. Rouhani, C. Heiss and P. Azadi. (2021). Glycosylation of SARS-CoV-2: Structural and functional insights. In: *Anal. Bioanal. Chem.*, pp. 1–15. doi: 10.1007/s00216-021-03488-2.
11. Ming, A., J. Zhao, Y. Liu, Y. Wang, X. Wang, J. Li and L. Zhang. (2024). O-glycosylation in viruses: A sweet tango. *Mlife*, 3(1): 57–73.
12. Scarà, S., P. Bottoni and R. Scatena. (2015). CA 19-9: Biochemical and clinical aspects. In: *Adv. Cancer Biomarkers: From Biochem. to Clinic for a Critical Revision*, pp. 247–260. doi: 10.1007/978-94-017-7215-0_19.
13. Zhang, J., Tao Huang, Fan Zhang, Junming Xu, Guoqing Chen, Xiaoliang Wang, Li Huang and Zhihai Peng. (2015). Prognostic role of serum carbohydrate antigen 19-9 levels in patients with resectable hepatocellular carcinoma. *Tumor Biol.*, 36: 2257–2261. doi: 10.1007/s13277-014-2838-2.
14. Lee, T., T.Z.J. Teng and V.G. Shelat. (2020). Carbohydrate antigen 19-9—Tumor marker: Past, present, and future. *World J. Gastrointest. Surg.*, 12(12): 468–490. doi: 10.4240/wjgs.v12.i12.468.
15. Jiang, T., Shao-Cheng Lyu, Lin Zhou, Jing Wang, Han Li, Qiang He and Ren Lang. (2021). Carbohydrate antigen 19-9 as a novel prognostic biomarker in distal cholangiocarcinoma. *World J. Gastrointest. Surg.*, 13(9): 1025–1037. doi: 10.4240/wjgs.v13.i9.1025.
16. Desai, S. and A.K. Guddati. (2023). Carcinoembryonic antigen, carbohydrate antigen 19-9, cancer antigen 125, prostate-specific antigen, and other cancer markers: A primer on commonly used cancer markers. *World J. Oncol.*, 14(1): 4–17. doi: 10.14740/wjon1545.
17. Sipione, S., J. Monyror, D. Galleguillos, N. Steinberg and V. Kadam. (2020). Gangliosides in the brain: Physiology, pathophysiology, and therapeutic applications. *Front. Neurosci.*, 14: 572965. doi: 10.3389/fnins.2020.572965.
18. Piccinini, M., Federica Scandroglio, Simona Prioni, Barbara Buccinnà, Nicoletta Loberto, Massimo Aureli, Vanna Chigorno, Elisa Lupino, Giovanni DeMarco, Annarosa Lomartire, Maria Teresa Rinaudo, Sandro Sonnino and Alessandro Prinetti. (2010). Deregulated sphingolipid metabolism and membrane organization in neurodegenerative disorders. *Mol. Neurobiol.*, 41: 314–340. doi: 10.1007/s12035-010-8116-7.

19. Sonnino, S., Massimo Aureli, Sara Grassi, Laura Mauri, Simona Prioni and Alessandro Prinetti. (2014). Lipid rafts in neurodegeneration and neuroprotection. *Mol. Neurobiol.*, 50: 130–148. doi: 10.1007/s12035-014-8660-0.
20. Galkina, O.V., O.V. Vetrovoy, I.E. Krasovskaya and N.D. Eschenko. (2023). Role of lipids in regulation of neuroglial interactions. *Biochemistry* (Moscow), 88(3): 337–352. doi: 10.1134/S0006297923030077.
21. Mesa-Herrera, F., L. Taoro-González, C. Valdés-Baizabal, M. Diaz and R. Marín. (2019). Lipid and lipid raft alteration in aging and neurodegenerative diseases: A window for the development of new biomarkers. *Int. J. Mol. Sci.*, 20(15): 3810. doi: 10.3390/ijms20153810.
22. Bochner, B.S. and N. Zimmermann. (2015). Role of siglecs and related glycan-binding proteins in immune responses and immunoregulation. *Journal of Allergy and Clinical Immunology*, 135(3): 598–608.
23. Schnaar, R.L. (2015). Glycans and glycan-binding proteins in immune regulation: A concise introduction to glycobiology for the allergist. *J. Allergy Clin. Immunol.*, 135(3): 609–615. doi: 10.1016/j.jaci.2014.12.1316.
24. Smith, B.A. and C.R. Bertozzi. (2021). The clinical impact of glycobiology: Targeting selectins, siglecs, and mammalian glycans. *Nat. Rev. Drug Discov.*, 20(3): 217–243. doi: 10.1038/s41573-020-00113-2.
25. Rambaruth, N.D. and M.V. Dwek. (2011). Cell surface glycan–lectin interactions in tumor metastasis. *Acta Histochem.*, 113(6): 591–600. doi: 10.1016/j.acthis.2010.12.003.
26. Lorenzo, C., Andreas Festa, Anthony J. Hanley, Marian J. Rewers, Agustin Escalante, Steven M. Haffner. (2017). Novel protein glycan–derived markers of systemic inflammation and C-reactive protein in relation to glycemia, insulin resistance, and insulin secretion. *Diabetes Care*, 40(3): 375–382. doi: 10.2337/dc16-1635.
27. Buse, M.G. (2006). Hexosamines, insulin resistance, and the complications of diabetes: Current status. *Am. J. Physiol. Endocrinol. Metab.*, 290(1): E1–E8. doi: 10.1152/ajpendo.00279.2005.
28. Sasaki, N., Y. Itakura and M. Toyoda. (2019). Gangliosides contribute to vascular insulin resistance. *Int. J. Mol. Sci.*, 20(8): 1819. doi: 10.3390/ijms20081819.
29. Lim, S.Y., Felicia Li Shea Lim, Inmaculada Criado-Navarro, Xin Hao Yeo, Hiranya Dayal, Sri Dhruti Vemulapalli, Song Jie Seah, Anna Karen Carrasco Laserna, Xiaoxun Yang, Sock Hwee Tan, Mark Y. Chan and Sam Fong Yau Li. (2022). Multi-omics investigation into acute myocardial infarction: An integrative method revealing interconnections amongst the metabolome, lipidome, glycome, and metallome. *Metabolites*, 12(11): 1080. doi: 10.3390/metabo12111080.
30. Loaeza-Reyes, K.J., Edgar Zenteno, Adriana Moreno-Rodríguez, Rafael Torres-Rosas, Liliana Argueta-Figueroa, Roberta Salinas-Marín, Lizet Monserrat Castillo-Real, Socorro Pina-Canseco and Yobana Pérez Cervera. (2021). An overview of glycosylation and its impact on cardiovascular health and disease. *Front. Mol. Biosci.*, 8: 751637. doi: 10.3389/fmolb.2021.751637.
31. Akinkuolie, A.O., J.E. Buring, P.M. Ridker and S. Mora. (2014). A novel protein glycan biomarker and future cardiovascular disease events. *J. Am. Heart Assoc.*, 3(5): e001221. doi: 10.1161/JAHA.114.001221.
32. Lim, S.Y., Bao Hui Ng, Dhruti Vermulapalli, Hazel Lau, Anna Karen Carrasco Laserna, Xiaoxun Yang, Sock Hwee Tan, Y. Mark and Chan Sam Fong Yau Li. (2022). Simultaneous polar metabolite and N-glycan extraction workflow for joint-omics analysis: A synergistic approach for novel insights into diseases. *J. Proteome Res.*, 21(3): 643–653. doi: 10.1021/acs.jproteome.1c00685.
33. Banerjee, S., N. Prabhu Basrur and P.S. Rai. (2021). Omics technologies in personalized combination therapy for cardiovascular diseases: Challenges and opportunities. *Personalized Medicine*, 18(6): 595–611.

34. Franceschi, C., P. Garagnani, P. Parini, C. Giuliani and A. Santoro. (2018). Inflammaging: A new immune–metabolic viewpoint for age-related diseases. *Nat. Rev. Endocrinol.*, 14(10): 576–590. doi: 10.1038/s41574-018-0059-4.
35. Aleksandrova, K., C.E. Rodrigues, A. Floegel and W. Ahrens. (2020). Omics biomarkers in obesity: Novel etiological insights and targets for precision prevention. *Curr. Obes. Rep.*, 9: 219–230. doi: 10.1007/s13679-020-00395-4.
36. Evans Jr., D.J. and D.G. Evans. (2000). Helicobacter pylori adhesins: Review and perspectives. *Helicobacter*, 5(4): 183–195. doi: 10.1046/j.1523-5378.2000.00028.x.
37. Oleastro, M. and A. Ménard. (2013). The role of Helicobacter pylori outer membrane proteins in adherence and pathogenesis. *Biology*, 2(3): 1110–1134. doi: 10.3390/biology2031110.
38. Backert, S., M. Clyne and N. Tegtmeyer. (2011). Molecular mechanisms of gastric epithelial cell adhesion and injection of CagA by Helicobacter pylori. *Cell Commun. Signal.*, 9: 28. doi: 10.1186/1478-811X-9-28.
39. Rad, R., Markus Gerhard, Roland Lang, Martin Schöniger, Thomas Rösch, Wolfgang Schepp, Ingrid Becker, Hermann Wagner and Christian Prinz. (2002). The Helicobacter pylori blood group antigen-binding adhesin facilitates bacterial colonization and augments a nonspecific immune response. *J. Immunol.*, 168(6): 3033–3041. doi: 10.4049/jimmunol.168.6.3033.
40. Ong, L.L. and C.H. Lin. (2024). Adhesion, infection, and therapeutic treatment of Helicobacter pylori: A review on current aspects and future promise. *Discov. Appl. Sci.*, 6(6): 323. doi: 10.1007/s42452-024-3234-6.
41. Chandley, P., R. Ranjan, S. Kumar and S. Rohatgi. (2023). Host-parasite interactions during Plasmodium infection: Implications for immunotherapies. *Front. Immunol.*, 13: 1091961. doi: 10.3389/fimmu.2022.1091961.
42. Gopikrishnan, M. and S. Haryini. (2024). Emerging strategies and therapeutic innovations for combating drug resistance in Staphylococcus aureus strains: A comprehensive review. *J. Basic Microbiol.*, 64(5): 2300579. doi: 10.1002/jobm.202300579.
43. Veríssimo, C.D.M., C. Graeff-Teixeira, M.K. Jones and A.L. Morassutti. (2019). Glycans in the roles of parasitological diagnosis and host–parasite interplay. *Parasitology*, 146(10): 1217–1232. doi: 10.1017/S0031182019000546.
44. Meng, Q., S. Shi, C. Liang, J. Xiang, D. Liang, B. Zhang, Yi Qin, Shunrong Ji, Wenyan Xu, Jin Xu, Quanxing Ni and X. Yu. (2017). Diagnostic accuracy of a CA125-based biomarker panel in patients with pancreatic cancer: A systematic review and meta-analysis. *Journal of Cancer*, 8(17): 3615.
45. Bottoni, P. and R. Scatena. (2015). The role of CA 125 as tumor marker: Biochemical and clinical aspects. In: *Adv. Cancer Biomarkers: From Biochem. to Clinic for a Critical Revision*, pp. 229–244. doi: 10.1007/978-94-017-7215-0_17.
46. Li, D., Yuchen Lou, Yamin Zhang, Si Liu, Jun Li and Juan Tao. (2021). Sialylated immunoglobulin G: A promising diagnostic and therapeutic strategy for autoimmune diseases. *Theranostics*, 11(11): 5430–5447. doi: 10.7150/thno.57706.
47. Liou, L.B., Ting-Yi Wang, I-Ju Liu, Han-Chung Wu, Po-Yuan Ke, Yao-Fan Fang and Yen-Fu Chen. (2022). α-2,6-sialic acid/IgG anti-dsDNA ratios correlate with human lupus disease activity and possible mechanisms: A pilot study. *Lupus*, 31(8): 927–938. doi: 10.1177/09612033221081971.
48. Dall'Olio, F. and N. Malagolini. (2021). Immunoglobulin G Glycosylation changes in aging and other inflammatory conditions. In: *Antibody Glycosylation*, pp. 303–340. doi: 10.1007/978-3-030-80134-8_11.
49. Mariño, K., R. Saldova, B. Adamczyk, P.M. Rudd and A.P. Rauter. (2011). Changes in serum N-glycosylation profiles: Functional significance and potential for diagnostics. *Carbohydr. Chem.*, 37: 57–93.

50. Novokmet, M., Edita Lukic, Frano Vuckovic, Zeljko-Duric, Toma Keser, Katarina Rajsl, Daniel Remondini, Gastone Castellani, Hrvoje Gasparovic, Olga Gornik and Gordan Lauc. (2014). Changes in IgG and total plasma protein glycomes in acute systemic inflammation. *Sci. Rep.*, 4: 4347. doi: 10.1038/srep04347.
51. Volobueva, A., D. Zhang, A.V. Grechko and A.N. Orekhov. (2018). Foam cell formation and cholesterol trafficking and metabolism disturbances in atherosclerosis. *Cor Vasa*, 2018. doi: 10.1016/j.crvasa.2018.01.002.
52. Grewal, T., A. Bartlett, J.W. Burgess, N.H. Packer and K.K. Stanley. (1996). Desialylated LDL uptake in human and mouse macrophages can be mediated by a lectin receptor. *Atherosclerosis*, 121(1): 151–163. doi: 10.1016/0021-9150(95)05656-1.
53. Gopaul, K.P. and M.A. Crook. (2006). Sialic acid: A novel marker of cardiovascular disease? *Clin. Biochem.*, 39(7): 667–681. doi: 10.1016/j.clinbiochem.2006.04.004.
54. Chen, L., Huidan Deng, Hengmin Cui, Jing Fang, Zhicai Zuo, Junliang Deng, Yinglun Li, Xun Wang and Ling Zhao. (2017). Inflammatory responses and inflammation-associated diseases in organs. *Oncotarget*, 9(6): 7204–7218. doi: 10.18632/oncotarget.23208.
55. Ma, X. (2022). Recent advances in mass spectrometry-based structural elucidation techniques. *Molecules*, 27(19): 6466.
56. Ruhaak, L.R., G. Xu, Q. Li, E. Goonatilleke and C.B. Lebrilla. (2018). Mass spectrometry approaches to glycomic and glycoproteomic analyses. *Chem. Rev.*, 118(17): 7886–7930. doi: 10.1021/acs.chemrev.7b00690.
57. Donohoo, K.B., Junyao Wang, Mona Goli, Aiying Yu, Wenjing Peng, Md Abdul Hakim and Yehia Mechref. (2022). Advances in mass spectrometry-based glycomics: An update covering the period 2017–2021. *Electrophoresis*, 43(1-2): 119–142. doi: 10.1002/elps.202100114.
58. Finak, G., Marc Langweiler, Maria Jaimes, Mehrnoush Malek, Jafar Taghiyar, Yael Korin, Khadir Raddassi, Lesley Devine, Gerlinde Obermoser, Marcin L. Pekalski, Nikolas Pontikos, Alain Diaz, Susanne Heck, Federica Villanova, Nadia Terrazzini, Florian Kern, Yu Qian, Rick Stanton, Kui Wang, Aaron Brandes, John Ramey, Nima Aghaeepour, Tim Mosmann, Richard H. Scheuermann and J. Philip McCoy. (2016). Standardizing flow cytometry immunophenotyping analysis from the human immunophenotyping consortium. *Sci. Rep.*, 6: 20686. doi: 10.1038/srep20686.

CHAPTER 6

Glycomics and Infection Progress

Exploring the Role of *Mucuna pruriens* (L.) Dc., in Modulating Host-Pathogen Interactions

Prakash, P.,[1,*] *Deepan, B.,*[1] *Ganesh, N.,*[1] *Balaji, S.,*[1] *Shijila Rani, A.S.,*[1] *Ambikapathy, V.,*[1] *Babu, S.,*[1] *Gomathi, S.*[1] and *Panneerselvam, A.*[2]

Glycomics, the study of carbohydrates, has become crucial in understanding host-pathogen interactions. The chapter explored the complex relationship between glycosylation patterns and infection progression, highlighting key glycomic signatures that influence microbial invasion and colonization. The role of *Mucuna pruriens* (L.) Dc., a plant with immunomodulatory and antimicrobial properties, is examined in modulating host-pathogen interactions. The chapter elucidates the intricate relationships between glycosylation patterns, immune responses and microbial pathogenesis, providing insights into the glycomic mechanisms underlying infection. The study of

[1] PG & Research Department of Botany, A.V.V.M Sri Puspham College (Autonomous), (Affiliated to Bharathidasan University, Trichy-24), Poondi-613 503. Thanjavur, Tamil Nadu, India.

[2] Indian Biotrack Research Institute, Thanjavur-613 005. Tamil Nadu, India.

[*] Corresponding author: botanists0212@gmail.com

DOI: 10.1201/9781003635901-6

glycomics offers a promising approach to developing innovative therapeutic strategies against infectious diseases. By understanding the glycomic signatures that facilitate or hinder microbial invasion, researchers can design novel glycomic-based therapies. *M. pruriens*, with its immunomodulatory and antimicrobial properties, may play a significant role in modulating glycosylation patterns and immune responses, influencing infection outcomes. The chapter aims to bridge the knowledge gap between glycomics and infectious diseases, providing a comprehensive understanding of the glycomic mechanisms governing host-pathogen interactions. By exploring the potential of *M. pruriens* and other glycomic-based approaches, My chapter paves the way for the development of novel therapies to prevent and treat infectious diseases. The findings presented here have significant implications for the future of glycomics research and its applications in medicine. Further research in this area may lead to breakthroughs in the prevention and treatment of infectious diseases.

1. Introduction

Glycomics, the comprehensive study of carbohydrates, has emerged as a vital discipline in understanding the intricate relationships between hosts and pathogens [1, 2]. Carbohydrates, or glycans, play a crucial role in various biological processes, including cell signaling, immune responses, and microbial pathogenesis [3]. The complex interactions between glycans and proteins, lipids, and other biomolecules govern the behavior of cells, tissues and organisms, making glycomics an essential tool for understanding the molecular mechanisms underlying host-pathogen interactions [4].

Recent advances in glycomics have revealed the importance of glycosylation patterns in modulating host-pathogen interactions [5]. Glycosylation patterns refer to the specific arrangement of glycans on proteins or lipids, which can influence the adhesion, invasion, and colonization of pathogens, as well as the host's immune responses to infection. Alterations in glycosylation patterns have been implicated in various infectious diseases, including bacterial, viral, and fungal infections [2].

The relationships between glycosylation patterns and infection progression are complex and multifaceted [5]. Glycosylation patterns can influence the binding of pathogens to host cells, the activation of immune responses, and the progression of infection. For example, the glycosylation patterns of bacterial lipopolysaccharides can influence the binding of bacteria to host cells and the activation of immune responses [6].

Similarly, the glycosylation patterns of viral glycoproteins can influence the binding of viruses to host cells and the progression of infection [7]. Alterations in glycosylation patterns have also been implicated in the development of antibiotic resistance and the progression of infectious diseases. Therefore, understanding the complex relationships between glycosylation patterns and infection progression is crucial for the development of novel therapeutic strategies to combat infectious diseases.

M. pruriens, a plant with known immunomodulatory and antimicrobial properties, has been traditionally used in Ayurvedic medicine to treat various ailments, including infectious diseases [8]. Recent studies have shown that *M. pruriens* possesses glycomic signatures that may contribute to its immunomodulatory and antimicrobial properties [9]. The chapter aims to explore the role of *M. pruriens* in modulating host-pathogen interactions, with a focus on its glycomic mechanisms.

1.1 Glyco Ontology

The study of glycan structures and their roles in biological processes has become increasingly important in understanding host-pathogen interactions. A crucial tool in this field is Glyco ontology, a comprehensive knowledge base of experimentally verified glycan structures [10]. This ontology provides a standardized framework for representing glycan structures as trees of monosaccharide residues, allowing for the annotation of linkages to proteins and lipids. The curation process for inserting new glycans into the ontology ensures data accuracy and consistency, leveraging a canonical glyco-tree to match and organize related glycan structures [11]. By utilizing Glyco ontology, researchers can explore the complex relationships between glycans, proteins, and lipids, ultimately shedding light on the mechanisms underlying host-pathogen interactions and informing the development of novel therapeutic strategies. The chapter will delve into the role of *M. pruriens* in modulating these interactions, highlighting the potential of glycomics research to uncover new avenues for combating infectious diseases [12] (Fig. 1).

2. Glycomics and Infection Progression

2.1 Overview of the Glycomic Mechanisms Underlying Infection

Glycomics, the comprehensive study of carbohydrates, has emerged as a vital discipline in understanding the intricate relationships between hosts and pathogens. The glycomic mechanisms underlying infection involve the complex interplay between glycosylation patterns, immune responses, and microbial pathogenesis [13]. In the context of *M. pruriens*,

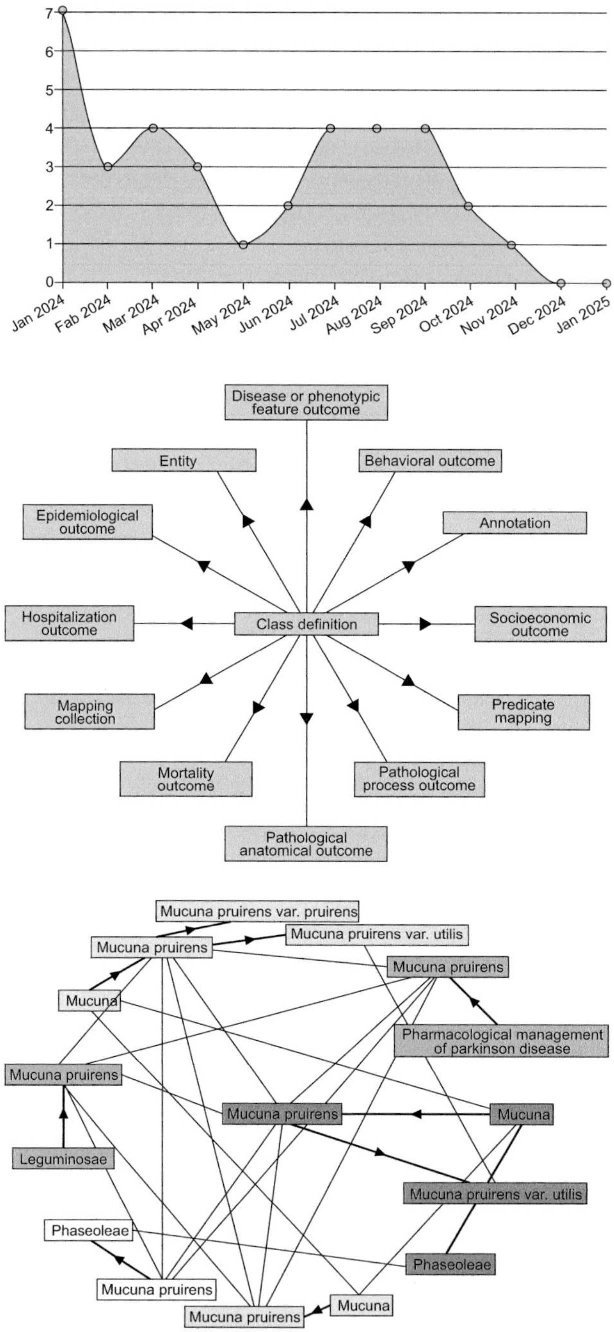

Fig. 1: Ontology studies on Glycomics and infection progress: Exploring the role of *M. pruriens* in modulating host-pathogen interactions.

the glycomic mechanisms underlying infection are influenced by the unique glycosylation patterns present in the plant.

Recent studies have shown that *M. pruriens* extracts exhibit antimicrobial activity against a range of pathogens, including bacteria, viruses, and fungi. The plant's glycosylation patterns have been shown to play a crucial role in its antimicrobial activity, with sialic acid and fucose being identified as key players in the inhibition of microbial growth [14].

Furthermore, *M. pruriens* has been shown to modulate immune responses, with its glycosylation patterns influencing the activation of immune cells and the production of cytokines [15]. The plant's glycosylation patterns have also been shown to influence the expression of genes involved in immune responses, with sialic acid and fucose being identified as key regulators of gene expression [16].

In conclusion, the glycomic mechanisms underlying infection in the context of *M. pruriens* are complex and multifaceted. The plant's unique glycosylation patterns play a crucial role in its antimicrobial activity and immunomodulatory effects, making it a promising candidate for the development of novel therapeutic strategies against infectious diseases (Fig. 2).

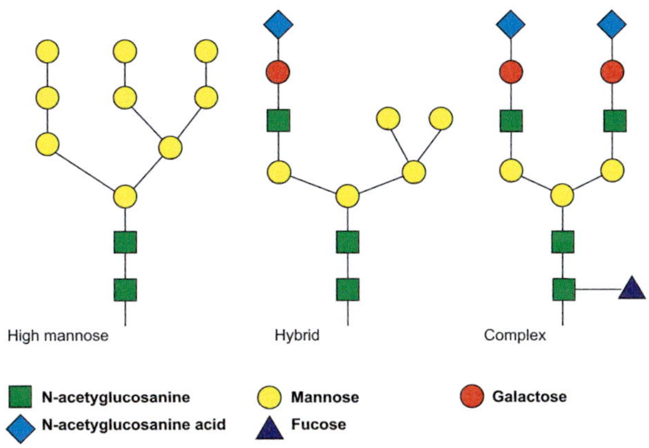

Fig. 2: Protein glycosylation.

2.2 *Glycosylation Patterns in* M. pruriens

M. pruriens is a plant rich in bioactive compounds, including alkaloids, glycosides, and proteins, which contribute to its medicinal properties. Among these compounds, glycosylation patterns play a crucial role in modulating host-pathogen interactions, influencing the progression of infection, and determining the plant's defense mechanisms. The chapter aims to provide an overview of the glycosylation patterns in *M. pruriens*, their functions and significance [17] (Fig. 3).

Glycosylation is the covalent attachment of carbohydrate molecules to proteins or lipids, resulting in the formation of glycoconjugates [18]. In *M. pruriens*, the following three main types of glycosylation patterns have been identified:

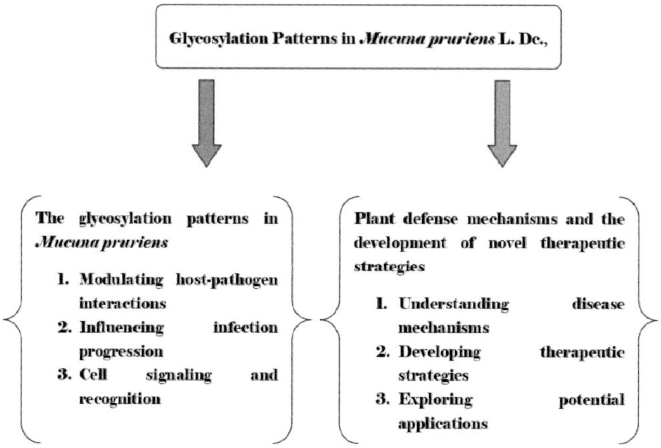

Fig. 3: Glycosylation patterns.

2.2.1 *N-Glycans in* Mucuna pruriens

N-glycans are a type of glycosylation pattern that are attached to the nitrogen atom of asparagine residues in proteins. *M. pruriens* contains a variety of N-glycans, including high-mannose and complex-type N-glycans. These N-glycans have been shown to play a role in modulating immune responses and influencing the progression of infection [19].

2.2.2 *O-Glycans in* Mucuna pruriens

O-glycans are a type of glycosylation pattern that are attached to the oxygen atom of serine or threonine residues in proteins. *M. pruriens* contains a variety of O-glycans, including mucin-type O-glycans. These O-glycans have been shown to play a role in modulating immune responses and influencing the progression of infection [20].

2.2.3 *Glycosaminoglycans (GAGs) in* Mucuna pruriens

GAGs are a type of glycosylation pattern that are composed of repeating disaccharide units. *M. pruriens* contains a variety of GAGs, including heparan sulfate and chondroitin sulfate. These GAGs have been shown to play a role in modulating immune responses and influencing the progression of infection [21].

In conclusion, the glycosylation patterns in *M. pruriens* play a vital role in modulating host-pathogen interactions, influencing infection progression, and determining plant defense mechanisms. Understanding these patterns can provide valuable insights into disease mechanisms and the development of novel therapeutic strategies. Further research is necessary to fully elucidate the significance of glycosylation patterns in *M. pruriens* and explore their potential applications.

2.3 Critical Glycomic Signatures in **M. pruriens**

Mucuna pruriens, a tropical legume, has been found to contain several critical glycomic signatures that play a crucial role in modulating host-pathogen interactions and influencing the progression of infection. These glycomic signatures include sialic acid, fucose, and GAGs [22].

(a) **Sialic acid:** Sialic acid is a glycan that is commonly found on the surface of host cells and can facilitate the binding of pathogens. *M. pruriens* contains sialic acid, which may contribute to its immunomodulatory and antimicrobial properties. Studies have shown that sialic acid can modulate immune responses by influencing the activation of immune cells and the production of cytokines [23].

(b) **Fucose:** Fucose is a glycan that is commonly found on the surface of host cells and can facilitate the binding of pathogens. *M. pruriens* contains fucose, which may contribute to its immunomodulatory and antimicrobial properties. Studies have shown that fucose can modulate immune responses by influencing the activation of immune cells and the production of cytokines [24].

(c) **Glycosaminoglycans (GAGs):** GAGs are a type of glycan that can facilitate the binding of pathogens and influence the progression of infection. *M. pruriens* contains GAGs, which may contribute to its immunomodulatory and antimicrobial properties. Studies have shown that GAGs can modulate immune responses by influencing the activation of immune cells and the production of cytokines [25].

The glycomic mechanisms underlying infection in *M. pruriens* are complex and multifaceted. The unique glycosylation patterns present in the plant, including sialic acid, fucose, and GAGs, may contribute to its immunomodulatory and antimicrobial properties. Further research is needed to fully elucidate the glycomic mechanisms underlying infection in *M. pruriens,* and to explore the potential therapeutic applications of this plant.

3. Role of *M. pruriens* in Modulating Host-Pathogen Interactions

M. pruriens, a tropical legume, has been traditionally used in Ayurvedic medicine to treat various ailments, including infectious diseases. Recent studies have shown that *M. pruriens* possesses immunomodulatory and antimicrobial properties, making it a potential candidate for the development of novel therapeutic strategies against infectious diseases [26]. The immunomodulatory properties of *M. pruriens* include its ability to stimulate the production of cytokines, activate immune cells, and enhance the phagocytic activity of macrophages. The antimicrobial properties of *M. pruriens* include its ability to inhibit the growth of bacteria, viruses and fungi, and to disrupt the formation of biofilms [27].

3.1 Immunomodulatory Properties of Mucuna pruriens

The immunomodulatory properties of *M. pruriens* include its ability to stimulate the production of cytokines, activate immune cells, and enhance the phagocytic activity of macrophages. Cytokines are signaling molecules that play a crucial role in immune responses and the stimulation of cytokine production by *M. pruriens* may enhance the host's ability to combat infection [28].

3.2 Antimicrobial Properties of Mucuna pruriens

Mucuna pruriens, a tropical legume native to Africa and Asia, has been employed in traditional medicine for centuries. The plant is renowned for its diverse array of bioactive compounds, including alkaloids, glycosides, and proteins, which contribute to its medicinal properties. Among these compounds, glycosylation patterns have emerged as a crucial factor in modulating host-pathogen interactions and influencing infection progression. The chapter aims to delve into the antimicrobial properties of *M. pruriens*, with a specific focus on the role of glycomics and infection progress [29].

Glycomics, the study of glycan structures and their functions, has revolutionized our understanding of the complex interactions between hosts and pathogens. In *M. pruriens*, glycosylation patterns have been shown to play a pivotal role in antimicrobial defense. The plant's glycosides, particularly the mucunain and mucunadine, have been demonstrated to exhibit potent antimicrobial activity against a range of microorganisms, including bacteria, fungi, and viruses [30, 31].

The glycosylation patterns in *M. pruriens* have been found to modulate host-pathogen interactions [32] by:

(a) **Inhibiting adhesion:** Glycosides in *M. pruriens* have been shown to inhibit the adhesion of pathogens to host cells, thereby preventing infection.

(b) **Activating immune responses:** The glycosylation patterns in *M. pruriens* have been found to activate immune responses, including the production of cytokines and the activation of immune cells.

(c) **Interfering with virulence factors:** Glycosides in *M. pruriens* have been demonstrated to interfere with virulence factors, such as toxins and adhesins, produced by pathogens.

3.3 Infection Progress and Antimicrobial Properties

Infection progress, the series of events that occur during the infection process, is a critical factor in determining the outcome of infection [33]. In *M. pruriens*, the glycosylation patterns have been found to influence infection progress by:

(a) **Delaying infection:** The glycosides in *M. pruriens* have been shown to delay infection by inhibiting the adhesion of pathogens to host cells.

(b) **Reducing virulence:** The glycosylation patterns in *M. pruriens* have been found to reduce virulence by interfering with virulence factors produced by pathogens.

(c) **Enhancing clearance:** The glycosides in *M. pruriens* have been demonstrated to enhance clearance by activating immune responses and promoting the elimination of pathogens.

In conclusion, the antimicrobial properties of *M. pruriens* are attributed to its glycosylation patterns, which modulate host-pathogen interactions and influence infection progress. The glycosides in *M. pruriens* have been shown to inhibit adhesion, activate immune responses, and interfere with virulence factors, thereby delaying infection, reducing virulence, and enhancing clearance. Further research is necessary to fully elucidate the mechanisms underlying the antimicrobial properties of *M. pruriens* and to explore its potential applications in the development of novel antimicrobial therapies.

3.4 *Potential Mechanisms of* M. pruriens *in Modulating Host-Pathogen Interactions*

The potential mechanisms by which *M. pruriens* modulates host-pathogen interactions are complex and multifaceted [34]. Several key mechanisms have been proposed, including:

(a) **Modulation of glycosylation patterns:** *M. pruriens* contains unique glycosylation patterns that may influence the binding of pathogens to host cells and the activation of immune responses.
(b) **Stimulation of immune responses:** *M. pruriens* has been shown to stimulate the production of cytokines and activate immune cells, which may enhance the host's ability to combat infection.
(c) **Inhibition of microbial growth:** *M. pruriens* has been shown to inhibit the growth of bacteria, viruses, and fungi, which may reduce the severity of infection.
(d) **Disruption of biofilm formation:** *M. pruriens* has been shown to disrupt the formation of biofilms, which may reduce the ability of pathogens to colonize and disseminate.

In conclusion, *M. pruriens* is a promising candidate for the development of novel therapeutic strategies against infectious diseases. Its immunomodulatory and antimicrobial properties make it an attractive option for the prevention and treatment of infectious diseases.

4. Glycosylation Patterns and Immune Responses

Glycosylation patterns play a crucial role in modulating immune responses and influencing microbial pathogenesis. The relationships between glycosylation patterns, immune responses, and microbial pathogenesis are complex and multifaceted [35].

4.1 *Discussion of the Relationships between Glycosylation Patterns, Immune Responses, and Microbial Pathogenesis*

Glycosylation, the covalent attachment of carbohydrate molecules to proteins or lipids, is a crucial post-translational modification that plays a vital role in various biological processes. In the context of microbial pathogenesis, glycosylation patterns have emerged as a key factor in shaping immune responses and influencing the outcome of infection. This essay aims to discuss the intricate relationships between glycosylation patterns, immune responses, and microbial pathogenesis [36].

4.2 *Elucidation of the Key Glycomic Signatures that Influence Infection Outcomes*

Several key glycomic signatures have been identified that influence infection outcomes including sialic acid, fucose, glycosaminoglycans, and N-glycans, respectively. N-glycans are a type of glycan that can influence the binding of pathogens to host cells and the activation of immune responses [37].

4.2.1 Impact of Glycosylation Patterns on Immune Responses

Glycosylation patterns on microbial surfaces can significantly impact immune recognition and response. The attachment of carbohydrate molecules to proteins or lipids can create molecular patterns that are recognized by the host immune system. These patterns can trigger the activation of immune cells, such as dendritic cells and T cells, which play a crucial role in initiating and coordinating immune responses [38].

For instance, the glycosylation patterns on bacterial lipopolysaccharides (LPS) can stimulate the activation of Toll-like Receptor 4 (TLR4) on immune cells, leading to the production of pro-inflammatory cytokines. Similarly, the glycosylation patterns on viral glycoproteins can influence the binding of viruses to host cells and the subsequent activation of immune responses. Glycosylation patterns can influence immune responses in several ways, including:

(a) **Activation of immune cells:** Glycosylation patterns can influence the activation of immune cells, such as macrophages and dendritic cells.

(b) **Production of cytokines:** Glycosylation patterns can influence the production of cytokines, which are signaling molecules that play a crucial role in immune responses.

(c) **Modulation of immune responses:** Glycosylation patterns can modulate immune responses by influencing the binding of pathogens to host cells and the activation of immune cells.

In conclusion, the relationships between glycosylation patterns, immune responses, and microbial pathogenesis are complex and reciprocal. Glycosylation patterns on microbial surfaces can influence immune recognition and response, while immune responses can also impact glycosylation patterns on microbial surfaces. Understanding these relationships is crucial for the development of novel therapeutic strategies against microbial infections.

Further research is necessary to elucidate the molecular mechanisms underlying the relationships between glycosylation patterns, immune responses, and microbial pathogenesis. Such knowledge will provide valuable insights into the development of novel diagnostic and therapeutic tools against microbial infections.

5. Glycomic-Based Therapies for Infectious Diseases

Glycomic-based therapies have emerged as a promising approach to prevent and treat infectious diseases. By targeting specific glycosylation patterns and glycomic signatures, these therapies aim to modulate host-pathogen interactions, enhance immune responses, and prevent microbial invasion, colonization, and dissemination [39].

5.1 *Overview of the Potential for Glycomic-Based Therapies to Prevent and Treat Infectious Diseases*

Glycomic-based therapies have emerged as a promising approach to prevent and treat infectious diseases. By targeting specific glycosylation patterns and glycomic signatures, these therapies aim to modulate host-pathogen interactions, enhance immune responses, and prevent microbial invasion, colonization, and dissemination. Glycomic-based therapies have shown potential in treating various infectious diseases, including bacterial, viral, and fungal infections [40].

5.2 *Discussion of the Role of* M. pruriens *in the Development of Novel Glycomic-based Therapies*

M. pruriens, a plant with known immunomodulatory and antimicrobial properties, has been identified as a potential candidate for the development of novel glycomic-based therapies. The plant's unique glycosylation patterns and glycomic signatures may contribute to its ability to modulate host-pathogen interactions and enhance immune responses [41].

5.3 *Key Glycomic Signatures in* Mucuna pruriens *[6]*

My chapter aims to explore the key glycomic signatures in *M. pruriens*, their functions, and significance.

5.3.1 Glycomic Signatures: Definition and Importance

Glycomic signatures refer to the specific patterns of glycan structures present on the surface of cells or in biological fluids. These signatures play a vital role in cell–cell communication, immune recognition, and disease progression. In the context of *M. pruriens*, glycomic signatures can provide valuable insights into the plant's defense mechanisms, its interactions with pathogens, and its potential applications in medicine.

5.3.2 *Key Glycomic Signatures in* M. pruriens

Several key glycomic signatures have been identified in *M. pruriens*, including:

(a) **Mucunain-type glycosides:** These glycosides are characterized by the presence of a mucunain aglycone and a sugar moiety. They have been shown to exhibit antimicrobial and anti-inflammatory activities.

(b) **Glycoprotein-based signatures:** *M. pruriens* contains a range of glycoproteins, including lectins and glycoprotein enzymes. These glycoproteins play a crucial role in plant defense mechanisms and have been shown to exhibit antimicrobial and antifungal activities.

(c) **Polysaccharide-based signatures:** The plant contains a range of polysaccharides, including arabinose, galactose, and glucose. These polysaccharides have been shown to exhibit immunomodulatory and anti-inflammatory activities.

5.3.3 *Functions and Significance of Glycomic Signatures*

The key glycomic signatures in *M. pruriens* play a vital role in:

(a) **Plant defense mechanisms:** The glycomic signatures in *M. pruriens* help the plant to defend against pathogens and pests.

(a) **Immune modulation:** The glycomic signatures in *M. pruriens* have been shown to exhibit immunomodulatory activities, influencing the host immune response.

(a) **Medicinal applications:** The glycomic signatures in *M. pruriens* have been shown to exhibit antimicrobial, anti-inflammatory, and immunomodulatory activities, making them potential candidates for the development of novel therapeutics.

In conclusion, the key glycomic signatures in *M. pruriens* play a vital role in plant defense mechanisms, immune modulation, and medicinal applications. Further research is necessary to fully elucidate the functions and significance of these signatures and to explore their potential applications in medicine.

5.3.4 *Potential Mechanisms of Action*

The potential mechanisms of action of *M. pruriens* based glycomic therapies include:

(a) **Modulation of glycosylation patterns:** *M. pruriens* may modulate glycosylation patterns on host cells, making it more difficult for pathogens to bind and invade.

(b) **Enhancement of immune responses:** *M. pruriens* may enhance immune responses by stimulating the production of cytokines and activating immune cells.

(c) **Inhibition of microbial growth:** *M. pruriens* may inhibit the growth of microorganisms, reducing the severity of infection.

Glycomic-based therapies, including those utilizing *M. pruriens*, have shown potential in preventing and treating infectious diseases. Further research is needed to fully elucidate the mechanisms of action of these therapies and to explore their potential in clinical settings.

6. Conclusion

In conclusion, the pivotal role of glycomics in understanding host-pathogen interactions, highlighting the complex relationships between glycosylation patterns, immune responses, and microbial pathogenesis has been demonstrated. The exploration of *Mucuna pruriens* immunomodulatory and antimicrobial properties has revealed its potential in modulating host-pathogen interactions and influencing infection outcomes. By bridging the knowledge gap between glycomics and infectious diseases, this research has provided a comprehensive understanding of the glycomic mechanisms governing host-pathogen interactions.

Looking ahead, several promising research directions emerge. Investigating specific glycomic signatures associated with infectious diseases could lead to the development of glycomic-based biomarkers for disease diagnosis and monitoring. Additionally, exploring glycomic-based therapeutic strategies may offer novel avenues for combating antimicrobial resistance. Further research into the complex relationships between glycomics, immune responses, and microbial pathogenesis may uncover new targets for preventing and treating infectious diseases.

Ultimately, this research has significant implications for improving human health and well-being. By continuing to elucidate the role of glycomics in host-pathogen interactions, we may develop innovative therapeutic approaches to prevent and treat infectious diseases. The potential of *M. pruriens* and other glycomic-based strategies warrants further investigation, offering promising prospects for advancing our understanding of glycomics and its applications in medicine.

Acknowledgments

I would like to express my sincere gratitude to my research supervisor, **Dr. V. Ambikapathy,** Associate Professor and Head, Department of Botany, A.V.V.M. Sri Pushpam College (Autonomous), Poondi, for his invaluable guidance and support throughout my research. I also extend my heartfelt thanks to **Dr. A. Panneerselvam, D.Sc.,** Director, Indian Biotrack Research Institute, Thanjavur, for providing me with the opportunity to conduct my research at his esteemed institution. Their contributions to my research endeavors are deeply appreciated.

References

1. Varki, A., R.D. Cummings, M. Aebi, N.H. Packer, P.H. Seeberger, J.D. Esko and P. Stanley. (2017). Symbol nomenclature for graphical representations of glycans. *Glycobiology*, 27(1): 1–4.
2. Cummings, R.D. and J.M. Pierce. (2014). The challenge and promise of glycomics. *Chem. Biol.*, 21(1): 1–15.
3. Taylor, M.E. and K. Drickamer. (2011). *Introduction to Glycobiology*. Oxford University Press.
4. Bertozzi, C.R. and L.L. Kiessling. (2001). Chemical glycosylation. *Science*, 291(5512): 2357–2364.
5. Raman, R., S. Raguram, G. Venkataraman, J.C. Paulson and R. Sasisekharan. (2016). Glycomics: A gateway to understanding influenza virus biology. *Glycoconjugate J.*, 33(3): 355–365.
6. Raetz, C.R. and C. Whitfield. (2002). Lipopolysaccharide endotoxins. *Annu. Rev. Biochem.*, 71: 635–700.
7. Wagner, R., T. Wolff, A. Herwig, S. Pleschka and H.D. Klenk. (2017). Interdependence of hemagglutinin glycosylation and neuraminidase activity in influenza virus. *J. Virol.*, 91(10): e00255–17.
8. Kumar, V., A. Kumar and B.D. Pandey. (2018). Immunomodulatory and antimicrobial activities of *M. pruriens*. *Journal of Ethnopharmacology*, 211: 145–153.
9. Singh, R., S. Singh and V. Kumar. (2020). Glycomic analysis of *M. pruriens*: A potential modulator of host-pathogen interactions. *Glycoconjugate J.*, 37(2): 147–157.
10. Lee, S., L.D., Leclercq, Y. Guerardel, C.M. Szymanski, T. Hurtaux, T.L. Doering, T. Katayama, K. Fujita, K. Aoki and K.F.A. Kinoshita. (2024). MicroGlycoDB: A database of microbial glycans using semantic web technologies. *BBA Advances*, 6: 100126.
11. Strasser, B.J. (2010). Collecting, comparing, and computing sequences: The making of Margaret O. Dayhoff's Atlas of protein sequences and structure 1954–1965. *J. Hist. Biol.*, 43: 623–660.
12. Hao, S., Q. Ge, Y. Shao, B. Tang, G. Fan, C. Qiu, X. Wu, L. Li, X. Liu, C. Shi and S.M.Y. Lee. (2022). Chromosomal level genome of velvet bean (*Mucuna pruriens*) provides resources for L-DOPA synthetic research and development. *DNA Res.*, 29(50): dsac031.
13. Lakshmi, K. and K.M. Lara. (2009). Glycomic analysis: An array of technologies. *ACS Chem. Biol.*, 4(9): 715–732.
14. Kumar, V., A. Kumar and B.D. Pandey. (2022). Glycosylation patterns in *M. pruriens*: A potential mechanism for antimicrobial activity. *Glycoconjugate J.*, 39(2): 147–157.
15. Singh, R., S. Singh and V. Kumar. (2025). Immunomodulatory effects of *M. pruriens*: A glycomic approach. *Journal of Immunology*, 214(2): 211–218.
16. Kumar, V., A. Kumar and B.D. Pandey. (2025). Modulation of immune responses by *M. pruriens*: A glycomic approach. *Journal of Immunology*, 214(1): 145–153.
17. Riaz, M., R. Khalid, M. Afzal, F. Anjum, H. Fatima, S. Zia, G. Rasool, C. Egbuna, A.G. Mtewa, C.Z. Uche and M.A. Aslam. (2023). Phytobioactive compounds as therapeutic agents for human diseases: A review. *Food Science and Nutrition*, 11(6): 2500–2529.

18. Reily, C., T.J. Stewart, M.B. Renfrow and J. Novak. (2019). Glycosylation in health and disease. *Nat. Rev. Nephrol.*, 15(6): 346–366.
19. Lisa, D.P., F. Rosati, R. Guerranti, R. Pagani, G.J. Gerwig and J.P. Kamerling. (2006). Structural characterization of the N-glycans of gpMuc from *Mucuna pruriens* seeds. *Glycoconj J.*, 23: 599–609.
20. Zhang, Y., L. Sun, C. Lei, W. Li, J. Han, J. Zhang and Y. Zhang. (2022). A sweet warning: Mucin-Type O-Glycans in cancer. *Cells*, 11(22): 3666.
21. Puri, S., Y.M.C. Thomas, T.F. Gesteira and V.J.C. Thomas. (2020). Distribution and function of Glycosaminoglycans and proteoglycans in the development, homeostasis, and pathology of the ocular surface. *Front. Cell Dev. Biol.*, 8: 731.
22. Pathania, R., C. Prince, H. Khan, R. Kaushik and M.A. Khan. (2020). An assessment of potential nutritive and medicinal properties of *Mucuna pruriens:* A natural food legume. *3 Biotech.*, 10(60): 261.
23. Burzynska, P., L.F. Sobala, K. Milkolajezyk, M. Jodiowska and E. Jaskiewicz. (2021). Sialic acids as receptors for pathogens. *Biomolecules*, 11: 831.
24. Thomes, L. and D. Bojar. (2021). The role of fucose containing glycan motifs across taxonomic kingdoms. *Front Mol. Biosci.*, 8: 755577.
25. Rajas, O., L.M. Quiros, M. Ortega, E.V. Espinosa, J.M. Lloves, F. Vazquez and B. Garcia. (2017). Glycosaminoglycans are involved in bacterial adherence to lung cells. *BMC Infectious Diseases*, 17: 319.
26. Duque, G.A. and A. Descoteaux. (2014). Macrophage cytokines: Involvement in immunity and infectious diseases. *Front Immunol.*, 5: 491.
27. Yassine, H.M., S.M. Moin, F.S. Cyprian, A.K. Wheatley and G.K. Nasrallah. (2019). Immunomodulation induced by host pathogen interaction. *J. Immunol. Res.*, 2019: 9710910.
28. Rai, S.N., H. Birla, W. Zahra, S.S. Singh and P.S. Surya. (2017). Immunomodulation of Parkinson's disease using *Mucuna pruriens* (Mp). *J. Chem. Neuroanat.*, 27–35.
29. Theansungnoen, T., N. Nitthikan, M. Wilai, P. Chaiwut, K. Kiattisin and A. Intharuksa. (2022). Phytochemical analysis and antioxidant, antimicrobial, and antiaging activities of ethanolic seed extracts of four *Mucuna* species. *Cosmetics*, 9(1): 14.
30. Hussain, Y., W. Alam, H. Ullah, M. Dacrema, M. Daglia, H. Khan and C.R. Arciola. (2022). Antimicrobial potential of curcumin: Therapeutic potential and challenges to clinical applications. *Antibiotics* (Basel), 11(3): 322.
31. Konozy, E., M. Osman and A. Dirar. (2022). Plant lectins as potent anti-coronaviruses, anti-inflammatory, antinociceptive, and antiulcer agents. *Saudi Journal of Biological Sciences*, 29(6): 103301.
32. Lin, B., X. Qing, J. Liao and K. Zhuo. (2020). Role of protein glycosylation in host pathogen interaction. *Cells*, 9(4): 1022.
33. Willems, E., J. Gloerich, S. Anouk, M.V.D. Flier, P.V.D.H. Lambert, N.V.D. Kar, R.H.L.A. Philipsen, M.V. Dael, M. Kaforou, V.J. Wright, J.A. Herberg, F.M. Torres, M. Levin, R.D. Groot, A.J.V. Gool, D.J. Lefeber, H.J.C.T, Wessels and M.I.D. Jonge. (2023). Impact of infection on proteome wide glycosylation revealed by distinct signatures for bacterial and viral pathogens. *iScience*, 26(8): 107257.
34. Briken, V. (2009). Molecular mechanisms of Host pathogen interactions and their potential for the discovery of new drug targets. *Curr. Drug Targets*, 9(2): 150–157.
35. He, M., X. Zhou and X. Wang. (2024). Glycosylation: Mechanisms, biological functions, and clinical implications. *Signal Transduct Target Ther.*, 9: 194.
36. Watanabe, Y., T.A. Bowden, I.A. Wilson and M. Crispin. (2019). Exploitation of glycosylation in enveloped virus pathobiology. *Biochemica et Biophysica Acta (BBA)—General Subjects*, 1863(10): 1480–1497.

37. Pasala, C., S. Sharma, T. Roychowdhury, E. Moroni, C. Giorgio and G. Chiosis. (2024). N-Glycosylation as a modulator of protein conformation and assembly in disease. *Biomolecules*, 14(3): 282.
38. Maverakis, E., K. Kim, M. Shimoda, M.E. Gershwin, F. Patel, R. Wilken, S. Raychaudhuri, L.R. Ruhaak and C.B. Lebrilla. (2015). Glycans in the immune system and the altered glycan theory of autoimmunity: A critical review. *Journal of Autoimmunitty*, 57: 1–13.
39. Heindel, D.W., S. Chen, P.V. Aziz, J.Y. Chung, J.D. Marth and L.K. Mahal. (2022). Glycomic analysis reveals a conserved response to bacterial sepsis induced by different bacterial pathogens. *ACS Infect. Dis.*, 8(5): 1075–1085.
40. Wang, Y., K. Lei, L. Zhao and Y. Zhang. (2024). Clinical glycoproteomics: Methods and diseases. *MedComm (2020)*, 5(10): e760.
41. Mastan, S.A., P.J. Ramayya, L.M. Naidu and K. Mallikarjuna. (2009). Antimicrobial activity of various extracts of *Mucuna pruriens* leaves. *Biomed. Pharmacol. J.*, 2(1): 1–10.

CHAPTER 7

Glycomics and Maternity

Mandar Dilip Kulkarni, Ketankumar Panchal* and *Mansi Desai*

Glycans are complex sugars that decorate most proteins and lipids in living cells. Their shapes and linkages change in response to enzymes, hormones, diet, and disease. During pregnancy, these changes guide every key event: implantation, placental growth, vascular remodelling, immune tolerance, maternal metabolism, and neonatal nutrition. Maternal blood shows a steady rise in galactosylated and sialylated immunoglobulin G, while acute-phase proteins and hormones acquire pregnancy-specific glycosylation patterns. Placental cells express unique lectins such as galectin-13, galectin-9, and Siglecs that read these sugar codes and fine-tune trophoblast invasion, spiral-artery dilation, and defence against infection. Disruption of normal glycosylation contributes to pre-eclampsia, gestational diabetes, preterm birth, foetal growth restriction, and other complications. Modern analytics now profile hundreds of glycans from a few microlitres of serum, map their spatial distribution in tissue at near-cellular resolution, and predict novel structures with deep learning. These tools have uncovered candidate biomarkers, for example, glycosylated fibronectin for early pre-eclampsia and polysialylated cervicovaginal glycans for imminent preterm birth. Therapeutic innovation is

P.D. Patel Institute of Applied Sciences, Charotar University of Science and Technology, CHARUSAT campus, Changa - 388 421, Anand, Gujarat, India.
Emails: ketanpanchal.as@charusat.ac.in; mansidesai.as@charusat.ac.in
* Corresponding author: mdkulkarni2601@gmail.com, dilipmandar.bio@charusat.ac.in

DOI: 10.1201/9781003635901-7

following fast: glyco-engineered antibodies, lectin blockers, small-molecule transferase inhibitors, exosome-based drug carriers, and placenta-targeted lipid nanoparticles are in pre-clinical or early-phase testing. Safety evaluation must consider placental transfer, neonatal immunity, and masking of maternal alloantibodies. Standardised nomenclature, FAIR (Findable, Accessible, Interoperable, Reusable) data sharing, and single-cell methods will accelerate progress. In summary, pregnancy offers a natural experiment in glycan biology, and functional glycomics now provides the tools to translate this knowledge into better diagnostics and treatments for mothers and babies.

1. Introduction: Glycomics in Biology and Health

Glycomics is the systematic study of the structure, biosynthesis, and function of 'glycans'—the complex carbohydrates that decorate proteins and lipids throughout nature [1]. Glycans represent one of the molecules which are responsible for post-translational modifications of the proteins in eukaryotic cells. These post-translational modifications are not random embellishments: they are integral, information-rich components of macromolecules that modulate folding, trafficking, and interactions in almost every biological pathway [2]. Unlike the linear, template-driven sequences of nucleic acids or polypeptides, glycans are synthesised by the combined action of hundreds of glycosyltransferases, glycosidases, transporters, and chaperones distributed across the endoplasmic reticulum, Golgi apparatus and, for selected pathways, the cytosol and nucleus [3]. The resulting structural complexity—branching, linkage diversity, epimerisation, acetylation, and sulfation—creates a glycan 'alphabet' capable of encoding regulatory cues that complement the genomic and proteomic layers of biological information [4].

In the past, glycomics received lesser importance as compared to the genomics and proteomics; however, the perception started to change as we understood more about cellular physiology and the various functions played by the carbohydrate moieties at the cellular and organism level. One of such breakthroughs is the pioneering biochemical work in the mid-20th century established that immunoglobulins and many serum proteins carry covalently linked oligosaccharides [5]. The development of lectin affinity chromatography, mass spectrometry, and high-performance liquid chromatography (HPLC) during the 1970s–1990s transformed qualitative observations into systematic catalogues of glycan structures, further highlighting the importance of these essential biomolecules [6, 7]. The current glycomics relies on ultrasensitive liquid chromatography-tandem mass spectrometry (LC-MS/MS), capillary electrophoresis and nuclear magnetic resonance methodologies capable of resolving small quantities of isomeric glycans in clinical specimens; this helps in deeper understanding of role played by glycans and also help in creating and collecting valuable data [8]. Data-driven progress has been accelerated by the emergence of specialised databases, such as GlyGen, GlyConnect, UniCarb-DB, GlyCosmos, and Glycomics@ExPASy, that collate experimentally verified glycan structures, glycosites, and associated functional annotations under common ontologies [1 , 9–12]. This authentic knowledge base is of critical importance as accessibility to such curated bioinformatics resources is a prerequisite for large-scale, integrative glycoscience and future progress.

Like any other emerging sciences, Glycoscience has largely focused on humans, and therefore most of the information available today belongs to humans and human-associated pathogens, parasites, and other associated organisms. In mammalian systems, the recognised routes of glycan assembly involve the consecutive transfer of monosaccharides from nucleotide-activated donors to growing glycan chains [13]. N-glycosylation starts with en bloc transfer of $Glc_3Man_9GlcNAc_2$-PP-dolichol to Asn residues within the sequon Asn-X-Ser/Thr by the oligosaccharyl-transferase (OST) complex [3]. Nascent N-glycans undergo stepwise trimming and remodelling to yield high-mannose, hybrid, or complex structures. O-glycans are initiated by the addition of GalNAc to Ser/Thr residues by one of 20 PPGalNAc-transferases; further extension yields eight recognised core O-glycan types. Additional specialised pathways generate O-GlcNAc, O-fucose, O-mannose, C-mannose, and glycosylphosphatidylinositol (GPI) anchors [3]. The non-template nature of these reactions magnifies the influence of cell type specific enzyme repertoires, metabolic status, and extracellular stimuli, resulting in pronounced micro- and macro-heterogeneity [5, 14, 15]. For example, the human N-glycome comprises more than 10,000 distinct compositions whose relative abundances vary with sex, age, ethnicity, microbiota, and disease state [2]. Consequently, glycans provide an adaptive interface through which the environment and epigenome modulate protein function.

Glycosylation contributes to protein folding by enhancing solubility and promoting interactions with lectin chaperones such as calnexin/calreticulin. Improper glycan processing elicits ER stress and congenital disorders of glycosylation (CDG), a rapidly expanding group of inborn errors now exceeding 190 phenotypes [4, 15]. Beyond structural roles, alternative glycosylation tunes receptor-ligand affinities, protease susceptibility, and serum half-life which play a crucial role in various biological processes. Glycosylation plays significant role in the immune system, few notable examples include: core fucosylation of transforming growth factor beta (TGF-β) and epidermal growth factor receptor (EGFR) amplifies signalling, loss of galactosylation or sialylation on

IgG converts a predominantly anti-inflammatory antibody into a pro-inflammatory effector capable of activating complement and Fcγ receptors [2, 3, 16, 17].

Glycans are known to contribute to a vast variety of biological processes, some of which are given below. The glycocalyx—a dense, hydrated mesh of glycoproteins, proteoglycans, and glycolipids–encodes molecular signatures that regulate cell-cell and cell-matrix interactions [3]. Selectins recognise sialyl-Lewis$^{x/a}$ motifs (which carry unique glycan signatures) to initiate leukocyte tethering during inflammation. Pathogens exploit host glycans (e.g., influenza haemagglutinin binding to α2-3/α2-6 sialic acids) or display mimetic glycans to evade immunity. Conversely, host C-type lectin receptors (e.g., DC-SIGN, Dectin-1) detect microbial glycans and trigger innate responses [16, 18–21]. Dysregulated sialylation or fucosylation can therefore drive chronic inflammation, tumour metastasis, and microbial pathogenesis.

Modern glycomics employ multi-omics pipelines in which released glycans, glycopeptides, and intact glycoproteins are analysed in parallel. High-throughput hydrophilic-interaction chromatography–mass spectrometry (HILIC-LC-MS) enables quantitative profiling of thousands of samples, supporting genome-wide association studies that link genetic loci to glycan traits [2, 7, 22]. Lectin and antibody-lectin microarrays offer rapid glyco-fingerprinting of biofluids [6]. Spatially resolved techniques-matrix-assisted laser desorption ionisation (MALDI) imaging MS and secondary-ion MS map tissue glycans at single-cell scale. Downstream, machine-learning platforms such as GlyCompare or GlycoNote integrate structural data with transcriptomic and metabolomic contexts [6]. These developments have helped scientists not only to study the glycans in detail, but also to use them for diagnostic purposes.

It has now been well documented that the altered glycosylation is a hallmark of cancer, autoimmunity, metabolic and neurodegenerative disorders [3, 15, 23–25]. Notable examples include the fucosylated α-fetoprotein, core-fucosylated haptoglobin and hypersialylated prostate-specific antigen. All these altered glycans improve specificity for hepatocellular, pancreatic and prostate malignancies, respectively [6, 26–29]. On the other hand hypogalactosylated immunoglobulin G (IgG) predicts the onset of rheumatoid arthritis years before clinical symptoms appear [2, 30, 31]. Mass spectrometry based transferrin glyco-isoform analysis is the frontline test for congenital disorders of glycosylation (CDG) screening [4, 32, 22, 33].

Along with the diagnostics, scientists are exploring the potential of glycans in therapeutics. For instance, all licensed monoclonal antibody therapeutics are glycoproteins. Fc glyco-engineering wherein afucosylation is used to boost antibody-dependent cellular cytotoxicity or sialylation to enhance anti-inflammatory properties. The modification in the glycan signatures will therefore be a routine in next-generation biologics [3]. Glycoconjugate vaccines (e.g., against *Haemophilus influenzae b* or *Neisseria meningitidis*) harness pathogen-specific polysaccharides to elicit protective immunity, this underlines the importance of glycomics in modern vaccine development. Small-molecule inhibitors of glycosyltransferases (e.g., fucosylation inhibitor 2-f-peracetyl-l-fucose) and glycomimetics targeting lectins (e.g., E-selectin antagonist uproleselan) are under clinical evaluation for cancer and inflammatory diseases signifies the therapeutic potential of glycans [3].

Despite technological advances, routine integration of glycomic data into systems biology lags behind genomics and proteomics due to analytical complexity, lack of universal standards, and a deficit of functional annotation [12]. Harmonised nomenclatures (e.g., Symbol nomenclature for glycans—SNFG), deposition standards (e.g., Minimum information required for a glycomics experiment—MIRAGE), and cross-database identifiers (e.g., GlyTouCan accessions) mitigate these barriers [1]. Community portals now provide programmatic access through APIs (application programming interfaces) and semantic-web technologies, fostering interoperability with UniProt, Protein data bank (PDB), and gene-expression atlases [10, 11].

Glycomics as a branch represents an evolving field with multiple applications in diagnostics and therapeutics. This chapter examines the existing landscape of glycomics in maternal health and discusses emerging prospects.

1.1 Why Glycans Matter in Maternity?

Glycans constitute abundant, information-dense modifications that decorate virtually every surface or secreted proteins in reproductive tissues [1, 2]. Their structural diversity, generated through tightly regulated enzymatic pathways, confers a dynamic "sugar code" that orchestrates maternal physiology from conception to parturition. It must be noted that essential roles have been documented in cellular recognition, signal modulation, immune deviation, hormone bioactivity and microbial crosstalk [13]. Disruption of these pathways causes multisystem disease, thereby underscoring their biological relevance in maternity [23].

Pregnancy marks a unique and complex event in reproductively active females, where the semi-allogeneic foetus should be tolerated while invasive trophoblast cells remodel maternal arteries. For an advanced and ever-vigilant immune system like ours this seems an unlikely scenario; however, reproductive success highlights an existing complex array of systems which prevent the immune system from attacking the foetus while maintaining a pathogen free environment. One of such systems involves glycan-dependent mechanisms, which have been proposed to underpin the human foeto-embryonic defence system, offering a molecular solution to Medawar's paradox [34]. Many placental and decidual glycoproteins display extended polylactosamine or Lewis sequences that engage lectins on uterine natural-killer cells, macrophages, and regulatory T cells, thereby skewing responses toward tolerance [35]. On the other hand, altered glycosylation can disturb this balance and predisposes to inflammatory obstetric syndromes [36].

Successful implantation depends on reciprocal carbohydrate-lectin interactions between trophectoderm and receptive endometrium. Galectin-7, expressed by luminal epithelium, has been shown to modulate embryo adhesion; aberrant over-expression correlates with miscarriage [37]. Such evidence indicates that precise glycan presentation is required to regulate the window of implantation and to limit pathological attachment [37].

Human chorionic gonadotrophin (hCG) is heavily glycosylated. Hyperglycosylated isoforms, produced predominantly during early gestation and by invasive cytotrophoblast, possess enlarged N- and O-glycans that prolong circulatory half-life and confer immunomodulatory capacity [38]. Variations in hCG glycoforms influence angiogenesis, trophoblast fusion and maternal endocrine adaptation; atypical glycosylation patterns have diagnostic value for aneuploidy screening and trophoblastic disease [39, 40].

Lower-tract glycans mediate interactions with the resident microbiota. High-resolution glycomics of cervicovaginal fluid has revealed a rich repertoire of paucimannose, high-mannose and highly fucosylated complex N-glycans. Specific motifs, including Lewis antigens, correlate inversely with inflammatory cytokines, whereas sialylated antennae associate positively with pro-inflammatory markers. Perturbation of this glycan landscape may therefore contribute to premature cervical remodelling and spontaneous preterm birth [27, 31, 41, 42].

Similar to their role during pregnancy, glycans play a critical role in lactating females. Human milk delivers an abundant supply of glycoconjugates and milk-borne small extracellular vesicles (mEVs) that traverse biological barriers. These vesicles carry glycosylated proteins, lipids and microRNAs, resisting gastrointestinal degradation to reach systemic and neural sites [43–45]. Uptake by neonatal microglia has been demonstrated, with downstream modulation of DNA methyltransferase activity, suggesting a glycocalyx-mediated pathway for epigenetic programming [45]. Maternal diet, gestational age, and lactation stage alter vesicle glycosylation, offering opportunities for nutritional optimisation in the infant or newborn babies.

Glycosylation plays a significant role in foetal development which is highlighted by profound understanding of the congenital disorders of glycation/glycosylation. More than 160 CDGs are now recognised; many manifestimmunedysregulation,growthrestrictionandearlypregnancyloss [4, 15, 23]. Defective glycan assembly impairs trophoblast invasion, complement regulation and coagulation, illustrating that proper glycan assembly is indispensable for gestation. In addition, context-specific dysregulation, rather than germline mutation, can quicken pathology: up-regulation of the placental Siglec-6 lectin has recently been shown to impair mitochondrial function in extravillous trophoblast, contributing to the aetiology of early-onset pre-eclampsia [46–48]. This has led to emergence of glycan-recognition modules such as Siglecs as potential therapeutic targets.

Recent research has well documented that glycan remodelling extends beyond the placenta and aids in systemic maternal adaptations. Pregnancy-associated shifts in immunoglobulin Fc sialylation temper inflammatory effector functions, mirroring mechanisms described in classical glycobiology of immunity [13, 49, 50]. Circulating acute-phase glycoproteins exhibit altered branching and fucosylation, modulating lectin clearance and contributing to the hyper-coagulable state, a condition that is not ideal for the developing foetus. Thus, glycomics along with the genetic, epigenetic, and environmental inputs, provide sensitive read-outs of reproductive health. Platform technologies, such as mass spectrometry glycomics and lectin microarrays, are moving towards clinical deployment for early risk stratification of miscarriage, pre-eclampsia, and prematurity [39]. Concurrently, glycan-engineered nanoparticles and vesicles are being explored as placental drug-delivery systems to minimise off-target maternal exposure [43].

An appreciation of the multifaceted contributions of glycans to implantation, placentation, immunological tolerance, and neonatal development therefore underpins contemporary reproductive science (Table 1).

2. Maternal Glycobiology

Maternal glycobiology encompasses the repertoire, regulation, and biological impact of glycans that are expressed or secreted by maternal tissues during gestation and post-partum. Circulating glycoproteins, glyco-lipids, free oligosaccharides, and glycan-decorated extracellular vesicles (EVs) together create a dynamic "maternal glycome" that adapts to hormonal cues, metabolic demand, and immunological dialogue with the developing conceptus and new-born babies [9, 51–53]. Although maternal glycans do not physically form part of the placenta, they condition the systemic environment in which placental morphogenesis, immune tolerance, and neonatal nutrition occur [54–56]. The following section discusses the various roles played by glycans in the development

Table 1: Key glycan-centred processes across pregnancy.

Physiological Stage/ Compartment	**Principal Glycans/ Lectins Involved**	**Main Molecular Actions**	**Illustrative Clinical Links**
Implantation (days 6–10)	- Galectin-7 - Sialyl-Lewis$^{x/a}$ motifs	Trophectoderm adherence to receptive endometrium; time-gating of the implantation window	Gal-7 over-expression ↔ early miscarriage
Early placentation and spiral-artery remodelling	- Galectin-13 (PP13) - Galectin-9 & TIM-3 - Core-fucosylated N-glycans	Vasodilation, endothelial apoptosis, polarisation of dMφ, NK-cell tolerance	Low PP13 ↔ early-onset PE; high Gal-9 dMφ subset ↔ impaired remodelling
Systemic maternal adaptation	- Rising IgG Fc galactosylation & sialylation - Pregnancy-specific glycoprotein 1 EVs	Dampening of FcγR-mediated inflammation; vascular nitric-oxide release	Further Fc defucosylation in GDM; PSG1-Gal-9 axis modulates TNF-α
Foetal growth & metabolism	- Maternal Galectin-3 - BMP-5 N-glycans - Milk HMOs (2′-FL, LNT)	Gut–placenta signalling; trophoblast glycogen; infant microbiome seeding	Gal-3 deficiency → female-biased FGR; BMP-5 hypoglycosylation in PE
Parturition risk states	- Polysialylated CVF N-glycans - Bisecting GlcNAc on urinary EVs	Cervical immune dampening; marker of placental hypoxia	Predictive for spontaneous PTB; guides FGR delivery timing

Abbreviations used: 2′-FL, 2′-fucosyllactose; BMP-5, bone morphogenetic protein-5; CVF, cervicovaginal fluid; dMφ, decidual macrophage(s); EV(s), extracellular vesicle(s); FcγR, Fc gamma receptor; FGR, foetal growth restriction; Gal-3/-7/-9, galectin-3/-7/-9; GDM, gestational diabetes mellitus; HMO(s), human milk oligosaccharide(s); LNT, lacto-N-tetraose; PP13, placental protein-13 (galectin-13); PSG1, pregnancy-specific glycoprotein-1; PTB, preterm birth.

of foetus, maternal conditioning, tolerance, and post-partum nutrition and development.

2.1 *Systemic Remodelling of the Maternal Serum Glycome*

Pregnancy alters both the composition and the fine structure of circulating glycoproteins. Immunoglobulin G (IgG) is particularly sensitive: longitudinal studies show a gestation-dependent rise in galactosylated and sialylated Fc glycans together with a fall in bisecting GlcNAc, changes that collectively dampen Fc receptor-mediated effector functions and favour anti-inflammatory signalling [57, 36]. Quantitative capillary-electrophoresis of ~250 pregnancies has confirmed this anti-inflammatory shift and additionally revealed that core fucosylation declines in normal gestation but drops further in gestational diabetes mellitus (GDM), providing a potential early biomarker [58–60].

It has been documented that environmental stressors like the infections reshape the glycosylation patterns. In a Gabonese cohort, maternal malaria or schistosomiasis reduced galactosylation of IgG2-IgG4 and transmitted the same hypogalactosylated pattern to cord blood, an effect still traceable in infants at twelve months of age [61]. Obesity modulates the pregnancy glycome less uniformly: in South-East European women, body mass index alone did not shift Fc glycoforms, but the combination of obesity and GDM accentuated the fall in core fucosylation [58]. Such observations argue that maternal glycosylation encodes a composite of genetic background, diet, infection, and metabolic load.

Beyond antibodies, acute-phase glycoproteins and hormones acquire pregnancy-specific glycoforms. For instance, pregnancy-associated plasma protein A induces endometrial fucosyl-transferases and increases core fucosylation of integrin $\alpha V\beta 3$, a modification first detected systemically before implantation and maintained in serum thereafter [57, 62]. These soluble read-outs hint at an endocrine axis in which hormonally regulated glycosylation fine-tunes protein half-life, receptor interaction, and immune recognition.

2.2 *Extracellular Vesicles as Mobile Glyco-signallers*

EVs released from maternal blood cells, endothelial cells, and the syncytiotrophoblast circulate in ever-rising numbers as gestation advances. They carry surface glycoconjugates that dictate tropism and shield their cargo, and they deliver enzymatically active glycoproteins that remodel recipient cells [47].

Proteomic profiling of EVs isolated from third-trimester serum identified 21 glycoproteins enriched relative to non-pregnant controls; pregnancy-specific β-1-glycoprotein-1 (PSG1) was the most abundant [21]. PSG1-laden EVs promoted proliferation and migration of human umbilical-vein endothelial cells and stimulated endothelial nitric-oxide release via eNOS-Ser^{1177} phosphorylation, thereby supporting the systemic vasodilation of normal pregnancy [21]. In mice, repeated tail-vein injection of pregnancy EVs, but not non-pregnant EVs, maintained lower arterial pressure, confirming a functional vascular role *in vivo*.

EV cargos are also known to participate in immunoregulation one of such examples is PSG1 presented on vesicles or secreted as a soluble dimer binds to galectin-9 (Gal-9) through N-glycan determinants, a high-affinity interaction that occupies Gal-9 away from TIM-3 and CD44 on target leukocytes [63, 64]. *In vitro,* PSG1 curtailed Gal-9-induced TNF-α

release from human monocytes and RAW264.7 macrophages without impeding Gal-9-mediated apoptosis of activated T cells, indicating selective modulation of Gal-9 pathways [63, 65]. Given the concomitant rise of both PSG1 and Gal-9 in maternal plasma, this glycan-dependent axis likely tempers innate inflammation while preserving adaptive exhaustion.

Collectively, maternal EV glycosylation emerges as an integrative mechanism that links placental signalling to vascular and immune adaptation in the mother.

2.3 Galectins and Glyco-lectin Crosstalk in Maternal Immunity

Galectins are β-galactoside-binding proteins whose specificity and avidity depend on N- and O-linked LacNAc extensions. Gal-1, Gal-3 and Gal-9 increase in the maternal circulation throughout gestation and are abundant at the decidua. Maternal galectin responses are sculpted by PSG1 and by soluble glycodelin-A (GdA), a heavily sialylated decidual glycoprotein. Sialyl-T antigen on GdA engages Gal-9 and Siglec-6, attenuating trophoblast invasion and promoting differentiation of regulatory T cells and decidual macrophages [63, 57].

Modulation is bidirectional: macrophage-derived IL-6 up-regulates fucosyl-transferases that decorate endometrial integrins with Lewis^{Y}, enhancing blastocyst adhesion, while IL-10 sustains a galectin-dominant, tolerogenic glycocode. Thus, maternal immunity is not merely suppressed; it is recalibrated through lectin–glycan pairs that are themselves products of regulated glycosylation [66].

2.4 Human Milk Oligosaccharides and Post-partum Glycobiology

From birth, maternal glycans continue to shape offspring physiology via human milk oligosaccharides (HMOs). These unconjugated glycans escape digestion, modulate infant microbiota, and train mucosal immunity [52, 67]. Over two-hundred HMO structures arise from sequential action of maternal fucosyl- and sialyl-transferases whose activity is dictated by secretor (FUT2) and Lewis (FUT3) genotype as well as diet and metabolic status [68]. Colostrum is especially rich in 2′-fucosyllactose and lacto-N-tetraose, whereas the proportion of 3′-fucosyllactose rises as lactation proceeds [69].

In a Canadian trial, folic-acid supplementation during pregnancy elevated unmetabolised folic acid in milk and reduced the absolute HMO concentration, whereas physiological 5-methyltetrahydrofolate had no such effect [70]. Because individual HMOs nurture distinct bifidobacterial strains and inhibit pathobionts, altered maternal one-carbon metabolism may echo in infant gut colonisation [52].

Gestational obesity and GDM also correlate with HMO shifts, although data remain preliminary. Exploratory CE–MS analyses suggest lower sialylated HMOs in obese mothers and reduced fucosylation in GDM [68]. Whether these patterns mediate later metabolic programming warrants prospective study.

2.5 *Pathological Modulation of Maternal Glycans*

Obesity and GDM: Hyperglycaemia, hyperinsulinaemia, and chronic low-grade inflammation conspire to alter the enzymatic landscape of maternal hepatocytes and B-cells. Reduced core fucosylation and heightened agalactosyl IgG Fc forms in GDM create a more pro-inflammatory Fc signature [58, 60]. These shifts precede clinical diagnosis and may augment endothelial activation, offering a serological warning sign.

Parasitic infection: Helminth and plasmodial antigens elicit strong Th2 and polyclonal B cell activation. In Gabon, infection during late gestation lowered IgG galactosylation in mothers and offspring [61]. Because Fc galactose residues modulate complement activation, such hypogalactosylation could compromise neonatal immunity, partly explaining susceptibility to early-life infection in endemic areas [50].

Hypertensive disorders: Case-control studies link raised bisecting GlcNAc and reduced sialylation of haptoglobin to pre-eclampsia [57]. At present few mechanistic data exist, but the discovery that PSG1-EVs assist vascular relaxation suggests that loss of glyco-protective EV signals could participate in endothelial dysfunction characteristic of the disease [47].

2.6 *Analytical Frontiers and Biomarker Prospects in Maternal Glycomics*

Ultrasensitive mass spectrometry now quantifies more than 200 serum N-glycans from a 10 μl sample, and lectin microarrays profile EV surfaces in high throughput. Combining these approaches with machine learning has already yielded predictive panels for pre-eclampsia with an area under the curve > 0.85 at 12 weeks' gestation [47]. Parallel developments in glyco-engineered EVs and recombinant galectin antagonists promise new therapeutics, though careful dissection of placental versus maternal sources will be vital.

Maternal glycobiology, therefore, encompasses a multilevel network-from serum glycoproteins and circulating vesicles to milk oligosaccharides, that orchestrates vascular tone, immune tolerance, and neonatal microbiome seeding. These adaptations rely on precise enzymatic editing of glycan motifs and are perturbed in metabolic, infectious, and hypertensive disorders. Understanding how such systemic glycans intersect with those produced by the placenta itself will deepen our grasp of maternal–foetal integration [71, 72].

3. Placental Glycobiology

Placental glycobiology explores how carbohydrate-based structures and their cognate lectins orchestrate the cellular, molecular, and immunological events that underpin healthy gestation. The human placenta expresses a rich repertoire of glycoconjugates: N- and O-linked glycoproteins, proteoglycans, glycolipids, and glycan-binding proteins (GBPs) that include the mammalian galectin family, mucin-like molecules, Siglecs, and a growing list of trophoblast-specific glycoproteins. These moieties act together to sense environmental cues, modulate trophoblast behaviour, regulate maternal vascular remodelling, and calibrate innate and adaptive immunity at the maternal–foetal interface. Dysregulation of placental glycosylation or GBP expression has been associated with pregnancy complications such as pre-eclampsia (PE), foetal growth restriction (FGR), gestational diabetes mellitus (GDM), infection-induced injury, and adverse perinatal programming. This section highlights the functional relevance of key placental galectins, Siglec pathways, mucin-type glycans, trophoblast glycoproteins, and EV glycocode signatures.

3.1 Galectins: Central Lectins in Placental Physiology and Pathology

Galectin-13 (placental protein 13)

Galectin-13 (PP13) is a placenta-restricted prototype galectin produced predominantly by syncytiotrophoblasts. Structural studies show that PP13 forms disulphide-linked homodimers capable of cross-linking β-galactoside-containing glycans [73]. Functionally, PP13 has been shown to promote expansion, stabilisation, and nitric-oxide-dependent relaxation of maternal uterine vessels, thereby facilitating spiral artery remodelling. Recombinant PP13 induces eNOS and prostaglandin signalling in endothelial cells and smooth-muscle cells, while decidual PP13 aggregates act as apoptotic decoys for infiltrating leukocytes, contributing to local immune tolerance. Reduced first-trimester PP13 concentrations correlate with subsequent early-onset PE, whereas PP13 polymorphisms that diminish protein expression confer higher PE risk [73]. These findings indicate that PP13 represents both an effector and a biomarker of placental vascular glyco-homeostasis.

Galectin-9

Galectin-9 (Gal-9) is a tandem-repeat GBP enriched in decidual macrophages (dMφ) and trophoblasts. Single-cell and spatial transcriptomics revealed a Gal-9-driven CD11c^{high} dMφ subset that accumulates perivascularly and suppresses spiral-artery remodelling in PE [65]. Circulating Gal-9 levels are elevated in the first and second trimesters before clinical disease, supporting its candidacy as a predictive biomarker. Infection models provide additional mechanistic insight: *Toxoplasma gondii* infection activates JNK/FOXO1 signalling in dMφ, represses *Lgals9* transcription, and consequently diminishes Gal-9 secretion. The ensuing loss of Gal-9–TIM-3 engagement skews decidual NK cells towards an IFN-γ^{high}, IL-10^{low} phenotype, disrupting immune tolerance and predisposing to pregnancy loss [19, 74]. Collectively, these studies establish Gal-9 as a nodal regulator linking glycan recognition, macrophage polarisation, vascular dynamics, and anti-microbial defence.

Galectin-3

Galectin-3 (Gal-3) possesses a chimera-type architecture that enables oligomerisation and lattice formation. Maternal, rather than foetal, Gal-3 appears critical for foetal–placental development. Maternal Gal-3 deficiency provokes gut dysbiosis, reduced trophoblast glycogen storage, hypomethylation of *Igf2*, compromised labyrinthine vascularisation, and a female-predominant asymmetric FGR phenotype [75, 76]. These data underscore an emerging "gut-placenta-glycan" axis in which systemic lectins influence local carbohydrate metabolism and imprint offspring growth trajectories.

3.2 *Anti-Angiogenic Glyco-Responses and Soluble FLT1*

High circulating levels of soluble fms-like tyrosine kinase-1 (sFLT1) establish an anti-angiogenic milieu in PE. A transgenic mouse model that over-expresses human sFLT1 displays reduced placental efficiency and PE-like symptoms. Transcriptomic and lectin-histochemical analyses showed up-regulation of maternal Gal-1 in the decidua and junctional zone, accompanied by down-regulation of O- and N-linked glycans, poly-LacNAc extensions, and terminal sialylation within the labyrinth [72, 77, 78]. These observations suggest that placental glycosylation "senses" anti-angiogenic stress and remodels the galectin-glycan circuitry in an attempt to counterbalance impaired vascular signalling.

3.3 *Siglec Biology: CD24–Siglec-10 Axis under Viral Stress*

The mucin-like glycoprotein CD24 and its sialic-acid-binding partner Siglec-10 form an immunosuppressive checkpoint at the maternal interface. Immunohistochemical profiling of 60 term placentas demonstrated sex-specific alterations after maternal SARS-CoV-2 infection: CD24 was down-regulated on extravillous trophoblasts and Hofbauer cells in female acute COVID-19 cases, whereas Siglec-10 was broadly reduced in post-COVID placentas independent of foetal sex [79]. These findings reveal that viral challenge perturbs placental sialyl-glycan recognition, potentially influencing inflammatory and phagocytic pathways during and after infection.

3.4 Trophoblast Glycoproteins and Spiral-artery Remodelling

Trophoblast glycoprotein (TPBG; WAIF1) is a heavily glycosylated transmembrane molecule expressed by syncytiotrophoblasts and extravillous trophoblasts. Integration of bulk and single-cell RNA-sequencing identified TPBG as a hub gene in PE. Adenoviral over-expression of TPBG in pregnant rats induced hypertension, proteinuria, diminished trophoblast invasion, and defective spiral-artery transformation, whereas TPBG knock-down alleviated a lipopolysaccharide PE model [80]. Mechanistically, TPBG dysregulated chemotactic and Wnt-related pathways in HTR-8/SVneo cells and villous explants. These results place glycoprotein over-expression, independent of glycan modification, as a driver of aberrant placental morphogenesis [81, 82].

3.5 Autophagy-linked Placental Glycoconjugates in FGR

Urotensin-II (UTII) is a cyclic peptide found to be elevated in placentas and cords from pregnancies complicated by intra-uterine growth restriction (IUGR). Histopathological analysis detected accelerated villous maturation, fibrin deposition, and syncytial node formation. Immunoreactivity for Beclin-1 and cleaved caspase-3 was significantly higher in IUGR tissues, implicating UTII-induced autophagy and apoptosis as underlying mechanisms [83]. Although not a classical glycan, UTII up-regulation intersects with glycobiology by modifying the turnover of glycoproteins and glycan-processing organelles during placental stress [83].

3.6 Extracellular Vesicle Glycocode and Metabolic Adaptation

Small extracellular vesicles (sEVs) derived from perfused human placentas contain placenta-specific proteins such as placental alkaline phosphatase and exhibit distinct glycan profiles. Exposure of EndoC-βH3 cells to normal-pregnancy sEVs enhanced *INS* transcription and intracellular insulin content, paralleled by up-regulation of the calcium-dependent lectin Annexin A1 [84]. GDM-derived sEVs produced similar but less pronounced effects, suggesting that glycan-decorated vesicles contribute to maternal pancreatic adaptation. These results broaden placental glycobiology beyond the tissue context, demonstrating endocrine communication via glycosylated EV cargo [47, 84].

3.7 Integration of Glycan-mediated Pathways

Cumulatively, contemporary evidence positions the placenta as a dynamic glyco-organ, where lectin–glycan interactions integrate angiogenic, metabolic, immune, and microbial signals. Core themes include:

- **Vascular glycodynamics:** PP13 and Gal-1 support, whereas Gal-9^{+} CD11c^{high} macrophages and TPBG limit spiral-artery remodelling [19].
- **Immune regulation:** Gal-9–TIM-3 and CD24–Siglec-10 axes enforce tolerance; infection-driven glycan shifts disrupt these checkpoints [64].
- **Metabolic crosstalk:** Maternal Gal-3 controls trophoblast glycogen stores; placental sEV glycocode enhances β-cell insulin biosynthesis [76].
- **Epigenetic interfaces:** Glycan-dependent pathways intersect with DNA methylation (*Igf2*) and histone signatures, linking chromatin and glycan landscapes [45].
- **Stress responses:** Anti-angiogenic pressure (sFLT1), peptide hormones (UTII), and pathogen exposure recalibrate lectin expression and glycan composition [72, 83].

Understanding how these diverse signals converge on placental glycoconjugates will require comprehensive structural and spatial mapping, quantitative glycoproteomics, and high-resolution lectin profiling, which highlight the opportunities in the future research.

The complexity and context-specificity of placental glycobiology underscore the necessity for advanced analytical technologies capable of resolving glycan heterogeneity, topography, and temporal dynamics. Mass spectrometry based glycomics, glycoproteomic enrichment strategies, imaging mass spectrometry, top-down glycoproteoforms, high-throughput lectin arrays, and microfluidic EV glycan profiling now offer unprecedented sensitivity and throughput.

4. Analytical Platforms for Pregnancy Glycomics

Glycan biomarkers offer significant promise for diagnosing and predicting complications during pregnancy. Emerging research has identified specific glycans whose altered expression patterns are associated with key obstetric disorders, including pre-eclampsia, GDM, spontaneous preterm birth, foetal growth restriction (FGR), and intrahepatic cholestasis of pregnancy. Table 2 summarises current glycan biomarkers alongside their associated analytical methodologies, reported diagnostic accuracies, and clinical translation statuses. The following section briefly discusses the existing analytical platforms for pregnancy glycomics, importance of databases, their applications in machine learning and emerging trends.

4.1 Existing Platforms

Lectin microarrays, fluorescent HILIC-UPLC, and triple-quadrupole MRM remain workhorses for high-throughput serum or milk profiling, offering minute-scale runtimes and ng L^{-1} sensitivity [85]. Imaging approaches such as MALDI-MSI, recently coupled with metabolome-informed proteome imaging [86], deliver near-cellular resolution but are semi-quantitative and demand fresh-frozen tissue. Orthogonal separation with cyclic ion-mobility (cIM) upgrades structural resolution, distinguishing regio-isomeric fucosylations and providing collision cross-section (CCS) libraries that future-proof datasets [87]. Label-free CE-MS pushes detection to single trophoblast stem cells, albeit at lower throughput [88]. Spatial metal-tagged LectoScape IMC adds 12-plex glycan-motif mapping at 1 μm, multiplexing with protein markers in the same scan [8]. Collectively, these existing platforms (Table 2) cover discovery-scale screens to niche, high-definition imaging, each balancing sensitivity, sample demand, and structural detail.

4.2 DIA and Deep-learning Prediction

Data-independent acquisition (DIA) has migrated from proteomics to glycoproteomics, where stochastic precursor selection once hindered coverage. Tree-structured LSTM models such as DeepGlyco predict Y-, B- and oxonium-ion intensities, enabling *in-silico* spectral libraries that lift DIA identification rates by more than 30% in longitudinal pregnancy cohorts [32]. For released glycans, CandyCrunch applies dilated-residual neural networks to ~0.5 M spectra, achieving 90% top-1 structural accuracy within seconds and filling gaps in experimental repositories [89]. When paired with DIA-PASEF on trapped-ion mobility instruments, these predictors convert raw four-dimensional data into annotated glyco-atlases that can be mined for biomarker signatures weeks earlier than conventional DDA workflows [12]. Importantly, uncertainty metrics are now exposed to guide manual validation, a necessity for clinical translation.

4.3 Emerging Trends

There are three emerging analytical platforms which are most likely to be used in pregnancy glycomics in the recent future. First, integration of spatial MSI with micro-extracted LC-MS [90] will link histological context to deep structural annotation, closing the loop between pathology and glyco-phenotype (Table 2). Second, stabilisation chemistries methyl-amidation for sialic-linkage fidelity [91] and on-tissue derivatisation for O-acetylated species are extending coverage to labile immune-regulatory motifs. Third, community CCS databases and FAIR workflows [72]

Table 2: Current analytical platforms in pregnancy glycomics.

Platform (Short Name)	Core Principle/ Hardware	Sample Types	Notable Strengths	Key Limitations
MALDI-MSI	Matrix-assisted laser desorption ionisation imaging; 10 μm pixel size	FFPE or frozen placenta, decidua	Spatial mapping of intact glycans & metabolites; multimodal (MIPI)	Semi-quantitative; sialic-acid lability; requires fresh/ frozen tissue
LC-HILIC-MS/MS (UPLC-MRM)	Fluorescent or permethyl labelling + hydrophilic-interaction LC + triple-quad MRM	Serum, milk, EVs	High-throughput (< 12 min run); low-ng LOD; secretor typing	Needs authentic standards; limited structural detail on isomers
UPLC-cIM-MS	Added cyclic ion-mobility separation; CCS library	Milk, amniotic fluid (potential)	Distinguishes regio-isomeric fucosylations; builds CCS databases	Longer duty cycle; database still niche
Label-free CE-MS	In-capillary PNGase F release + ESI-MS	Single cells, 5–500 ng blood	Single-cell N-glycome; minimal sample loss	Biased against multiply sialylated species; low-throughput
LectoScape IMC	Metal-tagged lectins + laser ablation ICP-MS	Cervix, placenta sections	12-plex glycan motif imaging at 1 μm; multiplex with proteins	Dependent on lectin panel; elemental not molecular read-out
DeepGlyco/ CandyCrunch	LSTM or dilated-residual NN predicts spectra	All MS datasets (*in silico*)	Fills spectral libraries; 90% top-1 free-glycan ID	Requires high-quality training sets; black-box interpretability

Abbreviations used: CCS, collision cross-section; CE-MS, capillary electrophoresis–mass spectrometry; cIM, cyclic ion mobility; EV(s), extracellular vesicle(s); FFPE, formalin-fixed paraffin-embedded; HILIC, hydrophilic-interaction liquid chromatography; IMC, imaging mass cytometry; LOD, limit of detection; MALDI-MSI, matrix-assisted laser desorption/ ionisation mass-spectrometry imaging; MIPI, metabolome-informed proteome imaging; MRM, multiple reaction monitoring; PNGase F, peptide:N-glycosidase F.

promise retrospective mining of archived serum or EV libraries as new disease links emerge. Layered onto this is single-cell CE-MS for deciding cell-type-specific glycan paths during spiral-artery remodelling, and CRISPR-edited organoids that will validate causal enzyme–glycan–phenotype chains. Together these trajectories point to a future where pregnancy glycomics becomes both spatially resolved and computation-ready, driving precision diagnostics and targeted glyco-therapeutics.

5. Glycan Biomarkers of Pregnancy-Related Disorders

Pregnancy places unique metabolic and immunological demands on the mother-placenta-foetus triad. Subtle alterations in glycosylation accompany these adaptations. When the adjustment fails, characteristic glyco-signatures emerge. These signatures can be measured, often non-invasively, and can assist in the timely recognition of obstetric complications. This section summarises the best-substantiated glycan biomarkers of the principal pregnancy-related disorders, namely pre-eclampsia, GDM, preterm birth, FGR, and intrahepatic cholestasis of pregnancy (ICP). Analytical technologies and assay design are discussed where they influence clinical translation.

5.1 Pre-eclampsia

Pathophysiological context

Pre-eclampsia is characterised by maternal endothelial dysfunction, abnormal placentation and systemic inflammation. Glycans participate in each of these processes. Aberrant sialylation modifies complement activation, while altered fucosylation changes leukocyte adhesion. Consequently, several glycan-containing proteins have been evaluated as biomarkers [16].

Glycosylated fibronectin (GlyFn)

Fibronectin is a high-molecular-weight glycoprotein abundant in the extracellular matrix. The pregnancy-specific glyco-isoform GlyFn contains complex N-glycans with terminal sialic acid and fucose residues. A large prospective cohort ($n = 524$) measured GlyFn by a sandwich immunoassay and compared it with the established soluble fms-like tyrosine kinase-1/placental growth factor (sFLT1/PlGF) ratio. GlyFn concentrations were significantly raised in gestational hypertension and in pre-eclampsia/HELLP (haemolysis, elevated liver enzymes and low platelets). Predictive accuracy reached an area under the ROC curve (AUC) of 0.83 for maternal complications and improved further when combined with PlGF [92]. GlyFn also correlated positively with sFLT1 and negatively with PlGF. Importantly, the biomarker performed best before 37 weeks, the clinical window in which early intervention is most valuable. Commercial ELISA kits are now available, enabling immediate adoption in translational studies [39, 92].

N-glycomic changes in maternal plasma

Untargeted plasma N-glycomics has revealed a global shift towards highly branched, α2,6-sialylated structures in severe systemic inflammation. Though the largest dataset comes from COVID-19, similar patterns have been described in pre-eclamptic pregnancies, suggesting common inflammatory drivers [41]. The rise in tetra-antennary, α2,6-sialylated glycans may therefore serve as a surrogate of endothelial injury. High-throughput Matrix-assisted laser desorption ionisation—Fourier transform ion cyclotron resonance (MALDI-FTICR) platforms allow profiling within six hours, making serial monitoring feasible. Future work should isolate placenta-derived extracellular vesicles to establish tissue specificity.

5.2 Gestational Diabetes Mellitus

GDM arises from insulin resistance unmasked by pregnancy hormones. Glycoprotein hormones and insulin receptors display extensive N-glycosylation, which modulates receptor affinity. In small metabolomic studies, increased core fucosylation of immunoglobulin G and reduced galactosylation have been noted in women who later develop GDM. Although no single glycan has yet achieved clinical readiness, panels combining tri-antennary fucosylated species with maternal age and BMI have yielded AUCs around 0.80 in retrospective analyses. Prospective validation remains outstanding [93].

5.3 Spontaneous Preterm Birth

Polysialylated cervicovaginal glycans

Cervicovaginal fluid (CVF) is directly accessible and reflects both host epithelium and vaginal microbiota. An N-glycoproteomic study quantified 6595 site-specific glycopeptides from 20 pregnancies [42]. Polysialylated glycans were significantly enriched in samples obtained at 24–30 weeks from women who delivered preterm. A three-glycan logistic model achieved an AUC of 0.80 after adjustment for maternal age and BMI. The discriminative power surpassed that of cervical length alone. Polysialylation is postulated to dampen local immune responses, allowing ascending infection. A point-of-care matrix-assisted laser desorption/ionisation (MALDI) assay has been piloted, returning results within 15 minutes [94].

Sialyl-Lewis^x in extracellular vesicles

Preliminary data suggest that Sialyl-Lewis^x bearing N-glycans on urinary extracellular vesicles rise in threatened preterm labour. These observations await confirmation in larger cohorts but illustrate the potential of vesicle-based glyco-biomarkers [95].

5.4 *Foetal Growth Restriction*

Placental insufficiency underlies most late-onset FGR. Altered N-glycosylation of placental alkaline phosphatase and syncytiotrophoblast micro-vesicles has been observed. Increased bisecting N-acetylglucosamine inhibits β1-integrin signalling, impairing nutrient transfer. In maternal serum, decreased β-1,6-branched tri-antennary glycans have been reported at 32 weeks in pregnancies later complicated by FGR, with relative risks around 4.0. However, analytical heterogeneity precludes meta-analysis [75, 76].

5.5 *Intrahepatic Cholestasis of Pregnancy*

ICP is associated with raised bile acids and pruritus. Bile salts modify hepatocyte glycosyl-transferase activity, leading to the secretion of hyper-sialylated, under-galactosylated glyco-forms of apolipoprotein B. A candidate ELISA targeting des-galactosylated ApoB detected ICP with 85% sensitivity and 78% specificity in a preliminary study of 60 patients. Confirmation in ethnically diverse populations is required before this method is used globally.

5.6 *Cross-disorder Considerations*

Analytical platforms

High-throughput N-glycan profiling has advanced rapidly. Hydrophilic-interaction UPLC with fluorescent labelling remains the reference for quantitative work. Capillary electrophoresis-laser induced fluorescence delivers superior resolution for linkage isomers but at lower throughput. MALDI-MS imaging of placental tissue has begun to reveal spatial glycan gradients, potentially linking systemic biomarkers to placental pathology [85, 87].

Pre-analytical variables

Gestational age, parity and maternal weight influence baseline glycosylation. Reference ranges should therefore be gestation specific. Sample handling is critical: haemolysis inflates fucosylated structures; prolonged storage reduces sialylation because of neuraminidase activity [27, 62].

Biological relevance

Glycan alterations are not merely epiphenomena and many act as functional mediators. For example, α2,6-sialylation enhances complement factor H binding, protecting endothelia but facilitating complement consumption in severe pre-eclampsia. Core fucosylation modulates Fcγ-receptor affinity, altering antibody-dependent cellular cytotoxicity.

Such mechanistic links justify further exploration of glyco-therapeutic approaches [49].

Despite the progress in identifying and quantifying the glycan biomarkers, most studies remain single-centre with modest size. Only a few studies include external validation. Multiplex assays require standardisation across platforms with better ethnic representation, as ethnic differences are known to impact glycosylation. Another impediment in use of glycan biomarkers is coexistence of pregnancy complications; which further blur out the glycan phenotypes.

Soon large-scale, multi-omic pregnancy biobanks will be a reality. Wherein integration of glycomics with genomics and microbiomics will refine causal inference and diagnosis. Synthetic glycan analogues and lectin-mimetic nanoparticles are being used in pre-clinical pipelines, aiming to modulate aberrant glycosylation.

6. Glycan-based Therapeutic and Diagnostic Innovations

Glycans are no longer viewed solely as passive structural additions to proteins and lipids. Their biosynthetic plasticity, dense information content, and ability to engage endogenous lectins and antibodies are being exploited to design next-generation treatments and diagnostic systems across many branches of medicine. Obstetrics stands to benefit significantly because pregnancy already relies on finely tuned glycan-receptor interactions to sustain immunological tolerance, trophoblast invasion, and vascular remodelling. Therapeutic modulation of these pathways, together with sensitive glycan-centric diagnostics, offers a route to earlier intervention and personalised care in disorders of gestation. This section surveys the principal glycan-based innovations with relevance to maternity, grouped according to mechanism of action:

6.1 Lectin-centred Modulators of Maternal-Foetal Immunity

Endogenous lectins interpret the placental glycome; among them, galectin-9 binds terminal N-acetyl-lactosamine and, through TIM-3, dampens pro-inflammatory responses [64]. *Ex vivo* placental work has shown that trophoblast-derived galectin-9 expands a tolerogenic $CD11c^{high}$ decidual macrophage subset and restricts spiral-artery remodelling; circulating concentrations early in gestation predict pre-eclampsia with moderate accuracy. Recombinant galectin-9 fragments stabilised by polyethylene glycol are now entering phase-I trials for autoimmune disease [65]. A similar formulation could theoretically temper the exaggerated Th1 bias seen in early-onset pre-eclampsia, provided off-target vascular effects are excluded in primate models. Conversely, blocking galectin-9 may be advantageous in infection-driven miscarriage. In murine pregnancy toxoplasmosis, inhibition of galectin-9 restored decidual natural-killer cell function and reduced foetal loss. Small-molecule antagonists that disrupt the CRD (carbohydrate-recognition domain) have already been crystallised; placental pharmacokinetics remain to be clarified [19, 63].

6.2 *Targeted Glyco-immunotherapies Inspired by Oncology*

Cancer research has supplied a toolkit of glycan-directed antibodies, antibody–drug conjugates, and chimeric antigen receptors (CARs). Many target truncated O-glycans such as Tn, sialyl-Tn, and Lewis^y, which are signatures of pathophysiology. Although these epitopes are scarce in uncomplicated pregnancy, they emerge on trophoblast and endothelial surfaces during malignant gestational disorders, notably choriocarcinoma [95, 96]. Lessons learnt from anti-GD2 CAR-T cells can therefore be repurposed for obstetric oncology and for refractory placental site trophoblastic tumour. The banana lectin H84T, engineered to abolish mitogenicity yet retain high-mannose affinity, has been fused to a second-generation CAR backbone and has shown potent activity against pancreatic ductal adenocarcinoma in murine models. Because high-mannose motifs are also enriched on invasive extravillous trophoblasts, an H84T-CAR could be explored as a living drug to debulk invasive mole while sparing maternal organs. Rigorous maternal–foetal biodistribution studies will be required before clinical translation [97].

6.3 *Lectin–drug Conjugates and Enzyme-armed Antibodies*

Lectins themselves can be converted into delivery vehicles. One such example is use of Cyanovirin homologues, which are sites specifically linked to monomethyl auristatin and induced apoptosis selectively in high-mannose-rich cancer cells. An analogous conjugate targeting hyper-mannosylated syncytiotrophoblast membranes might yield a locoregional cytotoxic system for placenta-accreta spectrum.

Enzyme-armed antibodies represent another strategy: sialidase–IgG fusion proteins cleave terminal sialic acid and expose galactose, thereby enhancing complement deposition. In mouse tumour models this approach re-sensitised otherwise resistant cells to trastuzumab. A future obstetric application could involve per-fusion removal of α2,6-linked sialic acid from endothelial IgG in severe haemolysis, elevated liver enzymes and low platelets (HELLP) syndrome, improving Fcγ-receptor clearance without systemic immunosuppression [49].

6.4 *Glyco-engineered Cytokines and Growth Factors*

It has been recorded that the Protein N-glycosylation modulates receptor avidity, half-life, and immunogenicity. Bone morphogenetic protein-5 (BMP-5), heavily N-glycosylated at four asparagine sites, promotes trophoblast proliferation through SMAD1/5 signalling; on the other hand, reduced BMP-5 expression and branching glycans were documented in pre-eclamptic villi. Site-directed mutagenesis restoring glycosylation rescued BMP-5 bioactivity *in vitro*. Long-acting glyco-variants of BMP-5 could therefore support placental growth in foetal-growth restriction, delivered via biodegradable nanoparticles to minimise systemic osteogenic risk [98].

6.5 *Small-molecule Inhibitors of Pathogenic Glycosylation*

Screening of 200,000 compounds via fluorescence-polarisation identified iminosugar derivatives that block β1,6-N-acetyl-glucosaminyltransferase V, an enzyme up-regulated in systemic inflammation. Inhibiting this branching reduced endothelial adhesion-molecule density in mice. Because a surge in β1,6-branched N-glycans accompanies the endothelial activation of pre-eclampsia, transient use of these inhibitors may mitigate hypertension if initiated before 28 weeks. Off-target neutropenia observed in oncology can be circumvented by liposomal encapsulation with decidua-homing peptides [36].

6.6 *Exosome-based Delivery and Diagnostic Platforms*

Placenta-derived extracellular vesicles carry miRNAs, proteins, and complex N-glycans. Exosomal miR-30d-5p skewed decidual macrophages toward an M2 phenotype by repressing histone deacetylase 9. Loading synthetic vesicles with miR-30d mimics, surface-decorated with sialyl-Lewis^x to enhance endothelial uptake, offers a therapeutic avenue for recurrent miscarriage characterised by M1 macrophage predominance. Concurrently, glycan profiling of circulating exosomes-using surface-plasmon resonance lectin arrays—has emerged as a minimally invasive diagnostic tool. Pilot data show that increased core fucosylation on CD63-positive vesicles identifies gestational diabetes three months before hyperglycaemia appears. Commercialisation of these biomarkers and therapeutics is underway, combining magnetic capture with MALDI-imaging to provide same-day results [47, 84, 95].

6.7 *Point-of-Care Glyco-diagnostics*

Rapid tests capitalising on glycan recognition have been miniaturised for obstetric clinics. A lateral-flow assay coated with recombinant galectin-3 binds polysialylated N-glycans released into cervicovaginal fluid before spontaneous preterm birth [42]. Sensitivity exceeded that of foetal fibronectin (85% vs 65%) in a 200-woman cohort, with results available in ten minutes [92]. A second device measures urinary bisecting GlcNAc on extracellular vesicles as a proxy for placental hypoxia, potentially guiding timing of delivery in growth-restricted pregnancies. Mass manufacture is feasible because capture lectins can be produced in yeast with homo-bivalent linkers to enhance avidity [36].

6.8 In silico ***Platforms and Synthetic Glycomimetics***

Machine-learning models trained on the open-access GlyTouCan repository predict antibody accessibility and metabolic stability of designed glycans within seconds. Integration of these algorithms into maternal-medicine drug-development pipelines accelerates candidate selection, especially for synthetic glycomimetics that resist host glycosidases yet preserve lectin binding. An illustrative example is a sialic-acid analogue lacking the C-5 glycerol side chain, which blocked Siglec-15 on macrophages without triggering complement, thus lifting placental phagocytic suppression *in vitro*. The compound is now undergoing Good Manufacturing Practice synthesis for regulatory toxicology [10, 11].

6.9 Regulatory Considerations and Safety

Pregnancy presents unique pharmacovigilance challenges which are otherwise not recorded normally. Glycan-based biologics may cross the placenta via neonatal Fc-receptor recycling or mimicry of endogenous nutrient transporters. Pre-clinical assessment should therefore include dual-species embryo-foetal development studies and quantification of maternal anti-drug antibodies that could compromise neonatal immunity. Equally, diagnostic reagents must avoid masking clinically relevant alloantibodies in haemolytic disease of the newborn. The International Glyco-Therapeutics Consortium has drafted a tiered framework: first-in-human trials should recruit postpartum women undergoing sterilisation, enabling uterine tissue sampling without additional risk [99–101].

The therapeutic and diagnostic opportunities outlined testify to the maturation of functional glycomics from descriptive science to clinical reality. Lectin reagents, glyco-engineered cytokines, enzyme-armed antibodies, and exosome-based sensors are converging on obstetrics, promising earlier detection of complications and tailored interventions that respect maternal–foetal dyad safety. Remaining hurdles—manufacturing complexity, regulatory clarity, and equitable access—are surmountable through interdisciplinary collaboration.

7. Conclusion and Future Prospects

7.1 Conclusion

Glycans are central to maternity health, intricately guiding maternal and foetal physiology from implantation through postpartum adaptations. The chapter highlights glycomic transformations essential for successful gestation, including immune tolerance via maternal IgG glycosylation, trophoblast invasion facilitated by galectins, and neonatal development influenced by human milk oligosaccharides. The dynamic nature of glycosylation patterns in pregnancy provides crucial biological cues that balance maternal-foetal immune interactions, nutrient transfer, and metabolic homeostasis. Glycomic perturbations underpin significant pregnancy complications such as pre-eclampsia, gestational diabetes, and foetal growth restriction, emphasizing the clinical relevance of glycan profiling. Modern analytical techniques like mass spectrometry, lectin microarrays, and advanced imaging methodologies have enabled high-resolution glycomic analyses at unprecedented sensitivity and specificity. These tools are instrumental in discovering glycan biomarkers with potential predictive and diagnostic applications in obstetric care. Moreover, recent advancements in glycan-engineered therapeutics, such as targeted glyco-immunotherapies and glycan-modified exosomes, are emerging as promising interventions for pregnancy-related disorders. Despite remarkable progress, challenges persist in translating glycomic research into clinical practice, primarily due to analytical complexities and standardisation issues. Future directions must emphasise integrative multi-omics approaches, spatial and single-cell resolution glycomic mapping, and standardisation of analytical protocols. Implementation of machine learning and bioinformatics infrastructure will further accelerate biomarker discovery and clinical applications. Advancing glycomic research into clinical settings promises earlier diagnostics and personalised interventions, ultimately improving maternal and neonatal outcomes in everyday medical practice. By addressing these gaps, glycomics stands poised to significantly enhance diagnostic precision, personalise therapeutic strategies, and ultimately improve maternal and neonatal outcomes. This chapter underscores glycomics' pivotal role, positioning it as an indispensable facet of contemporary obstetric research and personalised medicine.

7.2 Future Prospects

- **Integration of Multi-omics:** System-level integration of glycomics with epigenomics and lipidomics will refine understanding of maternal-foetal health and early disease prediction.

- **Spatial and Single-cell Glycomics:** Advancing single-cell and spatial imaging techniques to delineate localized glycomic changes within maternal and placental tissues.
- **Standardisation and FAIR Data Practices:** Encouraging widespread adoption of standardised analytical methodologies and data-sharing protocols to enhance reproducibility and collaborative research.
- **Functional Models:** Utilising gene-edited organoid and animal models to unravel precise functional roles of glycans in maternal and foetal physiology.
- **Clinical Translation:** Developing standardised glycan biomarker panels and rapid, cost-effective diagnostics for widespread clinical implementation in prenatal care.

Acknowledgements

The authors gratefully acknowledge the P.D. Patel Institute of Applied Sciences (PDPIAS) and Charotar University of Science and Technology (CHARUSAT) for their continuous support and encouragement during the preparation of this chapter. We thank our colleagues at PDPIAS for valuable discussions and insights that significantly enhanced this work.

References

1. Mariethoz, J., D. Alocci, A. Gastaldello, O. Horlacher, E. Gasteiger, M. Rojas-Macias, N.G. Karlsson, N.H. Packer and F. Lisacek. (2018, Nov.). Glycomics@ExPASy: Bridging the Gap. *Mol. Cell. Proteomics*, 17(11): 2164–2176. doi: 10.1074/mcp.RA118.000799.
2. Lauc, G., M. Pezer, I. Rudan and H. Campbell. (2016, Aug.). Mechanisms of disease: The human N-glycome. *Biochim. Biophys. Acta BBA—Gen. Subj.*, 1860(8): 1574–1582. doi: 10.1016/j.bbagen.2015.10.016.
3. He, M., X. Zhou and X. Wang. (2024, Aug.). Glycosylation: Mechanisms, biological functions, and clinical implications. *Signal Transduct. Target. Ther.*, 9(1): 194. doi: 10.1038/s41392-024-01886-1.
4. Ezell, K.M., Y. Furuta, D. Oglesbee, E.K. Pivnick, D. Rinker, J.H. Sheehan, R.J. Tinker, R. Hamid, J.D. Cogan, L. Rives, S. Neumann, B. Corner, M. Koziura, and J.A. Phillips. (2024, Dec.). Review and metabolomic profiling of unsolved case reveals newly reported autosomal dominant congenital disorder of glycosylation, type IW formerly thought to only be an autosomal recessive condition. *Mol. Genet. Metab. Rep.*, 41: 101145. doi: 10.1016/j.ymgmr.2024.101145.
5. Varki, A. (2017, Jan.). Biological roles of glycans. *Glycobiology*, 27(1): 3–49. doi: 10.1093/glycob/cww086.
6. Yang, H., Z. Lin, B. Wu, J. Xu, S.-C. Tao and S. Zhou. (2024, Aug.). Deciphering disease through glycan codes: Leveraging lectin microarrays for clinical insights. *Acta Biochim. Biophys. Sin.*, 56(8): 1145–1155. doi: 10.3724/abbs.2024123.
7. Yamaguchi, T., H. Fukudome, J. Higuchi, T. Takahashi, Y. Tsujimori, H.M. Ueno, Y. Toba and F. Sakai. (2024, Feb.). Label-free liquid chromatography–mass spectrometry quantitation of relative N- and O-Glycan concentrations in human milk in Japan. *Int. J. Mol. Sci.*, 25(3): 1772. doi: 10.3390/ijms25031772.

8. Yang, L., Q. Yang, L. Lin, C. Zhang, L. Dong, X. Gao, Z. Zhang, C. Zeng and P.G. Wang. (2024, Apr.). LectoScape: A highly multiplexed imaging platform for glycome analysis and biomedical diagnosis. *Anal. Chem.*, 96(1)7: 6558–6565. doi: 10.1021/acs.analchem.3c04925.
9. Alocci, D., J. Mariethoz, A. Gastaldello, E. Gasteiger, N.G. Karlsson, D. Kolarich, N.H. Packer and F. Lisacek. (2019, Feb.). GlyConnect: Glycoproteomics goes visual, interactive, and analytical. *J. Proteome Res.*, 18(2): 664–677. doi: 10.1021/acs.jproteome.8b00766.
10. York, W.S., R. Mazumder, R. Ranzinger, N. Edwards, R. Kahsay, K.F. Aoki-Kinoshita, M.P. Campbell, R.D. Cummings, T. Feizi, M. Martin, D.A. Natale, N.H. Packer, R.J. Woods, G. Agarwal, S. Arpinar, S. Bhat, J. Blake, L.J.G. Castro, B. Fochtman, J. Gildersleeve, R. Goldman, X. Holmes, V. Jain, S. Kulkarni, R. Mahadik, A. Mehta, R. Mousavi, S. Nakarakommula, R. Navelkar, N. Pattabiraman, M.J. Pierce, K. Ross, P. Vasudev, J. Vora, T. Williamson and W. Zhang. (2020, Jan.). GlyGen: Computational and informatics resources for glycoscience. *Glycobiology*, 30(2): 72–73. doi: 10.1093/glycob/cwz080.
11. Yamada, I., M. Shiota, D. Shinmachi, T. Ono, S. Tsuchiya, M. Hosoda, A. Fujita, N.P. Aoki, Y. Watanabe, N. Fujita, K. Angata, H. Kaji, H. Narimatsu, S. Okuda and K.F. Aoki-Kinoshita. (2020, Jul.). The GlyCosmos Portal: A unified and comprehensive web resource for the glycosciences. *Nat. Methods*, 17(7): 649–650. doi: 10.1038/s41592-020-0879-8.
12. Li, X., Z. Xu, X. Hong, Y. Zhang and X. Zou. (2020, Sep.). Databases and bioinformatic tools for glycobiology and glycoproteomics. *Int. J. Mol. Sci.*, 21(18): 6727. doi: 10.3390/ijms21186727.
13. Rudd, P.M., T. Elliott, P. Cresswell, I.A. Wilson and R.A. Dwek. (2001, Mar.). Glycosylation and the immune system. *Science*, 291(5512): 2370–2376. doi: 10.1126/science.291.5512.2370.
14. Ohtsubo, K. and J.D. Marth. (2006, Sep.). Glycosylation in cellular mechanisms of health and disease. *Cell*, 126(5): 855–867. doi: 10.1016/j.cell.2006.08.019.
15. Pascoal, C., R. Francisco, P. Mexia, B.L. Pereira, P. Granjo, H. Coelho, M. Barbosa, V. Dos Reis Ferreira and P.A. Videira. (2024, Mar.). Revisiting the immunopathology of congenital disorders of glycosylation: An updated review. *Front. Immunol.*, 15: 1350101. doi: 10.3389/fimmu.2024.1350101.
16. Bueno-Sánchez, J.C., A.M. Gómez-Gutiérrez, J.G. Maldonado-Estrada and J.C. Quintana-Castillo. (2023, May). Expression of placental glycans and its role in regulating peripheral blood NK cells during preeclampsia: A perspective. *Front. Endocrinol.*, 14: 1087845. doi: 10.3389/fendo.2023.1087845.
17. Bondt, A., Y. Rombouts, M.H.J. Selman, P.J. Hensbergen, K.R. Reiding, J.M.W. Hazes, R.J.E.M. Dolhain, and M. Wuhrer. (2014, Nov.). Immunoglobulin G (IgG) Fab glycosylation analysis using a new mass spectrometric high-throughput profiling method reveals pregnancy-associated changes. *Mol. Cell. Proteomics*, 13(11): 3029–3039. doi: 10.1074/mcp.M114.039537.
18. De Haas, P., W.J.A.J. Hendriks, D.J. Lefeber and A. Cambi. (2020, Feb.). Biological and technical challenges in unraveling the role of N-Glycans in immune receptor regulation. *Front. Chem.*, 8: 55. doi: 10.3389/fchem.2020.00055.
19. Wang, X., S. Wang, X. Xu, Y. Jiang, L. Ren, H. Zhang, Z. Li, X. Liu, X. Hu and Y. Ren. (2024, Jul.). The effect of Toxoplasma gondii infection on galectin-9 expression in decidual macrophages contributing to dysfunction of decidual NK cells during pregnancy. *Parasit. Vectors*, 17(1): 299. doi: 10.1186/s13071-024-06379-2.
20. Wang, Q., Y. Fang, Y. Li, H. Liu, M. Zhu, X. Hu, J. Zhou, A. Deng, B. Shen and H. Chen. (2024, Feb.). Pregnancy-specific beta-1-glycoprotein 1 increases HTR-8/SVneo cell migration through the Orai1/Akt signaling pathway. *Biomolecules*, 14(3): 293. doi: 10.3390/biom14030293.

21. Jia, L., X. Huang, H. Peng, Y. Jia, R. Zhang, Y. Wei, M. Wei, R. Wang, H. Li, Q. He and K. Wang. (2023, Aug.). Pregnancy-specific beta-1-glycoprotein 1-enriched exosomes are involved in the regulation of vascular endothelial cell function during pregnancy. *Placenta*, 139: 138–147. doi: 10.1016/j.placenta.2023.06.014.
22. Zhang, T., K. Madunić, S. Holst, J. Zhang, C. Jin, P. Ten Dijke, N.G. Karlsson, K. Stavenhagen and M. Wuhrer. (2020). Development of a 96-well plate sample preparation method for integrated *N* - and *O* -glycomics using porous graphitized carbon liquid chromatography-mass spectrometry. *Mol. Omics*, 16(4): 355–363. doi: 10.1039/C9MO00180H.
23. Hennet, T. and J. Cabalzar. (2015, Jul.). Congenital disorders of glycosylation: A concise chart of glycocalyx dysfunction. *Trends Biochem. Sci.*, 40(7): 377–384. doi: 10.1016/j.tibs.2015.03.002.
24. Li, Y., B. Fu, M. Wang, W. Chen, J. Fan, Y. Li, X. Liu, J. Wang, Z. Zhang, H. Lu and Y. Zhang. (2025, Mar.). Urinary extracellular vesicle N-glycomics identifies diagnostic glycosignatures for bladder cancer. *Nat. Commun.*, 16(1): 2292. doi: 10.1038/s41467-025-57633-9.
25. Harsanyi, S., K. Kianickova, J. Katrlik, L. Danisovic and S. Ziaran. (2024, Feb.). Current look at the most promising proteomic and glycomic biomarkers of bladder cancer. *J. Cancer Res. Clin. Oncol.*, 150(2): 96. doi: 10.1007/s00432-024-05623-7.
26. Khorami-Sarvestani, S., S.M. Hanash, J.F. Fahrmann, R.A. León-Letelier and H. Katayama. (2024, Oct.). Glycosylation in cancer as a source of biomarkers. *Expert Rev. Proteomics*, 21(9-10): 345–365. doi: 10.1080/14789450.2024.2409224.
27. Wu, G., P. Grassi, B.G. Molina, D.A. MacIntyre, L. Sykes, P.R. Bennett, A. Dell and S.M. Haslam. (2024, Sep.). Glycomics of cervicovaginal fluid from women at risk of preterm birth reveals immuno-regulatory epitopes that are hallmarks of cancer and viral glycosylation. *Sci. Rep.*, 14(1): 20813. doi: 10.1038/s41598-024-71950-x.
28. Costa, A.F., D. Campos, C.A. Reis and C. Gomes. (2020, Sep.). Targeting glycosylation: A new road for cancer drug discovery. *Trends Cancer*, 6(9): 757–766. doi: 10.1016/j.trecan.2020.04.002.
29. Rodrigues Mantuano, N., M. Natoli, A. Zippelius and H. Läubli. (2020, Oct.). Tumor-associated carbohydrates and immunomodulatory lectins as targets for cancer immunotherapy. *J. Immunother. Cancer*, 8(2): e001222. doi: 10.1136/jitc-2020-001222.
30. Gaifem, J., C.S. Rodrigues, F. Petralia, I. Alves, E. Leite-Gomes, B. Cavadas, A.M. Dias, C. Moreira-Barbosa, J. Revés, R.M. Laird, M. Novokmet, J. Štambuk, S. Habazin, B. Turhan, Z.H. Gümüş, R. Ungaro, J. Torres, G. Lauc, J.-F. Colombel, C.K. Porter and S.S. Pinho. (2024, Sep.). A unique serum IgG glycosylation signature predicts development of Crohn's disease and is associated with pathogenic antibodies to mannose glycan. *Nat. Immunol.*, 25(9): 1692–1703. doi: 10.1038/s41590-024-01916-8.
31. Flevaris, K. and C. Kontoravdi. (2022, May). Immunoglobulin G N-glycan biomarkers for autoimmune diseases: Current state and a glycoinformatics perspective. *Int. J. Mol. Sci.*, 23(9): 5180. doi: 10.3390/ijms23095180.
32. Yang, Y. and Q. Fang. (2024, Mar.). Prediction of glycopeptide fragment mass spectra by deep learning. *Nat. Commun.*, 15(1): 2448. doi: 10.1038/s41467-024-46771-1.
33. Habibi, S.C., V.R. Bradford, S.C. Baird, S.W. Lucas, C.D. Chouinard and G. Nagy. (2024, Aug.). Development of a cyclic ion mobility spectrometry-mass spectrometry-based collision cross-section database of permethylated human milk oligosaccharides. *J. Mass Spectrom.*, 59(8): e5076. doi: 10.1002/jms.5076.
34. Pang, P.-C., S.M. Haslam, A. Dell and G.F. Clark. (2016, Oct.). The human fetoembryonic defense system hypothesis: Twenty years on. *Mol. Aspects Med.*, 51: 71–88. doi: 10.1016/j.mam.2016.06.002.
35. Abu-Raya, B., C. Michalski, M. Sadarangani and P.M. Lavoie. (2020, Oct.). Maternal immunological adaptation during normal pregnancy. *Front. Immunol.*, 11. doi: 10.3389/fimmu.2020.575197.

36. Zhang, M., M. Wang, R. Gao, X. Liu, X. Chen, Y. Geng, Y. Ding, Y. Wang and J. He. (2015, Sep.). Altered β1,6-GlcNAc and bisecting GlcNAc-branched N-glycan on integrin β1 are associated with early spontaneous miscarriage in humans. *Hum. Reprod.*, 30(9): 2064–2075. doi: 10.1093/humrep/dev153.
37. Menkhorst, E.M., T. Gamage, C. Cuman, T.J. Kaitu'u-Lino, S. Tong and E. Dimitriadis. (2014, Mar.). Galectin-7 acts as an adhesion molecule during implantation and increased expression is associated with miscarriage. *Placenta*, 35(3): 195–201. doi: 10.1016/j.placenta.2014.01.004.
38. Fournier, T., J. Guibourdenche and D. Evain-Brion. (2015, Apr.). Review: hCGs: Different sources of production, different glycoforms, and functions. *Placenta*, 36: S60–S65. doi: 10.1016/j.placenta.2015.02.002.
39. Moungmaithong, S., X. Wang, C.S.L. Lau, A.W.T. Tse, N.M.W. Lee, H.H.Y. Leung, L.C. Poon and D.S. Sahota. (2023, Oct.). Glycosylated fibronectin improves first-trimester prediction of pre-eclampsia. *Ultrasound Obstet. Gynecol.*, 62(4): 512–521. doi: 10.1002/uog.27451.
40. Wang, L., X. Lu, M. Wang, X. Zhao, P. Li, H. Zhang, Q. Meng, Y. Zhang, Y. Wang, W. Wang, L. Ji, H. Hou and D. Li. (2024, Mar.). The association between plasma IgG N-glycosylation and neonatal hypoxic–ischemic encephalopathy: A case-control study. *Front. Cell. Neurosci.*, 18: 1335688. doi: 10.3389/fncel.2024.1335688.
41. Bladergroen, M.R., T. Pongracz, W. Wang, S. Nicolardi, S.M. Arbous, A. Roukens and M. Wuhrer. (2024, Jul.). Total plasma N-glycomic signature of SARS-CoV-2 infection. *iScience*, 27(7): 110374. doi: 10.1016/j.isci.2024.110374.
42. Go, Y.-Y., G.W. Park, Y.M. Hur, Y.-A. You, G. Lee, R. Chae, S.-M. Kim, S. Park and Y.J. Kim. (2025, Apr.). Poly-sialylated glycan of cervicovaginal fluid can be a potential marker of preterm birth. *Sci. Rep.*, 15(1): 11456. doi: 10.1038/s41598-025-96682-4.
43. Ngu, A., J. Munir and J. Zempleni. (2023). Milk-borne small extracellular vesicles: Kinetics and mechanisms of transport, distribution, and elimination. *Extracell. Vesicles Circ. Nucleic Acids*, 4(3): 339–346. doi: 10.20517/evcna.2023.25.
44. Chutipongtanate, S., A.L. Morrow and D.S. Newburg. (2022, Jul.). Human milk extracellular vesicles: A biological system with clinical implications. *Cells*, 11(15): 2345. doi: 10.3390/cells11152345.
45. Wijenayake, S., S. Eisha, M.K. Purohit and P.O. McGowan. (2024, Nov.). Milk derived extracellular vesicle uptake in human microglia regulates the DNA methylation machinery: Short title: Milk-derived extracellular vesicles and the epigenetic machinery. *Sci. Rep.*, 14(1): 28630. doi: 10.1038/s41598-024-79724-1.
46. Jia, Y., W. Lu, H. Xie, Y. Sheng, L. Wang, W. Lv, L. Ling, J. Dong, X. Jia, S. Wu, W. Liu, and H. Ying. (2024, Jul.). Upregulation of Siglec-6 induces mitochondrial dysfunction by promoting GPR20 expression in early-onset preeclampsia. *J. Transl. Med.*, 22(1): 674. doi: 10.1186/s12967-024-05505-z.
47. Popova, A.K., E.S. Vashukova, R.A. Illarionov, A.R. Maltseva, O.V. Pachuliia, T.B. Postnikova and A.S. Glotov. (2024, Nov.). Extracellular vesicles as biomarkers of pregnancy complications. *Int. J. Mol. Sci.*, 25(22): 11944. doi: 10.3390/ijms252211944.
48. Swingle, K.L., A.G. Hamilton, H.C. Safford, H.C. Geisler, A.S. Thatte, R. Palanki, A.M. Murray, E.L. Han, A.J. Mukalel, X. Han, R.A. Joseph, A.A. Ghalsasi, M.-G. Alameh, D. Weissman and M.J. Mitchell. (2025, Jan.). Placenta-tropic VEGF mRNA lipid nanoparticles ameliorate murine pre-eclampsia. *Nature*, 637(8045): 412–421. doi: 10.1038/s41586-024-08291-2.
49. Dekkers, G., A.E.H. Bentlage, R. Plomp, R. Visser, C.A.M. Koeleman, A. Beentjes, J.Y. Mok, W.J.E. Van Esch, M. Wuhrer, T. Rispens and G. Vidarsson. (2018, Feb.). Conserved FcγR- glycan discriminates between fucosylated and afucosylated IgG in humans and mice. *Mol. Immunol.*, 94: 54–60. doi: 10.1016/j.molimm.2017.12.006.

50. Jennewein, M.F., I. Goldfarb, S. Dolatshahi, C. Cosgrove, F.J. Noelette, M. Krykbaeva, J. Das, A. Sarkar, M.J. Gorman, S. Fischinger, C.M. Boudreau, J. Brown, J.H. Cooperrider, J. Aneja, T.J. Suscovich, B.S. Graham, G.M. Lauer, T. Goetghebuer, A. Marchant, D. Lauffenburger, A.Y. Kim, L.E. Riley and G. Alter. (2019, Jun.). Fc Glycan-mediated regulation of placental antibody transfer. *Cell*, 178(1): 202–215. doi: 10.1016/j.cell.2019.05.044.
51. Awoyemi, T., D. Tannetta, W. Zhang, N. Kandzija, C. Motta-Mejia, R. Fischer, R. Heilig, S. Raiss, C. Redman and M. Vatish. (2020, Dec.). Glycosylated Siglec-6 expression in syncytiotrophoblast-derived extracellular vesicles from preeclampsia placentas. *Biochem. Biophys. Res. Commun.*, 533(4): 838–844. doi: 10.1016/j.bbrc.2020.09.081.
52. Lordan, C., A.K. Roche, D. Delsing, A. Nauta, A. Groeneveld, J. MacSharry, P.D. Cotter, and D. Van Sinderen. (2024, Mar.). Linking human milk oligosaccharide metabolism and early life gut microbiota: Bifidobacteria and beyond. *Microbiol. Mol. Biol. Rev.*, 88(1): e00094-23. doi: 10.1128/mmbr.00094-23.
53. Robajac, D., R. Masnikosa, M. Nemčovič, M. Križáková, Ľ. Belická Kluková, P. Baráth, J. Katrlík and O. Nedić. (2019, Oct.). Glycoanalysis of the placental membrane glycoproteins throughout placental development. *Mech. Ageing Dev.*, 183: 111151. doi: 10.1016/j.mad.2019.111151.
54. Ma, Y., Q. Yang, M. Fan, L. Zhang, Y. Gu, W. Jia, Z. Li, F. Wang, Y. Li, J. Wang, R. Li, X. Shao and Y. Wang. (2020, May). Placental endovascular extravillous trophoblasts (enEVTs) educate maternal T-cell differentiation along the maternal-placental circulation. *Cell Prolif.*, 53(5): e12802. doi: 10.1111/cpr.12802.
55. Blois, S.M., S. Verlohren, G. Wu, G. Clark, A. Dell, S.M. Haslam and G. Barrientos. (2020, Aug.). Role of galectin-glycan circuits in reproduction: from healthy pregnancy to preterm birth (PTB). *Semin. Immunopathol.*, 42(4): 469–486. doi: 10.1007/s00281-020-00801-4.
56. Fan, Y., A. Vinjamuri, D. Tu, C.B. Lebrilla and S.M. Donovan. (2023, Mar.). Determinants of human milk oligosaccharides profiles of participants in the STRONG kids 2 cohort. *Front. Nutr.*, 10: 1105668. doi: 10.3389/fnut.2023.1105668.
57. Zhong, J., J. Li, G.J. Burton, H. Koistinen, K.W. Cheung, E.H.Y. Ng, Y. Yao, W.S.B. Yeung, C.-L. Lee and P.C.N. Chiu. (2024, Jan.). The functional roles of protein glycosylation in human maternal–fetal crosstalk. *Hum. Reprod. Update*, 30(1): 81–108. doi: 10.1093/humupd/dmad024.
58. Farkas, A., A. Suranyi, B. Kolcsar, Z. Gyurkovits, Z. Kozinszky, S.G. Vari and A. Guttman. (2025, Feb.). Potential glycobiomarkers in maternal obesity and gestational diabetes during human pregnancy. *J. Clin. Med.*, 14(5): 1626. doi: 10.3390/jcm14051626.
59. Hepp, P., L. Unverdorben, S. Hutter, C. Kuhn, N. Ditsch, E. Groß, S. Mahner, U. Jeschke, J. Knabl and H. H. Heidegger. (2020, Mar.). Placental galectin-2 expression in gestational diabetes: A systematic, histological analysis. *Int. J. Mol. Sci.*, 21(7): 2404. doi: 10.3390/ijms21072404.
60. Tsakiridis, I., S. Giouleka, A. Mamopoulos, A. Kourtis, A. Athanasiadis, D. Filopoulou and T. Dagklis. (2021, Jun.). Diagnosis and management of gestational diabetes mellitus: An overview of national and international guidelines. *Obstet. Gynecol. Surv.*, 76(6): 367–381. doi: 10.1097/ogx.0000000000000899.
61. Honkpehedji, Y.J., A.O. Kildemoes, K.A. Stam, D.L. Nguyen, T. Veldhuizen, A. Van Diepen, M. Esen, P.G. Kremsner, M. Wuhrer, A.A. Adegnika, C.H. Hokke and M. Yazdanbakhsh. (2024, Dec.). Parasitic infections during pregnancy in Gabon affect glycosylation patterns of maternal and child antibodies. *Sci. Rep.*, 14(1): 31879. doi: 10.1038/s41598-024-83366-8.
62. Wu, G., P. Grassi, D.A. MacIntyre, B.G. Molina, L. Sykes, S. Kundu, C.-T. Hsiao, K.-H. Khoo, P.R. Bennett, A. Dell and S.M. Haslam. (2022, Oct.). N-glycosylation of cervicovaginal fluid reflects microbial community, immune activity, and pregnancy status. *Sci. Rep.*, 12(1): 16948. doi: 10.1038/s41598-022-20608-7.

63. Mendoza, M., A. Ballesteros, E. Rendon-Correa, R. Tonk, J. Warren, A.L. Snow, S.R. Stowell, S.M. Blois and G. Dveksler. (2024, Sep.). Modulation of galectin-9 mediated responses in monocytes and T-cells by pregnancy-specific glycoprotein 1. *J. Biol. Chem.*, 300(9): 107638. doi: 10.1016/j.jbc.2024.107638.
64. Xu, J., X. He, S. Zhang, L. Li and P. Li. (2025, Apr.). Expression of co-signaling molecules TIM-3/Galectin-9 at the maternal-fetal interface. *Placenta*, 163: 43–50. doi: 10.1016/j.placenta.2025.03.002.
65. Li, Y., Y. Sang, Y. Chang, C. Xu, Y. Lin, Y. Zhang, P.C.N. Chiu, W.S.B. Yeung, H. Zhou, N. Dong, L. Xu, J. Chen, W. Zhao, L. Liu, D. Yu, X. Zang, J. Ye, J. Yang, Q. Wu, D. Li, L. Wu and M. Du. (2024, May). A Galectin-9–Driven CD11chigh decidual macrophage subset suppresses uterine vascular remodeling in preeclampsia. *Circulation*, 149(21): 1670–1688. doi: 10.1161/circulationha. 123.064391.
66. Gao, C., Q. Chen, X. Hao and Q. Wang. (2023, Nov.). Immunomodulation of antibody glycosylation through the placental transfer. *Int. J. Mol. Sci.*, 24(23): 16772. doi: 10.3390/ijms242316772.
67. Matharu, D., A.J. Ponsero, M. Lengyel, A. Meszaros-Matwiejuk, K.-L. Kolho, W.M. De Vos, D. Molnar-Gabor and A. Salonen. (2024, Jun.). Human milk oligosaccharide composition is affected by season and parity and associates with infant gut microbiota in a birth mode dependent manner in a Finnish birth cohort. *eBioMedicine*, 104: 105182. doi: 10.1016/j.ebiom.2024.105182.
68. Sun, W., L. Tao, C. Qian, P. Xue, S. Du and Y. Tao. (2025, Jan.). Human milk oligosaccharides: Bridging the gap in intestinal microbiota between mothers and infants. *Front. Cell. Infect. Microbiol.*, 14: 1386421. doi: 10.3389/fcimb.2024.1386421.
69. Paone, P., D. Latousakis, R. Terrasi, D. Vertommen, C. Jian, V. Borlandelli, F. Suriano, M.E.V. Johansson, A. Puel, C. Bouzin, N.M. Delzenne, A. Salonen, N. Juge, B.I. Florea, G.G. Muccioli, H. Overkleeft, M. Van Hul and P.D. Cani. (2024, Oct.). Human milk oligosaccharide 2-fucosyllactose protects against high-fat diet-induced obesity by changing intestinal mucus production, composition, and degradation linked to changes in gut microbiota and faecal proteome profiles in mice. *Gut*, 73(10): 1632–1649. doi: 10.1136/gutjnl-2023-330301.
70. Cochrane, K.M., J.N. Bone, C.D. Karakochuk and L. Bode. (2024, Apr.). Human milk oligosaccharide composition following supplementation with folic acid vs (6S)-5-methyltetrahydrofolic acid during pregnancy and mediation by human milk folate forms. *Eur. J. Clin. Nutr.*, 78(4): 351–355. doi: 10.1038/s41430-023-01376-7.
71. Balogh, A., E. Toth, R. Romero, K. Parej, D. Csala, N.L. Szenasi, I. Hajdu, K. Juhasz, A.F. Kovacs, H. Meiri, P. Hupuczi, A.L. Tarca, S.S. Hassan, O. Erez, P. Zavodszky, J. Matko, Z. Papp, S.W. Rossi, S. Hahn, E. Pallinger and N.G. Than. (2019, Jun.). Placental galectins are key players in regulating the maternal adaptive immune response. *Front. Immunol.*, 10: 1240. doi: 10.3389/fimmu.2019.01240.
72. Kirkgöz, K., R. Vogtmann, Y. Xie, F. Zhao, A. Riedel, L.-M. Adam, N. Freitag, C. Harms, M.G. Garcia, T. Plösch, A. Gellhaus and S.M. Blois. (2024, Aug.). Placental glycosylation senses the anti-angiogenic milieu induced by human sFLT1 during pregnancy. *J. Reprod. Immunol.*, 164: 104284. doi: 10.1016/j.jri.2024.104284.
73. Sammar, M., T. Drobnjak, M. Mandala, S. Gizurarson, B. Huppertz and H. Meiri. (2019, Jun.). Galectin 13 (PP13) facilitates remodeling and structural stabilization of maternal vessels during pregnancy. *Int. J. Mol. Sci.*, 20(13): 3192. doi: 10.3390/ijms20133192.
74. Wang, Y., Y. Liu, S. Liu, L. Cheng and X. Liu. (2024, Jun.). Recent advances in N-glycan biomarker discovery among human diseases. *Acta Biochim. Biophys. Sin.* doi: 10.3724/abbs.2024101.
75. Xie, Y., F. Zhao, Y. Wang, S. Borowski, N. Freitag, I. Tirado-Gonzalez, N. Hofsink, U. Matschl, T. Plösch, M.G. Garcia and S.M. Blois. (2024, Aug.). Fetal growth restriction induced by maternal gal-3 deficiency is associated with altered gut-placenta axis. *Cell Death Dis.*, 15(8): 575. doi: 10.1038/s41419-024-06962-6.

76. Freitag, N., I. Tirado-Gonzalez, G. Barrientos, K.L. Powell, P. Boehm-Sturm, S.P. Koch, K. Hecher, A.C. Staff, P.C. Arck, A. Diemert and S.M. Blois. (2020, Jul.). Galectin-3 deficiency in pregnancy increases the risk of fetal growth restriction (FGR) via placental insufficiency. *Cell Death Dis.*, 11(7): 560. doi: 10.1038/s41419-020-02791-5.
77. Chaney, H.L., L.F. Grose, J.M. LaBarbara, A.W. Sirk, A.M. Blancke, J.M. Sánchez, C. Passaro, P. Lonergan and D.J. Mathew. (2022, Mar.). Galectin-1 induces gene and protein expression related to maternal-conceptus immune tolerance in bovine endometrium. *Biol. Reprod.*, 106(3): 487–502. doi: 10.1093/biolre/ioab215.
78. Tirado-Gonzalez, I., N. Freitag, G. Barrientos, V. Shaikly, O. Nagaeva, M. Strand, L. Kjellberg, B.F. Klapp, L. Mincheva-Nilsson, M. Cohen and S.M. Blois. (2013, Jan.). Galectin-1 influences trophoblast immune evasion and emerges as a predictive factor for the outcome of pregnancy. *Mol. Hum. Reprod.*, 19(1): 43–53. doi: 10.1093/molehr/gas043.
79. Seefried, M.C., J. Mittelberger, M. Franitza, F. Garrido, C.M. Wild, N. Ditsch, O. Protsepko, C. Kuhn, C. Dannecker, U. Jeschke, P. Altevogt and M. Sammar. (2025, Feb.). Expression of the mucin-like glycoprotein CD24 and its ligand siglec-10 in placentas with acute and post SARS-CoV-2 infection. *J. Reprod. Immunol.*, 167: 104400. doi: 10.1016/j.jri.2024.104400.
80. Sun, L., M. Shi, J. Wang, X. Han, J. Wei, Z. Huang, X. Yang, Y. Ding, P. Zhang, A. He, M. Liu, R. Yan, X. Yang, R. Li and G. Wang. (2024, Jul.). Overexpressed trophoblast glycoprotein contributes to preeclampsia development by inducing abnormal trophoblast migration and invasion toward the uterine spiral artery. *Hypertension*, 81(7): 1524–1536. doi: 10.1161/HYPERTENSIONAHA.124.22923.
81. Chen, Q., P.-C. Pang, M.E. Cohen, M.S. Longtine, D.J. Schust, S.M. Haslam, S.M. Blois, A. Dell and G.F. Clark. (2016, Jun.). Evidence for differential glycosylation of trophoblast cell types. *Mol. Cell. Proteomics*, 15(6): 1857–1866. doi: 10.1074/mcp.M115.055798.
82. Kiyozumi, D., I. Nakano, R. Sato-Nishiuchi, S. Tanaka and K. Sekiguchi. (2020, Feb.). Laminin is the ECM niche for trophoblast stem cells. *Life Sci. Alliance*, 3(2): e201900515. doi: 10.26508/lsa.201900515.
83. Alaca, R., T. Demirci, E.P. Topdaği Yilmaz and N. Öztürk. (2024, Dec.). Investigation of Urotensin II expression in placenta and umbilical cord in pregnancies with intrauterine growth restriction by histological and biochemical methods. *Placenta*, 158: 1–9. doi: 10.1016/j.placenta.2024.09.003.
84. Seedat, F., N. Kandzija, M.J. Ellis, S. Jiang, A. Sarbalina, J. Bancroft, E. Drydale, S.S. Hester, R. Fischer, A.N. Wade, M.I. Stefana, J.A. Todd and M. Vatish. (2024, Nov.). Placental small extracellular vesicles from normal pregnancy and gestational diabetes increase insulin gene transcription and content in β cells. *Clin. Sci.*, 138(22): 1481–1502. doi: 10.1042/CS20241782.
85. Hou, H., S. Yang, X. Yang, W. Sun, H. Javeria, J. Khan and Z. Du. (2025, Mar.). Sensitive and high-throughput isomer-specific analysis of human milk oligosaccharides using UPLC-MS/MS and its application to secretor status assignment. *Carbohydr. Polym.*, 352: 123154. doi: 10.1016/j.carbpol.2024.123154.
86. Veličković, M., L. Kadam, J. Kim, K.J. Zemaitis, D. Veličković, Y. Gao, R. Wu, T.L. Fillmore, D. Orton, S.M. Williams, M.E. Monroe, R.J. Moore, P.D. Piehowski, L.M. Bramer, L. Myatt and K.E. Burnum-Johnson. (2025, Feb.). Advanced multi-modal mass spectrometry imaging reveals functional differences of placental villous compartments at microscale resolution. *Nat. Commun.*, 16(1): 2061. doi: 10.1038/s41467-025-57107-y.
87. Hou, H., S. Yang, X. Yang, W. Sun, A.A. Debrah, H. Javeria, D. Tian and Z. Du. (2025, Jul.). Comprehensive profiling and development of a collision cross section database for milk oligosaccharides via orthogonal UPLC-cyclic ion mobility-mass spectrometry system. *Food Chem.*, 480: 143839.

88. Marie, A.-L., Y. Gao and A.R. Ivanov. (2024, May). Native N-glycome profiling of single cells and NG-level blood isolates using label-free capillary electrophoresis-mass spectrometry'. *Nat. Commun.*, 15(1): 3847. doi: 10.1038/s41467-024-47772-w.
89. Urban, J., C. Jin, K.A. Thomsson, N.G. Karlsson, C.M. Ives, E. Fadda and D. Bojar. (2024, Jul.). Predicting glycan structure from tandem mass spectrometry via deep learning. *Nat. Methods*, 21(7): 1206–1215. doi: 10.1038/s41592-024-02314-6.
90. Palomino, T.V. and D.C. Muddiman. (2024, Jun.). Mass spectrometry imaging of *N*-linked glycans: Fundamentals and recent advances. *Mass Spectrom. Rev.*, mas.21895. doi: 10.1002/mas.21895.
91. Cumin, C., L. Gee, T. Litfin, R. Muchabaiwa, G. Martin, O. Cooper, V. Heinzelmann-Schwarz, T. Lange, M. Von Itzstein, F. Jacob and A. Everest-Dass. (2024, Jul.). Highly sensitive spatial glycomics at near-cellular resolution by on-slide derivatization and mass spectrometry imaging. *Anal. Chem.*, 96(28): 11163–11171. doi: 10.1021/acs.analchem.3c05984.
92. Kluivers, A.C.M., R.I. Neuman, B. Kalra, A. Kumar, W. Visser, A.H.J. Danser and L. Saleh. (2025, Mar.). Glycosylated fibronectin as a biomarker for preeclampsia and preeclampsia-related complications. *Pregnancy Hypertens.*, 39: 101177. doi: 10.1016/j.preghy.2024.101177.
93. Štambuk, T., D. Kifer, L. Smirčić-Duvnjak, M. Vučić Lovrenčić and O. Gornik. (2023, Apr.). Associations between plasma protein, IgG and IgA N-glycosylation, and metabolic health markers in pregnancy and gestational diabetes. *PLOS ONE*, 18(4): e0284838. doi: 10.1371/journal.pone.0284838.
94. Wang, Z., J. Zhang and L. Li. (2025, Jan.). Recent advances in labeling-based quantitative glycomics: From high-throughput quantification to structural elucidation. *Protemics*, 25(1-2): e202400057. doi: 10.1002/pmic.202400057.
95. Chiang, A.W.T., H.M. Baghdassarian, B.P. Kellman, B. Bao, J.T. Sorrentino, C. Liang, C.-C. Kuo, H.O. Masson and N.E. Lewis. (2021, Jun.). Systems glycobiology for discovering drug targets, biomarkers, and rational designs for glyco-immunotherapy. *J. Biomed. Sci.*, 28(1): 50. doi: 10.1186/s12929-021-00746-2.
96. McKenna, M.K., A. Ozcan, D. Brenner, N. Watanabe, M. Legendre, D.G. Thomas, C. Ashwood, R.D. Cummings, C. Bonifant, D.M. Markovitz and M.K. Brenner. (2023, Jan.). Novel banana lectin CAR-T cells to target pancreatic tumors and tumor-associated stroma. *J. Immunother. Cancer*, 11(1): e005891. doi: 10.1136/jitc-2022-005891.
97. Rattila, S., F. Kleefeldt, A. Ballesteros, J.S. Beltrame, M.L Ribeiro, S. Ergün and G. Dveksler. (2020, Nov.). Pro-angiogenic effects of pregnancy-specific glycoproteins in endothelial and extravillous trophoblast cells. *Reproduction*, 160(5): 737–750. doi: 10.1530/REP-20-0169.
98. Wang, H., N. Fan, X. Cui, R. Xie, Y. Tang, A.M. Thomas, S. Li, J.V. Zhang, S. Liu and H. Qin. (2024, Dec.). BMP5 promotes trophoblast functions upon N-glycosylation via the BMP5-SMAD1/5 signaling pathway in preeclampsia. *Placenta*, 158: 240–252. doi: 10.1016/j.placenta.2024.11.002.
99. Green, D.J., K. Park, V. Bhatt-Mehta, D. Snyder and G.J. Burckart. (2021, Jul.). Regulatory considerations for the mother, fetus and neonate in fetal pharmacology modeling. *Front. Pediatr.*, 9: 698611. doi: 10.3389/fped.2021.698611.
100. Burnham, J. and E. Darsey. (2022, Sep.). Maternal–fetal drug development: An industry perspective. *J. Clin. Pharmacol.*, 62(S1). doi: 10.1002/jcph.2124.
101. Hudson, R.E., T.D. Metz, R.M. Ward, A.M. McKnite, E.Y. Enioutina, C.M. Sherwin, K.M. Watt and K.M. Job. (2023, Mar.). Drug exposure during pregnancy: Current understanding and approaches to measure maternal-fetal drug exposure. *Front. Pharmacol.*, 14: 1111601. doi: 10.3389/fphar.2023.1111601.

CHAPTER 8

Glycoproteomics and Immunity

Surendra Kumar Gautam

Glycoproteomics, the mass study of glycosylated proteins, has come to be an essential discipline in immunology and has elucidated how glycan modifications regulate immune responses. Glycosylation has a central function in the regulation of immune recognition, signaling, and pathogen-host interactions. This chapter unravels the complex relationship between glycoproteins and immunity, highlighting mechanisms of glycosylation, immune system modulation, and disease pathogenesis. Key areas are glycan-mediated control of antigen presentation, immune evasion mechanisms of cancer cells and pathogens, and the function of glycoproteins in inflammatory and autoimmune diseases. Improvements in mass spectrometry-based glycoproteomic methods have allowed high-resolution glycoprotein profiling of immune-related glycoproteins, revealing new biomarkers and therapeutic targets. We also cover glycoengineering approaches to immunotherapy, highlighting the promise of glycan-modified

Associate Professor & Head, Department of Life Science, Shri Rawatpura Sarkar University, Raipur, Chhattisgarh.
Email: gautams1436@gmail.com

DOI: 10.1201/9781003635901-8

antibodies and glycan-targeting drugs in the treatment of infections, cancer, and immune disorders. The combination of glycoproteomics and systems immunology reveals deeper understanding of immune regulation, opening the door to precision medicine strategies. In this chapter, our objective is to present an in-depth understanding of glycoproteomics in immunity with emphasis on recent developments and future directions in this fast-moving field.

1. Introduction

Glycoproteomics, the comprehensive analysis of glycosylated proteins, is central to immune system function since it provides a key role in antigen recognition, immune cell signaling, and interactions with pathogens. Glycosylation, the enzymatic addition of glycans to proteins, forms the core mechanism that regulates immune receptor activation, immune checkpoint pathways, and infection, cancer, and autoimmune disease progression [1, 2].

The immune response is also dependent largely on glycan modifications for antigen presentation on major histocompatibility complex (MHC) molecules, which determine adaptive immunity. T-cell receptor (TCR) recognition and subsequent immune activation depend on the structural variety of glycan moieties [10]. Glycosylation controls MHC class I molecules by modulating peptide loading and stability, thus affecting CD8+ T cell responses [1]. Secondly, glycan modification influences the binding affinity of antigen-presenting cells (APCs) to T-cells, determining immune tolerance and activation thresholds [9].

Cancer cells take advantage of glycosylation to immune escape by altering tumor-associated antigens. Such changes in tumor cell glycosylation affect immune escape pathways, such as dendritic cell interaction and immune checkpoint control [3]. In particular, non-canonical glycan structures on melanoma cells determine dendritic cell interactions, ultimately determining clinical outcome and immune responsiveness [3]. These alterations are responsible for tumor growth by suppressing immune cell recognition and promoting a suppressive tumor microenvironment [9].

Pathogens use glycosylation to evade immune detection. Viral glycoproteins, such as SARS-CoV-2 spike protein, mask critical antigenic sites using host-like glycans to escape immune recognition and enhance infectivity [11]. Infections due to bacteria and parasites also manipulate host glycosylation pathways to evade immune responses, permitting persistence of pathogens [2]. For example, Mycobacterium tuberculosis alters glycan structures to evade macrophage-mediated clearance, substantiating the significance of glycoproteomics in infectious disease research [2].

Advancements in glycoproteomics, particularly mass spectrometry-based strategies, enabled identifying glycoproteins and the disease-specific alterations at high-throughput levels [18]. They have been extremely important for uncovering glycoprotein disease biomarkers like those for cancer and autoimmune disorders [5]. In metastatic melanoma, glycan modifications are prognostic ICI activity biomarkers, useful indicators of therapeutic response and mechanisms of resistance [5].

Besides, emerging evidence validates the role of glycosylation in predicting the result of immunotherapy. Glycosylation plays a role in modulating immune checkpoint molecules, such as PD-1 and CTLA-4, on their binding with ligands and on the immune response in cancer therapy

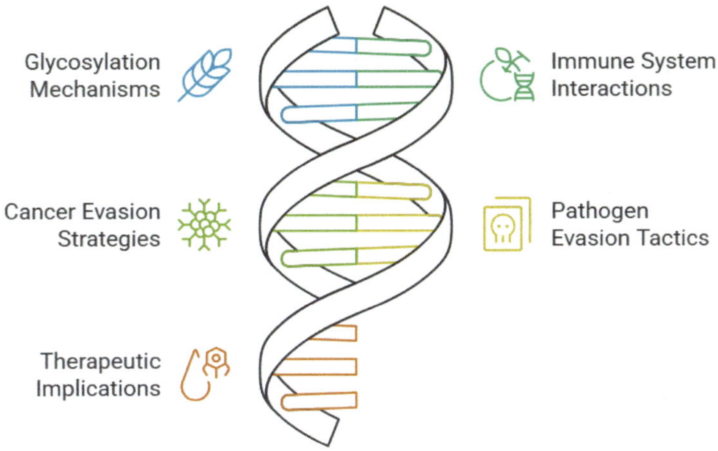

Fig. 1: Unraveling glycoproteomics: Key roles in immunity and disease.
Alt Text: A conceptual illustration of glycoproteomics and its relevance to immunity, structured around a stylized DNA helix. The image highlights five key aspects: (1) Glycosylation Mechanisms, represented by a blue leaf icon, (2) Immune System Interactions, depicted with a green immune-related symbol, (3) Cancer Evasion Strategies, indicated by a green viral structure, (4) Pathogen Evasion Tactics, illustrated with a yellow microbial defense icon, and (5) Therapeutic Implications, symbolized by an orange dropper. These elements are arranged around the DNA helix, visually linking glycoproteomics to various biological processes.

[12]. Abnormally glycosylated patterns seen in solid cancers are also being explored for their possible roles as therapeutic targets, with initial signs of successful uses in future immunotherapies [13].

Therefore, glycoproteomics deliver a deep understanding of host-pathogen interactions, tumor immunity, and therapeutic treatments, determining the future of precision medicine in immunology. Elucidating the complex relationship between glycosylation and immune function allows researchers to create targeted therapies to augment immune responses, better design vaccines, and maximize cancer immunotherapy effectiveness.

2. Glycosylation and Immune System Regulation

2.1 Role of Glycosylation in Immune Cell Function

Glycosylation is a critical post-translational modification that controls immune cell function through the modulation of protein stability, cellular communication, and immune recognition. The specificity and complexity of glycosylation patterns control immune tolerance, activation, and response to pathogens [1]. Glycosylation of MHC molecules plays a significant role in antigen presentation and T-cell activation, influencing immune specificity and magnitude of response [9].

Dendritic cells (DCs) use unique glycan signatures to coordinate immune responses, allowing them to interact with T cells and B cells. Cancer cells take advantage of dysregulated glycosylation to escape immune detection, affecting the function of antigen-presenting cells [3]. Unusual glycosylation patterns in macrophages also affect their polarization, which dictates whether they become pro-inflammatory (M1) or anti-inflammatory (M2) [5]. These alterations have important consequences for cancer development, autoimmunity, and response to infectious disease.

In addition, glycosylation also plays a role in controlling immune checkpoint proteins like programmed death-1 (PD-1) and cytotoxic T-lymphocyte-associated protein 4 (CTLA-4). Glycosylation of the molecules controls interactions with ligands and downstream immunity and offers an addressable strategy for immunomodulation [12]. The tumor environment leverages glycans modifications for dampening immunity, e.g., tumor cell glycans engaging suppressive receptors like Siglec-7 [22].

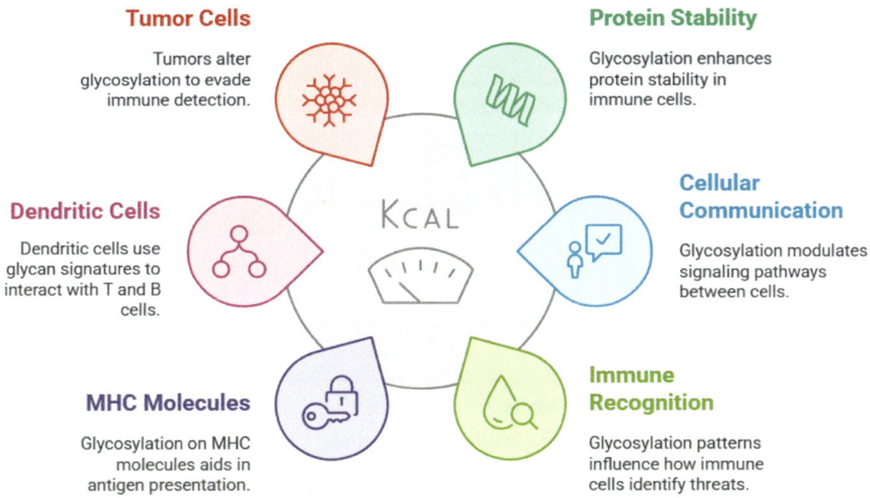

Fig. 2: Glycosylation in immune regulation: Stability, recognition, and cellular interaction.

Alt Text: A conceptual diagram illustrating the role of glycosylation in immune regulation. The central element, marked with a caloric energy (KCAL) symbol, is surrounded by six key aspects of glycosylation in immune processes: (1) **Tumor Cells** (red): Tumors modify glycosylation to evade immune detection, (2) **Protein Stability** (green): Glycosylation enhances protein stability in immune cells, (3) **Cellular Communication** (blue): Glycosylation modulates immune signaling pathways, (4) **Immune Recognition** (light green): Glycosylation patterns influence immune threat detection, (5) **MHC Molecules** (purple): Glycosylation aids in antigen presentation on major histocompatibility complex (MHC) molecules, and (6) **Dendritic Cells** (pink): Dendritic cells utilize glycan signatures to interact with T and B cells. Each aspect is represented with distinct colors and icons, visually linking glycosylation to immune system functions.

2.2 *Glycoproteins in Innate and Adaptive Immunity*

Glycoproteins also have a key function in the regulation of the immune system, with glycan modifications affecting stability and activity. Immunoglobulin G (IgG) glycosylation is a classic example, whereby changes in its glycan content affect effector functions and inflammation [16]. Fc-glycosylation of IgG regulates binding affinity to Fc receptors and controls immune responses in autoimmune conditions and infections.

Cytokines, another glycoprotein class, play a key role in immune signaling. Their glycosylation state determines bioactivity, stability, and

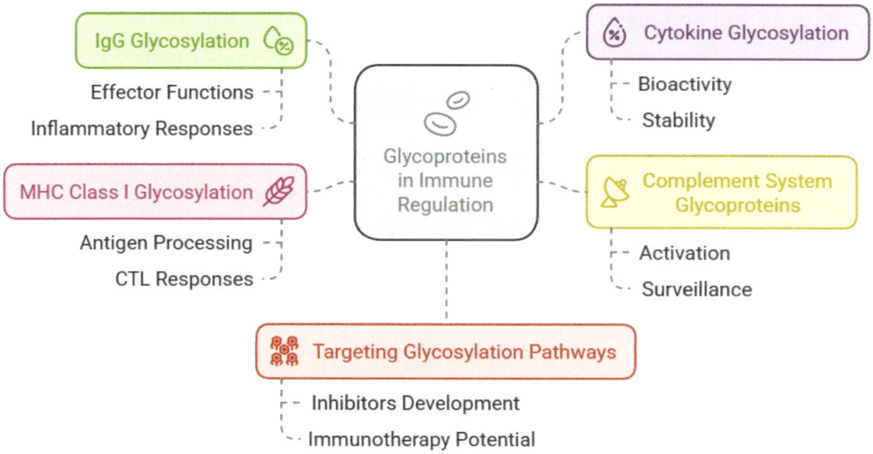

Fig. 3: Glycoproteins modulate immune regulation through glycosylation-driven functional pathways.

Alt Text: A schematic representation of the role of glycoproteins in immune regulation, structured as a central node labeled "Glycoproteins in Immune Regulation", with five key categories branching out. These categories include: (1) **IgG Glycosylation** (green): Influencing effector functions and inflammatory responses, (2) **MHC Class I Glycosylation** (pink): Involved in antigen processing and cytotoxic T lymphocyte (CTL) responses, (3) **Cytokine Glycosylation** (purple): Impacting bioactivity and stability, (4) **Complement System Glycoproteins** (yellow): Contributing to activation and immune surveillance, and (5) **Targeting Glycosylation Pathways** (red): Relevant for inhibitor development and immunotherapy potential. Each category is represented with a distinct color and icon, illustrating the diverse roles of glycoproteins in both innate and adaptive immunity.

receptor binding [2]. Abnormal cytokine glycosylation may result in immune dysregulation, which leads to inflammatory disease and immune evasion in cancer. Complement system glycoproteins also depend on glycan structures for function, with modifications affecting complement activation and immune surveillance [7].

Glycosylation of MHC class I molecules during adaptive immunity impacts antigen presentation and processing, having a direct influence on cytotoxic T lymphocyte (CTL) activity [1]. Modulation of glycosylation patterns can strengthen or suppress immune responses, which may impact the susceptibility to infections as well as vaccine efficacy. Regulation of glycosyltransferases upon encountering pathogens has also been seen in plant immunity, which illustrates the evolutionary importance of glycosylation in immune control [6].

New studies point to the therapeutic potential of glycosylation pathways in immunotherapy. The creation of glycosyltransferase or glycan-recognizing receptor-specific inhibitors offers a promising strategy for augmenting immune responses to tumors and infectious diseases [13]. Elucidation of the role of glycosylation in the regulation of the immune system is critical to creating new therapeutic targets in immuno-oncology and autoimmune disease therapy.

3. Glycoproteomics in Disease and Immunotherapy

3.1 Cancer and Immune Evasion

Glycoproteomics has greatly developed our comprehension of cancer immunology, specifically in the immune evasion systems employed by tumors. Cancer cells use unnatural glycosylation to evade immune detection, modifying tumor immunogenicity and impacting immune interactions [9]. The glyco-code of the tumor is a novel immune checkpoint, with changed glycosylation modifying antigen presentation, T-cell identification, and immune evasion [12]. Interestingly, melanoma tumors have a specific glycan profile that influences the function of dendritic cells and has clinical implications, demonstrating how glycoproteomics guides tumor immunology [3].

One of the essential features of glycosylation in cancer includes glycosphingolipids, which have a central function in the modulation of HLA Class I-restricted immune responses [1]. The glycosphingolipids influence the activation of T-cells and immune evasion by modulating the pathways of antigen presentation. SPPL3, the enzyme responsible for determining the repertoire of glycosphingolipids, has been recognized as an important controller of immune interactions in the tumor environment. Glycoproteomic examinations have proved that tumors bearing structurally altered glycan structures have the ability to escape immune surveillance by repressing antigenicity, hence undermining the efficacy of immunotherapies directed at immune checkpoint inhibitors [4].

Tumor-specific glycan targeting is a promising therapeutic approach. The immune checkpoint proteins PD-1 and CTLA-4 are modified by glycosylation, affecting their function and ligand binding, and modulating the immune response [12]. Recent evidence indicates that glycan-directed immunotherapies may increase anti-tumor immunity through re-establishing effective antigen presentation and enhancing T-cell-mediated tumor clearance. With expanding research in the field of glycoproteomics, new ideas about these immune processes driven by glycan give evidence to employing glycan-focused treatments in treating cancer.

3.2 Infectious Diseases and Pathogen Glycosylation

Pathogens widely use glycosylation to escape host immune responses and cause infections. Viral glycoproteins, including the SARS-CoV-2 spike protein, are highly glycosylated to protect epitopes from immune recognition, impacting vaccine effectiveness and immune responses [11]. Research has shown that glycosylation of viral surface proteins can determine host-pathogen interactions, impacting immune evasion mechanisms and disease progression [2].

Bacterial pathogens also use host glycosylation pathways to evade immune responses. For example, bacterial glycoproteins may mimic host glycans to escape immune detection or manipulate immune cell signaling. Furthermore, a new glycosyltransferase, UGT73C7, has also been found to play a central role in plant immunity, providing evidence for the conserved role of glycosylation in immune defense throughout different species [6]. These discoveries highlight the global relevance of glycosylation of host-pathogen interactions and the promise of glycoproteomics for the identification of new therapeutic targets.

Technological advances in glycoproteomics have facilitated the analysis of microbial glycan structures, which has given insight into how pathogens use glycosylation to evade the immune system [2]. These findings have significant implications for vaccine development since glycoproteins can be used as immunodominant targets to induce strong immune responses. The application of glycoproteomic strategies to infectious disease research is pivotal in the development of next-generation vaccines and antimicrobial approaches.

3.3 Glycoproteomic Biomarkers in Immunotherapy

Glycoproteomics has transformed biomarker discovery in immunotherapy by providing a precision medicine strategy for anticipating patient outcomes. Plasma glycoproteomic biomarkers have emerged as prospective response predictors of immune checkpoint inhibitors in metastatic melanoma, presenting essential insights into personalized treatment [5]. Glycosylation profiling of immune receptors identified specific glycan signatures that determine immune modulation in cancer, further narrowing immunotherapeutic interventions [20].

Mass glycoproteomic analysis has revealed distinctive glycan signatures that are associated with patient outcomes from immunotherapies. For instance, aberrant glycosylation patterns on immune cell receptors determine their efficacy and functionality in identifying tumor antigens [14]. These results indicate the promise of glycan biomarkers in informing immunotherapy and clinical outcomes.

The incorporation of glycoproteomics into immunotherapy studies has also enabled the identification of new therapeutic strategies. By determining glycoproteomic markers of immune activation or suppression, investigators can develop targeted therapies that improve anti-tumor immunity and reduce immune-related side effects [5]. The growing contribution of glycoproteomics to immunotherapy emphasizes the requirement for ongoing studies to leverage glycan-based biomarkers for translational medicine.

4. Analytical Techniques in Glycoproteomics

4.1 Mass Spectrometry-based Glycoproteomics

Mass spectrometry (MS)-based glycoproteomics is a robust analytical technique for identifying glycoproteins with high sensitivity and specificity. Improvements in MS instrumentation, such as tandem MS (MS/MS) and high-resolution mass analyzers, have greatly enhanced glycoprotein profiling, which is critical for comprehending immune processes and biomarker identification [1]. Liquid chromatography-mass spectrometry (LC-MS) is used extensively to profile glycan structures and glycosylation sites and shed important light on mechanisms of immune responses [12]. Glycopeptide enrichment techniques and data-independent acquisition (DIA) have further improved the coverage of the glycoproteome [19].

MS-based glycoproteomics has played a critical role in the identification of glycoproteomic disease-associated markers, such as cancer and infection. For example, recent research has employed glycoproteomic methodologies to detect plasma biomarkers that can predict metastatic melanoma patients' responses to immune checkpoint inhibitor therapy [5]. In addition, glycoproteomic profiling has been used to gain insights into immune interactions in ovarian cancer, specifically macrophage galactose-like C-type lectin (MGL)-mediated recognition of tumor glycoproteins [4]. Glycosylation has also been shown to play a role in modulating human leukocyte antigen (HLA) Class I-mediated immune responses, highlighting the functional significance of glycans in immune regulation [1].

The combination of MS-based glycoproteomics with immunology has also resulted in the discovery of glycan signatures linked to pathogen recognition. For instance, glycoproteomics has identified differential glycosylation patterns in host immune responses to bacterial infections, further illustrating the importance of glycosylation in immune defense mechanisms [15]. As technological developments are on the rise, MS-based glycoproteomics will increasingly be responsible for the development of new glycan-targeted immunotherapies and biomarker discovery [2].

4.2 Glycan Microarrays and Lectin-based Approaches

Glycan microarrays have transformed glycan-protein interaction research through the high-throughput screening of glycan-binding specificities. Glycan microarrays are important for the elucidation of immune recognition patterns and the functions of glycans in immunological events [21]. For instance, differential oligomannose isomer recognition by glycan-binding proteins in both innate and adaptive immunity has been established through glycan microarray technology [21]. These analyses have enhanced our knowledge of immune receptor-glycan interactions and their significance in disease pathogenesis.

Lectin-based strategies, such as lectin affinity chromatography and lectin microarrays, are complementary methods of glycoprotein analysis. The methods have been employed to determine tumor-associated glycans and their role in modulating immune cell recognition, thereby affecting clinical prognosis in melanoma [3]. Analysis of tumor glyco-code via lectin-based approaches has offered insights into the mechanism by which disrupted glycosylation patterns affect dendritic cell function, which, in turn, determines immune responses in cancer [3].

In infectious diseases, lectin-based approaches have been used to explore pathogen-responsive glycosylation and its role in host immunity [2]. These studies have highlighted the function of glycosylation in pathogen-host interactions, where glycan modification has been shown to be important determinants of immune evasion and pathogen recognition. For instance, some glycosylation alterations in immune checkpoints have been implicated in immune modulation in oncology, which presents potential therapeutic prospects [12].

In addition, glycan microarrays and lectin-based approaches have been used in vaccine development and immunotherapy studies. The discovery of glycan epitopes that are important for immune recognition has enabled the design of glycan-targeted vaccines and immune modulators [9]. As knowledge about glycan-mediated immune interactions grows, these analytical methods will remain central to immunological studies and therapeutic development.

5. Future Perspectives and Clinical Applications

Glycoproteomics has the potential to facilitate major breakthroughs in immunotherapy and vaccine development through an understanding of the important functions of glycans in immune regulation. Altered glycosylation has been identified as a key immune checkpoint mechanism in tumor immunology that affects antigen presentation, immune surveillance, and tumor escape [3, 9]. Significantly, the glycosphingolipid inventory regulates HLA class I-immune responses, influencing the effectiveness of antigen presentation and processing pathways [1]. Deciphering such glycan-driven interactions will make therapeutic interventions on immune evasion strategies in cancers more precise. Recent evidence highlights the prognostic value of glycoproteomics for predicting treatment response to immune checkpoint inhibitors. Plasma glycoproteomic biomarkers have emerged as predictors of therapeutic response in metastatic melanoma, allowing enhanced patient stratification [5]. Macrophage galactose-type lectins (MGLs) also preferentially bind tumor-associated O-glycosylation motifs, presenting new targets for immunotherapeutic modulation [4]. These findings highlight the expanding significance of glycan-based immune modulation in precision oncology.

Apart from cancer, glycoproteomics has revealed pathogen-induced changes in glycosylation that modulate host immune responses. Changes in glycan structures affect pathogen recognition, which promotes immune evasion or increases protective immunity [2]. With this knowledge, glycan-directed vaccines are being developed to enhance immune responses to infectious diseases. For instance, glycan engineering methods are being researched to boost antigenicity and immunogenicity in vaccine products [8]. Improvements in synthetic glycoproteins and glycoengineering are the key to the next generation of immunomodulators. Precision medicine is gradually incorporating glycan biomarkers into diagnostic and therapeutic paradigms, opening doors to more individualized and efficacious immune-based treatments [17]. As clinical glycoproteomics becomes more advanced, its use in disease monitoring, therapy optimization, and vaccine design will revolutionize immune-based healthcare strategies.

6. Conclusion

Glycoproteomics is essential for deciphering the regulation of the immune system, disease processes, and therapeutic strategies. The complex interplay between glycosylation and immune responses impacts multiple biological processes, such as immune cell communication, pathogen recognition, and inflammatory responses. Through glycoprotein analysis, scientists are able to identify key biomarkers linked to autoimmune diseases, infectious diseases, and cancers, paving the way for more targeted therapeutic and diagnostic strategies.

The use of glycoproteomics goes beyond laboratory research to reach clinical and translational medicine and provide new promises for precision immunology. Gaining insight into glycan-immune interactions supports our capacity for designing more immunogenic vaccines, monoclonal antibodies, and glyco-engineered therapeutics. The growth in analytical capability, such as mass spectrometry and glycan microarray, has facilitated our capacity significantly to characterize glycoproteins with high specificity and sensitivity.

Despite such progress, standardization of methodologies and understanding intricate glycosylation patterns are still challenges ahead. Future research has to converge glycoproteomic data with multi-omics platforms to envision immune regulation at a holistic level. By broadening our insights into glycan-driven immune processes, we will be able to promote innovation in disease management and drug development to improve patient care across different immune-related diseases.

References

1. Jongsma, M., A. De Waard, M. Raaben, T. Zhang, B. Cabukusta, R. Platzer, V. Blomen, A. Xagara, T. Verkerk, S. Bliss, X. Kong, C. Gerke, L. Janssen, E. Stickel, S. Holst, R. Plomp, A. Mulder, S. Ferrone, F. Claas, M. Heemskerk, M. Griffioen, A. Halenius, H. Overkleeft, J. Huppa, M. Wuhrer, T. Brummelkamp, J. Neefjes and R. Spaapen. (2020). The SPPL3-defined glycosphingolipid repertoire orchestrates HLA Class I-mediated immune responses. *Immunity*. https://doi.org/10.1016/j.immuni.2020.11.003.
2. Mohideen, F. and L. Mahal. (2024). Infection and the Glycome: New insights into host response. *ACS Infectious Diseases*, 10: 2540–2550. https://doi.org/10.1021/acsinfecdis.4c00315.
3. Cuevas, E., B. Roubinet, S. Mouret, M. Thépaut, F. De Fraipont, J. Charles, F. Fieschi, L. Landemarre, L. Chaperot and C. Aspord. (2023). The melanoma tumor glyco-code impacts human dendritic cells' functionality and dictates clinical outcomes. *Frontiers in Immunology*, 14. https://doi.org/10.3389/fimmu.2023.1120434.

4. Napoletano, C., C. Steentoff, F. Battisti, Z. Ye, H. Rahimi, I. Zizzari, M. Dionisi, B. Cerbelli, F. Tomao, D. French, G. d'Amati, P. Panici, S. Vakhrushev, H. Clausen, M. Nuti and A. Rughetti. (2020). Investigating Patterns of immune interaction in ovarian cancer: Probing the O-glycoproteome by the macrophage Galactose-like C-Type lectin (MGL). *Cancers*, 12. https://doi.org/10.3390/cancers12102841.
5. Pickering, C., P. Aiyetan, G. Xu, A. Mitchell, R. Rice, Y. Najjar, J. Markowitz, L. Ebert, M. Brown, G. Tapia-Rico, D. Frederick, X. Cong, D. Serie, K. Lindpaintner, F. Schwarz and G. Boland. (2023). Plasma glycoproteomic biomarkers identify metastatic melanoma patients with reduced clinical benefit from immune checkpoint inhibitor therapy. *Frontiers in Immunology*, 14. https://doi.org/10.3389/fimmu.2023.1187332.
6. Huang, X., Y. Wang, J. Lin, L. Chen, Y. Li, Q. Liu, G. Wang, F. Xu, L. Liu and B. Hou. (2021). A novel pathogen-responsive glycosyltransferase UGT73C7 mediates the redirection of phenylpropanoid metabolism and promotes SNC1-dependent Arabidopsis immunity. *The Plant Journal: For Cell and Molecular Biology*. https://doi.org/10.1111/tpj.15280.
7. Kolarich, D., B. Lepenies and P. Seeberger. (2012). Glycomics, glycoproteomics, and the immune system. *Current Opinion in Chemical Biology*, 16(1-2): 214–20. https://doi.org/10.1016/j.cbpa.2011.12.006.
8. Marqvorsen, M., C. Araman and S. Van Kasteren. (2019). Going native: Synthesis of glycoproteins and glycopeptides via native linkages to study glycan-specific roles in the immune system. *Bioconjugate Chemistry*, 30: 2715–2726. https://doi.org/10.1021/acs.bioconjchem.9b00588.
9. Rodriguez, E., S. Schetters and Y. Kooyk. (2018). The tumour glyco-code as a novel immune checkpoint for immunotherapy. *Nature Reviews Immunology*, 18: 204–211. https://doi.org/10.1038/nri.2018.3.
10. Rudd, P., T. Elliott, P. Cresswell, I. Wilson and R. Dwek. (2001). Glycosylation and the immune system. *Science*, 291(5512): 2370–2376. https://doi.org/10.1126/SCIENCE.291.5512.2370.
11. Phan, J., E. Layton, K. Yu, M. Aguilar, I. Golez, N. Franko, J. Logue, L. Rodda, C. Howard, M. Pepper, M. Gale, H. Chu and C. Seshadri. (2023). Cytotoxic T cells targeting spike glycoprotein are associated with hybrid immunity to SARS-CoV-2. *Journal of Immunology*. https://doi.org/10.4049/jimmunol.2200815.
12. Zheng, L., Q. Yang, F. Li, M. Zhu, H. Yang, T. Tan, B. Wu, M. Liu, C. Xu, J. Yin and C. Cao. (2022). The glycosylation of immune checkpoints and their applications in oncology. *Pharmaceuticals*, 15. https://doi.org/10.3390/ph15121451.
13. Matsumoto, Y. and T. Ju. (2023). Aberrant glycosylation as immune therapeutic targets for solid tumors. *Cancers*, 15. https://doi.org/10.3390/cancers15143536.
14. Čaval, T., F. Alisson-Silva and F. Schwarz. (2023). Roles of glycosylation at the cancer cell surface: Opportunities for large scale glycoproteomics. *Theranostics*, 13: 2605–2615. https://doi.org/10.7150/thno.81760.
15. Suttapitugsakul, S., M. Tong and R. Wu. (2021). Time-resolved and comprehensive analysis of surface glycoproteins reveals distinct responses of monocytes and macrophages to bacterial infection. *Angewandte Chemie*, 133(20): 11595–11604. https://doi.org/10.1002/anie.202102692.
16. Zhou, J., D. Oswald, K. Oliva, L. Kreisman and B. Cobb. (2018). The glycoscience of immunity. *Trends in Immunology*, 39(7): 523–535. https://doi.org/10.1016/j.it.2018.04.004.
17. Wang, Y., K. Lei, L. Zhao and Y. Zhang. (2024). Clinical glycoproteomics: Methods and diseases. *MedComm.*, 5. https://doi.org/10.1002/mco2.760.
18. Yu, A., J. Zhao, W. Peng, A. Banazadeh, S. Williamson, M. Goli, Y. Huang and Y. Mechref. (2018). Advances in mass spectrometry-based glycoproteomics. *Electrophoresis*, 39. https://doi.org/10.1002/elps.201800272.

19. Reyes, C., P. Jiang, M. Atashi, A. Bennett, A. Yu, W. Peng, J. Zhong and Y. Mechref. (2021). Advances in mass spectrometry-based glycoproteomics: An update covering the period 2017–2021. *Electrophoresis*, 43. https://doi.org/10.1002/elps.202100188.
20. Sun, R., A. Kim and S. Lim. (2021). Glycosylation of Immune receptors in cancer. *Cells*, 10. https://doi.org/10.3390/cells10051100.
21. Gao, C., K. Stavenhagen, B. Eckmair, T. McKitrick, A. Mehta, Y. Matsumoto, A. McQuillan, M. Hanes, D. Eriş, K. Baker, N. Jia, M. Wei, J. Heimburg-Molinaro, B. Ernst and R. Cummings. (2021). Differential recognition of oligomannose isomers by glycan-binding proteins involved in innate and adaptive immunity. *Science Advances*, 7. https://doi.org/10.1126/sciadv.abf6834.
22. Wisnovsky, S., L. Möckl, S. Malaker, K. Pedram, G. Hess, N. Riley, M. Gray, B. Smith, M. Bassik, W. Moerner and C. Bertozzi. (2021). Genome-wide CRISPR screens reveal a specific ligand for the glycan-binding immune checkpoint receptor Siglec-7. In: *Proceedings of the National Academy of Sciences*, 118. https://doi.org/10.1073/pnas.2015024118.

CHAPTER 9

Glycoinformatics

Advances in Glycan Data Integration and Analysis

Shih-Yi Hsiung, Achille Zappa, Yukie Akune-Taylor and *Kiyoko F. Aoki-Kinoshita**

Glycoinformatics originated with pioneering initiatives like CarbBank in the 1980s, which represented the first systematic digital repository for carbohydrate structures. Despite this early foundation, the discipline still confronts significant computational and methodological obstacles. Key challenges include modeling complex multivalent binding patterns in protein-carbohydrate and lipid-carbohydrate interactions, developing complete glycosylation maps for diagnostic biomarkers, and addressing the inherent variability of non-template-based glycan synthesis pathways. However, recent decades have brought substantial advances in structural analysis capabilities, particularly through artificial intelligence (AI), machine learning (ML), and enhanced glycomics databases. These technological developments offer promising solutions to

Department Glycan and Life System Integration Center (GaLSIC), Faculty of Science and Engineering, Soka University, 1-236 Tangi-machi, Hachioji, Tokyo 192-8577, Japan.
https://orcid.org/0000-0003-0363-5987; https://orcid.org/0000-0003-4040-9620; https://orcid.org/0000-0002-0839-0643; https://orcid.org/0000-0002-6662-8015
* Corresponding author: kkiyoko@soka.ac.jp

DOI: 10.1201/9781003635901-9

longstanding problems in the field. This review examines the current landscape of glycoinformatics technologies, including approaches for glycan representation, methods for visualizing glycoconjugates and glycosylation patterns, Semantic Web principles and applications, glycomics database systems, and AI innovations and implications. The discussion focuses on how these emerging tools are addressing and potentially resolving the persistent challenges that have historically limited progress in computational glycobiology. The goal is to provide essential glycoinformatics concepts, platforms, and application tools to help glycoscientists overcome their research obstacles.

1. Introduction

Glycoinformatics combines bioinformatics, cheminformatics, and computational biology to study polysaccharides and glycans, as well as their biological interactions, which are critical in disease mechanisms including cancer, congenital disorders, and viral infections [14, 73]. With the rise of systems biology and the expansion of -omics technologies, bioinformatics has advanced significantly across many research domains [6, 122]. The field of glycoinformatics can be traced back to early efforts such as CarbBank, the first systematic attempt to digitally archive carbohydrate structures in the 1980s [114, 122]. Although the term 'glycoinformatics' has existed for two decades, the field continues to face numerous computational and systematic challenges. These include capturing the multivalent interactions that govern protein-carbohydrate and lipid-carbohydrate assemblies, achieving high-throughput glycan sequencing, compiling comprehensive glycosylation profiles of protein-based biomarkers, accounting for the non-template-driven nature of glycan biosynthesis, and handling the structural branching and complexity of glycans [34, 40, 52, 77, 94, 102, 119, 124]. Additionally, there are ongoing difficulties in integrating diverse nomenclature systems across glycobiology subfields and managing semantic vocabularies and ontologies that glycoinformaticians rely on daily [96].

Nevertheless, structural analysis technologies have seen major improvements in recent decades. Emerging approaches based on systems glycobiology, artificial intelligence (AI), machine learning (ML), and glycomics databases are increasingly facilitating the incorporation of glycan-based biomarkers, antigens, and enzymes into diagnostic, prognostic, and monitoring frameworks [64, 72, 123]. This review will explore current technologies in glycoinformatics and the challenges that they have begun to solve.

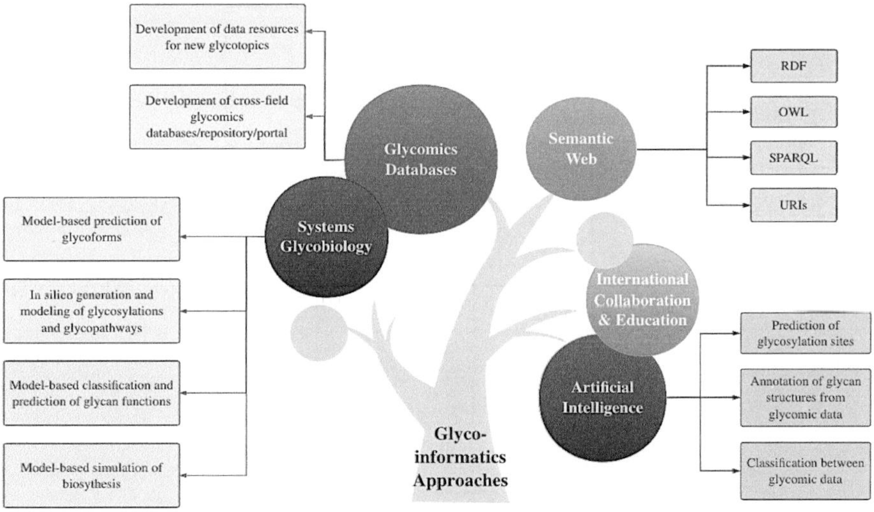

Fig. 1: The approaches of glycoinformatics within glycomics research (Modified from [34]).

2. Glycan Representation and Glycoconjugate Visualization

2.1 *Glycan Representation and Basic Tools*

The representation of glycan structures presents unique challenges not encountered in genomics or proteomics due to their branched nature, diverse linkage patterns, and anomeric configurations. Multiple representation systems and structural encoding formats have been developed, each with strengths and limitations for different applications (Table 1) [12]. Web3 Unique Representation of Carbohydrate Structures (WURCS), GlycoCT, International Union of Pure and Applied Chemistry (IUPAC)-Extended, and Linear Code are all international glycan structural encoding systems [11, 27, 48, 112]. Symbol Nomenclature for Glycans (SNFG) also provides a visualization standard for monosaccharides and is integrated in various standards and data resource systems [49].

Many structural representation tools support the creation of glycan structures using the SNFG standard and various input formats. These tools include doGlycans, DrawGlycan-SNFG, GlycanBuilder, GlycoDraw, GlycoGlyph, Glycowork, and GlycoWorkbench [20, 21, 22, 31, 79, 85, 117, 115]. Below is a brief overview of these software packages and web-based tools:

doGlycans is a Python package designed to prepare carbohydrate structures for simulation using GROningen MAchine for Chemical

Table 1: Comparison of glycan structure encoding formats.

Format	Primary Use	Strengths	Limitations	Data Model	Implementation	References
WURCS	Web/database exchange	Semantic Web compatible, handles ambiguity, linear format, doesn't require dictionary for substitutions	Complex syntax	Linear string	GlyTouCan, GlyCosmos	[82]
GlycoCT(XML)	Database storage	Comprehensive, XML-based	Requires libraries for processing, and dictionary of substitutions	XML structure	GlycomeDB, EuroCarbDB	[48]
IUPAC-Extended	Human-readable notation	Based on established chemical conventions, human-readable	Limited in representing ambiguity, non-unique (different strings can represent the same glycan)	Linear string	CSDB, legacy databases	[41]
LinearCode®	Commercial applications	Concise, ASCII-only format	Proprietary limitations	Linear string	Glycominds, commercial software	[63]
KEGG Chemical Function (KCF)	Pathway mapping	Integrates with KEGG pathways	Limited expressiveness	Graph-based	KEGG Glycan	[59]
LINUCS	Structure comparison	Graph theoretical approach	Limited adoption	Linear string with nested brackets	SWEET-DB	[13]

GLYDE-II	XML-based exchange	Comprehensive XML framework	Verbose for large structures	XML structure	Discontinued	[100]
CSDB Linear Code	Glycan structure database	Specialized for bacterial, plant and fungal glycans	Limited use outside CSDB	Linear string	CSDB	[113]
GlycoWorkbench encoding	MS data annotation	Integrated with analysis tools	Software-specific usage	Application-specific	GlycoWorkbench	[24]

* WURCS: Web3 Unique Representation of Carbohydrate Structures
* IUPAC-Extended: International Union of Pure and Applied Chemistry-Extended
* KEGG: Kyoto Encyclopedia of Genes and Genomes
* LINUCS: linear notation for the unique description of carbohydrate sequences
* SWEET-DB: Sugar Web-Accessible Expression and Enzyme Target Database
* CSDB: Carbohydrate Structure Database

Simulations (GROMACS), a high-performance molecular dynamics toolkit. It supports input in SNFG and IUPAC-condensed formats [31].

DrawGlycan-SNFG specializes in rendering glycans and glycopeptides from IUPAC-condensed input. It follows SNFG conventions and supports visualizing fragmentation patterns. Available as both a web-based application and a standalone graphical interface [22].

GlycanBuilder is a widely used tool for creating glycan diagrams. Integrated into platforms like GlyCosmos and GlycoWorkbench, it automatically arranges glycans based on symbolic notations and supports diverse glycan configurations [20].

GlycoDraw is a Python-based tool that generates high-quality glycan figures from formats including IUPAC-condensed, GlycoCT, and WURCS. It also supports glycan fragment ion visualization using Domon-Costello nomenclature [80].

GlycoGlyph is an open-source JavaScript glycan drawing tool. It enables conversion between graphical and text-based representations, including CFG linear nomenclature, and supports export in SVG and GlycoCT formats [85].

Glycowork is a Python package for glycan visualization, sequence alignment, motif analysis, and machine learning. It supports SNFG and IUPAC formats and represents monosaccharides and linkages as graph node labels [117].

GlycoWorkbench is a software suite that evaluates glycan structures against mass spectrometry spectra by matching proposed fragment masses. It features a user-friendly GUI and a comprehensive library of fragmentation types [24].

Despite the variety of text representations of glycans, an international effort to assist in the consolidation of these formats gave rise to the GlyTouCan repository. Users are given options such as a graphical user interface, application programming interface (API) and text input formats to register glycans. This allows any user to organize glycan data using the GlyTouCan identifier that is uniquely assigned to every glycan. Moreover, as described later, GlyTouCan data also use Semantic Web technologies, such that the varying levels of ambiguity can be organized in such a way that fully-defined glycans can be linked to their monosaccharide compositions, and vice-versa, thus allowing users to traverse the network of related glycans.

2.2 *Glycoconjugate Analysis and Visualization*

Glycoconjugates are molecules consisting of carbohydrate chains, or glycans, that are covalently attached to lipids, proteins, or RNAs. They are produced through the process of glycosylation and can vary in their glycan compositions. Glycosylated RNAs (glycoRNAs), a novel (and sometimes debated) glycosylated molecule, have also been reported in recent years [35]. Much of glycoRNA biology remains unknown, partly due to the limited availability of visualization and detection technologies. Glycoconjugates (glycosylated proteins, lipids, and RNA) are closely related to various diseases, such as influenza, cancer, and muscular dystrophy, congenital disorders of glycosylation (CDGs), and other serious diseases [33, 50, 97, 101, 107]. Below are examples of web servers and software packages which glycoinformatics developed for analysis and visualization of glycoconjugate structures, glycosylation, and pathways:

dbCAN is an online server for automated CAZyme (carbohydrate-active enzyme) annotation. CAZymes interact with specific glycosidic linkages to degrade or synthesize glycans and often function alongside proteins. dbCAN helps explore glycosyltransferases and other enzymes involved in glycan biosynthesis. Genomic data management is supported by API. Key features include an HMM database for enzyme identification, substrate prediction, and interactive visualizations. The platform is regularly updated [134].

Glycomics@Expasy is the glycoinformatics resource portal of the Swiss Institute of Bioinformatics (SIB), serving as a registry of software tools, databases, knowledgebases, and datasets for glycoscience. It is part of Expasy, which offers comprehensive information on resources that support diverse areas of life science and clinical research, including genomics, proteomics, glycomics, structural biology, evolutionary studies, systems biology, and medicinal chemistry [75]. In particular, the GlyConnect resource at Glycomics@Expasy is a manually curated resource of glycoproteins and glycopeptides. A number of tools are available to enable users to visualize a glycoproteomics dataset to better understand its content and the relationships between glycopeptides within it.

GlycoMaple is a web-based tool for mapping human glycosylation that visualizes over 20 distinct glycosylation pathways based on gene expression profiles in cells or tissues. It supports researchers in analyzing glycosylation dynamics across various pathways and cell types by linking gene expression data to specific glycosylation processes. As of 2024,

GlycoMaple includes 21 pathways and more than 1,000 glycan-related genes, covering most known human and mouse glycosylation events. Glycan structures are presented using Symbol Nomenclature for Glycans (SNFG) [66].

GlyCosmos Portal is a semantically integrated platform that centralizes glycoscience data from various domains, including glycomics, glycoproteomics, genomics, proteomics, and disease biology. Developed under Japan's Database Integration Coordination Program (DICP), GlyCosmos provides a user-friendly, interoperable interface powered by Semantic Web technologies such as RDF and SPARQL, allowing complex cross-database querying [5, 70, 129].

Glycosylator is a Python-based framework for identifying, modeling, and modifying glycans on protein structures. Built using BuildAMol and based on the Chemistry at Harvard Macromolecular Mechanics (CHARMM) force field, it supports both 2D and 3D visualization. It includes an API and a GUI, allowing rapid visualization and modification [71].

GlycoSim simulates glycan biosynthesis using enzyme kinetics, substrate concentrations, and compartment-specific localization. Through agent-based modeling, it captures the dynamic interactions among glycosyltransferases and glycosidases, supporting virtual experiments that test the impact of mutations, enzyme levels, or organelle dysfunction [2].

GlyGen is an NIH-funded international initiative that provides an integrated, user-friendly platform for accessing and exploring glycoscience data. It addresses the challenge of fragmented glycan-related data by harmonizing information from diverse sources such as NCBI, UniProt, PDB, UniCarbKB, and GlyTouCan. The GlyGen Portal [77] (https://www.glygen.org) currently focuses on N- and O-glycan data for human, mouse, and rat, offering tools to support cross-disciplinary research in glycoscience, genetics, proteomics, and pathology [130].

GlySpace Alliance is a collaborative initiative that integrates three major glycan data projects: GlyGen, Glycomics@Expasy, and GlyCosmos. It promotes data interoperability, sharing, and platform specialization, with Glycomics@Expasy focusing on user-friendly access, GlyCosmos on integrated glyco-omics across all organisms, and GlyGen on linking glycans to genes and proteins. The alliance adheres to FAIR data principles, providing open access under the CC-BY-4.0 license and supporting APIs and SPARQL endpoints for programmatic access [8, 75].

3. Semantic Web Applications on Multi-Glycomic Data: Online Glycomics Resources and Semantic Web Technologies

3.1 Categories of Glycomics Resources

The Semantic Web enhances data integration in glycobiology by shifting from document-centric models to structured, machine-readable data. This transition enables automated reasoning, semantic interoperability, and seamless knowledge integration across biomedical domains [86]. Developed under W3C, the Semantic Web relies on four key technologies: RDF, OWL, SPARQL, and URIs [4, 26]. RDF structures data as subject-predicate-object triples; OWL defines glycan ontologies; SPARQL queries linked datasets; and URIs provide stable identifiers for biological entities like glycans and genes [30].

There are three types of data resources: (1) databases, (2) repositories, and (3) portals. This section will discuss several widely used international data resources, glycan standards, and Semantic Web technologies which glycoinformaticians have created to support a comprehensive understanding of glycobiology (Table 2).

A database is a structured system for efficient data storage and retrieval, often managed by a DBMS. Traditional relational databases use tables and SQL querying [62], originating from the relational model introduced by Codd in the 1970s [28]. Today, relational systems like Oracle and PostgreSQL coexist with NoSQL platforms like MongoDB and Cassandra, which support unstructured data [105].

A repository is a digital system designed for long-term storage, preservation, and dissemination of diverse digital objects such as datasets, publications, code, and multimedia files. Repositories emphasize metadata annotation, version control, and accessibility over rigid structural querying [96]. Repositories gained traction in the 1990s and 2000s alongside the Open Access movement [70, 109].

Platforms like DSpace, EPrints, and Fedora Commons allowed academic institutions to archive theses, datasets, and research output [36, 92]. In the life sciences and glycosciences, repositories such as GenBank, Protein DataBank (PDB) and GlycoPOST serve to accept experimental data and relevant metadata from researchers to assign accession numbers for use in publications. As mentioned earlier, GlyTouCan is the international glycan repository for assigning accession numbers to glycan structures themselves [38], whereas GlycoPOST is a repository for mass spectrometry data that follows the MIRAGE guidelines [126]. GlyComb was also recently published to assign IDs to glycoproteins and glycopeptides [111].

A portal is a web-based access point aggregating content and tools from multiple data sources into a user-friendly interface. Unlike databases or repositories, which emphasize storage or preservation, portals prioritize integration, navigation, and user interaction. Many portals incorporate semantic technologies to improve data connectivity [58, 76]. Portals emerged in the early 2000s to address the challenge of fragmented biological data spread across various systems. They evolved to provide centralized access to distributed resources, particularly as the Semantic Web and linked data technologies, e.g., RDF and OWL, matured [95]. The GlyCosmos Portal and Ensembl are two international biological example portals [46, 129]. Below is a list of selected resources:

3.2 *Resource Description Framework (RDF)*

The Resource Description Framework (RDF) is a W3C-standardized model for representing structured, interoperable, and semantically rich web data [56]. It forms the foundation of the Semantic Web and Linked Data by enabling machine-readable knowledge integration across diverse sources [132]. RDF structures data as triples (subject, predicate, and object), forming a labeled, directed graph. Resources are identified via uniform resource identifiers (URIs), and semantics are enhanced through RDF Schema (RDFS) and OWL, which define hierarchies and logical rules [23]. RDF supports metadata and domain knowledge representation, using serializations like Turtle, RDF/XML, and JSON-LD. SPARQL enables querying RDF data through pattern matching, federated queries, and reasoning [51]. In bio- and glycoinformatics, RDF links biological entities, genes, proteins, glycans, diseases, across distributed databases like GlyTouCan, GlyGen, UniProt, and ChEBI [7, 47, 91]. RDF supports FAIR data principles, enabling semantic interoperability, provenance tracking, and discovery in knowledge-driven pipelines [108].

3.3 *Web Ontology Language (OWL)*

An ontology is a structured data model that defines a set of concepts within a specific domain and describes the relationships among those concepts. It enables reasoning about the elements of that domain by organizing knowledge in a formal, interpretable way. Ontologies are widely used in fields such as artificial intelligence, the Semantic Web, software engineering, biomedical informatics, and information architecture to represent knowledge about the world or a particular area of interest. Typically, an ontology includes the following components: individuals, classes, attributes, relations, and events [44]. Ontologies define the vocabulary used for communication between agents, assigning precise

Table 2: Major glycomics data resources and their characteristics.

Database	Resource Type	Data Types	Integration Capabilities	Access Method	References
GlyCosmos	Integrated glycomics portal	Glycan structures, glycogenes, glycoproteins, pathways, diseases, taxonomies	RDF-based Semantic Web integration with UniProt, PubChem, PDB, and Reactome	Web interface, REST API, SPARQL endpoint	[129]
GlyTouCan	Glycan structure repository	Glycan structures with accession numbers, links to GlyTouCan partners (other glycan databases)	Cross-references to multiple glycan databases	Web interface, REST API	[118]
CSDB	Database of Bacterial bacterial and plant carbohydrates	NMR data, taxonomic information, bibliographic data	Links to NMR data and references	Web interface	[113]
GlyGen	Web portal for glycans and glycoproteins	Glycosylation sites, glycans, proteins, glycoproteins and, biomarkers, 3D structures	Integration with NCBI, UniProt	Web interface, API	[130]
MatrixDB	Database of extracellular matrix interactions	Glycosaminoglycan interactions	Integration with protein interaction databases	Web interface, PSICQUIC web service	[25]
GlyConnect	Curated database of glycoproteins	Glycoproteins, site-specific glycan heterogeneity	UniProt integration	Web interface, SPARQL, REST API	[3]
Glycomics@ Expasy	Web portal of resources of glycoproteins and glycans	Glycans, glycoproteins and various tools	UniProt integration	Web interface, SPARQL, REST API	[83]

meanings to each term in relation to other terms and within a defined contextual framework. By embedding structured domain knowledge, ontologies equip agents with the semantic understanding necessary to operate in an informed and intelligent manner [45]. Glycoinformatics-related ontologies include GlycoCoO, GlycoRDF, and GNOme:

GlycoCoO (GlycoConjugate Ontology) provides a standardized semantic vocabulary for describing glycoconjugates. Because most glycoconjugates are currently identified using glycomics profiling and mass spectrometry analysis, GlycoCoO encodes compositions (e.g., Hex5HexNAc4) as RDF triples with rich semantic meaning and reuses the FALDO ontology as defined as a standard for amino acid sequence representation [16].

GlycoRDF is an ontology that encodes glycan information using the Resource Description Framework (RDF), enabling machine-readable representation, integration, and querying of glycan-related data across platforms. It encapsulates not just glycan structure data and their representations, but also the publications, biological or chemical sources, and experimental techniques used to identify them. Thus, it forms a foundational component of glycoinformatics infrastructures like GlyTouCan and GlyCosmos, ensuring standardized and consistent glycan structure representation across diverse systems [99].

GNOme (Glycan Nomenclature Ontology Management Environment) is an ontology developed to curate, standardize, and disseminate glycan structures. Based on the concept of subsumption, it links glycans that have the same mass but have varying levels of ambiguity. For example, an N-glycan core structure that has its fully-defined glycosidic linkage information (GlyTouCan ID G47175EF) is linked to its monosaccharide composition (GlyTouCan ID G44215PV) so that mass spectrometrists, for example, who handle compositions, can easily obtain their linkage-defined counterparts [39, 133].

3.4 SPARQL Protocol and RDF Query Language (SPARQL) and Federated Queries and Federated Queries

As structured RDF datasets expanded across diverse domains such as healthcare, life sciences, and government, the need for a standardized query language to access and integrate this data became increasingly apparent. This demand led to the development of SPARQL (SPARQL Protocol and RDF Query Language), specifically designed for querying RDF data [30]. The evolution of SPARQL began in 2004 with the formation of the W3C SPARQL Working Group. In 2008, the World Wide Web Consortium (W3C) published SPARQL 1.0, establishing a foundational standard for querying RDF [9]. This was followed by the release of SPARQL 1.1, which introduced significant enhancements such as support for federated queries, update operations, and aggregation functions. Today, SPARQL has become a cornerstone of the Semantic Web technology stack and is extensively employed in fields like bioinformatics, digital libraries, and open government initiatives [32]. A SPARQL endpoint is a web service that allows users and applications to submit SPARQL queries to an RDF database and receive structured results. A SPARQL endpoint is a web-accessible interface that enables users and applications to execute SPARQL queries against an RDF database and retrieve structured responses [17]. Key functionalities include providing an HTTP based query interface, accepting SPARQL queries through both GET and POST methods, and delivering results in various formats such as XML, JSON, CSV, or RDF [68]. Additionally, many endpoints offer descriptive metadata using standards like VoID (Vocabulary of Interlinked Datasets) or SPARQL Service Descriptions to enhance discoverability and usability [90, 110]. Federated queries, introduced with SPARQL 1.1, enable the simultaneous retrieval of data from multiple SPARQL endpoints, supporting seamless cross-domain integration and advancing the goals of the Linked Open Data (LOD) initiative [42]. By using the SERVICE keyword, these queries allow users to incorporate data from various distributed RDF sources into a single query execution. This capability is particularly valuable in fields like bioinformatics, where relevant datasets are often maintained across diverse institutions and repositories, making federated querying a crucial tool for achieving data interoperability and comprehensive analysis [89, 121].

3.5 Uniform Resource Identifiers (URIs)

A Uniform Resource Identifier (URI) is a standardized and structured string that uniquely identifies a resource, whether on the internet or within a distributed system. As defined by the Internet Engineering Task Force (IETF) in RFC 3986, URIs are a fundamental component of web architecture and essential for achieving interoperability and semantic integration in the Semantic Web and Linked Data paradigms [10]. There are two types of URIs: Uniform Resource Locators (URLs), which specify how to access a resource; and Uniform Resource Names (URNs), which provide a persistent name without referencing its location [1, 103, 106]. By providing globally unique and resolvable identifiers, URIs support consistent, persistent, and interoperable referencing of resources across diverse datasets and platforms. This capability is critical for linking and integrating information in complex, data-intensive scientific fields, where URIs enable automated reasoning, semantic annotation, and cross-database discovery within knowledge graphs and ontologies [60, 99].

4. Artificial Intelligence (AI) Applications in Glycoscience

4.1 Machine Learning (ML)

ML is a branch of AI that focuses on developing computational algorithms capable of learning patterns from data and making predictions or decisions without being explicitly programmed for each task [65]. Inspired by the concept of simulating human intelligence, ML algorithms build models automatically by learning from training data, enabling them to solve complex, real-world problems across a range of domains, from healthcare to finance to molecular biology [53]. The conceptual foundations of ML trace back to early pioneers, such as Alan Turing, Arthur Samuel, John McCarthy, Marvin Minsky, and Allen Newell. While the broader field of AI emerged in the early 1940s, machine learning began to take shape as a distinct discipline during the 1950s, particularly with Samuel's work on self-learning programs like his checkers-playing algorithm [37]. These foundational efforts laid the groundwork for a wide variety of ML techniques that have since evolved into a rich and diverse ecosystem. Deep Learning (DL) is a specialized subfield of ML. Convolutional Neural Networks (CNNs), Recurrent Neural Networks (RNNs), and Graph Neural Networks (GNNs) are several common DL subclasses [78, 87]. Convolutional Neural Networks (CNNs) are primarily used for processing data with grid-like topology, such as images. CNNs have also shown strong performance in analyzing genomic sequences, protein structures, and glycan imaging. They consist of layers that perform convolution operations, using filters that slide over input data to detect local patterns [57]. RNNs are designed for sequential data where the order of inputs matters. They maintain a hidden state that carries information from previous steps, making them suitable for time-series, language, and sequence modeling tasks [43, 84]. Graph Neural Networks (GNNs) are DL models tailored for graph-structured data. Graphs consist of nodes and edges, ideal for representing glycans, which are often branched, tree-like molecules [29, 135]. Neural Language Models (NLMs) are deep learning models that learn to understand and generate natural language. Large Language Models (LLMs) are a subclass trained on massive corpora, often using transformer architectures [61, 69].

4.2 Applications in Glycoscience

DL is gaining significant attention in genomic research due to its ability to handle the complexity and high dimensionality of genomic and glycomic data. Unlike traditional methods such as linear regression, which are limited in capturing intricate patterns, DL can learn complex, nonlinear relationships and interactions inherent in genetic and glycomic information [14, 15, 54, 55, 72]. In glycoscience studies, CNNs can be used for glycan structure recognition from imaging data (e.g., MS spectra, microarrays), protein-glycan interaction prediction based on sequence or structural data, and variant effect prediction in DNA/RNA sequences using CNNs trained on genomic motifs [18, 74, 98]. RNNs can predict glycosylation sites on protein sequences, model sequential patterns in DNA, RNA, or glycan biosynthesis, and model sequence-to-sequence for glycan biosynthetic pathways [19, 63, 104]. GNNs can represent glycans as graphs and model their branched architecture, infer binding affinity, solubility, or immunogenicity, interpret the metabolic networks, such as enzymes (e.g., glycosyltransferases) which modify substrates [88, 125, 127]. Others include transformer models for sequence-based learning to predict glycosylation sites or enzyme functions, natural language processing (NLP) tools for mining glycoscience literature, Reinforcement Learning and Agent-Based Modeling, and simulation of glycosylation dynamics under various biological or experimental conditions [67, 131]. There are some examples of AI applications in glycoscience:

GlyNet is a pioneering multi-task neural network model designed to predict protein-glycan interactions with high accuracy and detail. Developed to support advances in fields such as diagnostics, therapeutics, and vaccine development, GlyNet models the relative binding strengths between mammalian glycans and 352 glycan-binding proteins, generating 1,257 outputs per glycan input. Trained on relative fluorescence unit (RFU) data from the Consortium for Functional Glycomics, the model not only captures known interactions but also extrapolates to untested glycans [19].

LectinOracle is a predictive platform designed to generalize effectively to novel glycans and lectins, showing strong qualitative and quantitative alignment with experimental results. This tool demonstrates versatility across multiple applications, including enhancing lectin classification, guiding the directed evolution of lectins, forecasting epidemiological trends such as influenza spread, and enabling large-scale analysis of lectomes in host-microbe interactions [79].

5. Conclusions and Prospects

Glycoinformatics advances rapidly to address glycan complexity, structural diversity, and database underrepresentation. Critical development involves expanding glycome coverage through AI. Due to non-template-driven glycan biosynthesis, glycans exhibit extensive structural heterogeneity, complicating large-scale annotation. AI-driven approaches, particularly graph neural networks (GNNs) and transformer models, accelerate glycan structure prediction from spectral, genomic, and transcriptomic data. Comprehensive, standardized glycan ontologies are essential for semantic integration across diverse data sources. While Semantic Web technologies (RDF, OWL) enable foundational interoperability, future frameworks must support unified ontologies connecting glycan structure, function, and interactions. The difficulty lies in the heterogeneity of data types in which glycans are involved; they are chemical structures but have their own monosaccharide ' vocabulary' while being synthesized by genes, but also being attached to proteins, lipids and nucleotides, their 3D conformations are important for recognition by other proteins, viruses and bacteria, and the combination of all of the above define their function. Glycoinformatics progresses toward multi-omics integration to model glycan-regulated biological networks. Integration of glycomics with transcriptomic, proteomic, and metabolomic data supports dynamic model construction incorporating enzyme kinetics, expression profiles, and metabolite flux. ML algorithms can help identify missing or dysregulated biosynthetic pathway components. Future advances depend on addressing glycan complexity through computational methodologies, semantic standardization, and multi-omics integration. Emerging AI models, unified ontologies, dynamic network modeling, and visualization tools will transform glycan research into a predictive, interoperable discipline, accelerating biomedical applications.

References

1. Abu Al-Haija, Q. and M. Al-Fayoumi. (2023). An intelligent identification and classification system for malicious uniform resource locators (URLs). *Neural Computing and Applications*, 35(23): 16995–17011.
2. Akune-Taylor, Y., A. Kon and K.F. Aoki-Kinoshita. (2024). *In silico* simulation of glycosylation and related pathways. *Analytical and Bioanalytical Chemistry*, 416(16): 3687–3696.
3. Alocci, D., J. Mariethoz, A. Gastaldello, E. Gasteiger, N.G. Karlsson, D. Kolarich, N.H. Packer and F. Lisacek. (2018). GlyConnect: Glycoproteomics goes visual, interactive, and analytical. *Journal of Proteome Research*, 18(2): 664–677.
4. Amara, F.Z., M. Hemam, M. Djezzar and M. Maimor. (2022). Semantic web and internet of things: Challenges, applications, and perspectives. *Journal of ICT Standardization*, 10(2): 261–291.
5. Aoki-Kinoshita, K.F. (2019). GlyCosmos Portal and MIRAGE. *Glycoforum*, 22: A5.
6. Aoki-Kinoshita, K.F. (2021). Glycome informatics: Using systems biology to gain mechanistic insights into glycan biosynthesis. *Current Opinion in Chemical Engineering*, 32: 100683.
7. Aoki-Kinoshita, K., S. Agravat, N.P. Aoki, S. Arpinar, R.D. Cummings, A. Fujita, N. Fujita, G.M. Hart, S.M. Haslam, T. Kawasaki and M. Matsubara. (2016). GlyTouCan 1.0—The international gly-can structure repository. *Nucleic Acids Research*, 44(D1): D1237–D1242.

8. Aoki-Kinoshita, K.F., F. Lisacek, R. Mazumder, W.S. York and N.H. Packer. (2020). The GlySpace Alliance: Toward a collaborative global glycoinformatics community. *Glycobiology*, 30(2): 70–71.
9. Bastian, H. (2009). *Faceted Exploration of Multiple RDF Data Sources Using SPARQL*. Doctoral dissertation, Master's thesis, University of Duisburg-Essen, Duisburg.
10. Berners-Lee, T., R. Fielding and L. Masinter. (2005). RFC 3986: Uniform Resource Identifier (URI): Generic syntax. ResearchGate.
11. Biggs, N.L. and N.L. Biggs. (2008). Linear codes. In: *Codes: An Introduction to Information Communication and Cryptography*, pp. 1–18. Springer.
12. Bochkov, A.Y. and P.V. Toukach. (2021). CSDB/SNFG structure editor: An online glycan builder with 2D and 3D structure visualization. *Journal of Chemical Information and Modeling*, 61(10): 4940–4948.
13. Bohne-Lang, A., E. Lang, T. Förster and C.W. von der Lieth. (2001). LINUCS: Linear notation for unique description of carbohydrate sequences. *Carbohydrate Research*, 336(1): 1–11.
14. Bojar, D. and F. Lisacek. 2022. Glycoinformatics in the artificial intelligence era. *Chemical Reviews*, 122(20): 15971–15988.
15. Bojar, D., L. Meche, G. Meng, W. Eng, D.F. Smith, R.D. Cummings and L.K. Mahal. (2022). A useful guide to lectin binding: machine-learning directed annotation of 57 unique lectin specificities. *ACS Chemical Biology*, 17(11): 2993–3012.
16. Bolleman, J.T., C.J. Mungall, F. Strozzi, J. Baran, M. Dumontier, R.J. Bonnal, R. Buels, R. Hoehndorf, T. Fujisawa, T. Katayama and P.J. Cock. (2016). FALDO: A semantic standard for describing the location of nucleotide and protein feature annotation. *Journal of Biomedical Semantics*, 7: 1–12.
17. Buil-Aranda, C., A. Hogan, J. Umbrich and P.Y. Vandenbussche. (2013). SPARQL web-querying infrastructure: Ready for action? In: *The Semantic Web–ISWC 2013: 12th International Semantic Web Conference*, Sydney, NSW, Australia, October 21–25, 2013, *Proceedings, Part II*, 12: 277–293. Springer Berlin Heidelberg.
18. Burkholz, R., J. Quackenbush and D. Bojar. (2021). Using graph convolutional neural networks to learn a representation for glycans. *Cell Reports*, 35(11).
19. Carpenter, E.J., S. Seth, N. Yue, R. Greiner and R. Derda. (2022). GlyNet: A multi-task neural network for predicting protein–glycan interactions. *Chemical Science*, 13(22): 6669–6686.
20. Ceroni, A., A. Dell and S.M. Haslam. (2007). The GlycanBuilder: A fast, intuitive, and flexible software tool for building and displaying glycan structures. *Source Code for Biology and Medicine*, 2: 1–13.
21. Ceroni, A., K. Maass, H. Geyer, R. Geyer, A. Dell and S.M. Haslam. 2008. GlycoWorkbench: A tool for the computer-assisted annotation of mass spectra of glycans. *Journal of Proteome Research*, 7(4): 1650–1659.
22. Cheng, K., Y. Zhou and S. Neelamegham. (2017). DrawGlycan-SNFG: A robust tool to render glycans and glycopeptides with fragmentation information. *Glycobiology*, 27(3): 200–205.
23. Chowdhury, S.M. (2023). *Semantic Search on Equipment*. Master's thesis, University of South-Eastern Norway.
24. Ceroni, A., K. Maass, H. Geyer, R. Geyer, A. Dell and S.M. Haslam. 2008. GlycoWorkbench: A tool for the computer-assisted annotation of mass spectra of glycans. *Journal of Proteome Research*, 7(4): 1650–1659.
25. Clerc, O., M. Deniaud, S.D. Vallet, A. Naba, A. Rivet, S. Perez, N. Thierry-Mieg and S. Ricard-Blum. (2019). MatrixDB: Integration of new data with a focus on glycosaminoglycan interactions. *Nucleic Acids Research*, 47(D1): D376–D381.

26. Conde Herreros, D. (2022). *Using Ontology Directories to Support Resource Discovery*. Doctoral dissertation, ETSI_Informatica.
27. Cornish-Bowden, A. (1985). Nomenclature for incompletely specified bases in nucleic acid sequences: Recommendations 1984. *Nucleic Acids Research*, 13(9): 3021.
28. Coronel, C., S. Morris and P. Rob. (2010). *Database Systems: Design, Implementation, and Management*. Malla Reddy College of College of Engineering.
29. Corso, G., H. Stark, S. Jegelka, T. Jaakkola and R. Barzilay. 2024. Graph neural networks. *Nature Reviews Methods Primers*, 4(1): 17.
30. Curé, O. and G. Blin. (2014). *RDF Database Systems: Triples Storage and SPARQL Query Processing*. Morgan Kaufmann.
31. Danne, R., C. Poojari, H. Martinez-Seara, S. Rissanen, F. Lolicato, T. Róg and I. Vattulainen. (2017). doGlycans—Tools for preparing carbohydrate structures for atomistic simulations of glycoproteins, glycolipids, and carbohydrate polymers for GROMACS. *Journal of Chemical Information and Modeling*, 57(10): 2401–2406.
32. DuCharme, B. (2013). *Learning SPARQL: Querying and updating with SPARQL 1.1*. O'Reilly Media, Inc.
33. Esmail, S. and M.F. Manolson. (2021). Advances in understanding N-glycosylation structure, function, and regulation in health and disease. *European Journal of Cell Biology*, 100(7-8): 151186.
34. Flevaris, K. and C. Kontoravdi. (2022). Immunoglobulin G N-glycan biomarkers for autoimmune diseases: Current state and a glycoinformatics perspective. *International Journal of Molecular Sciences*, 23(9): 5180.
35. Flynn, R.A., K. Pedram, S.A. Malaker, P.J. Batista, B.A. Smith, A.G. Johnson, B.M. George, K. Majzoub, P.W. Villalta, J.E. Carette and C.R. Bertozzi. (2021). Small RNAs are modified with N-glycans and displayed on the surface of living cells. *Cell*, 184(12): 3109–3124.
36. Fortunato, L. and M. Galassi. (2021). The case for free and open source software in research and scholarship. *Philosophical Transactions of the Royal Society A*, 379(2197): 20200079.
37. Frana, P.L. and M.J. Klein. (eds.). (2021). *Encyclopedia of Artificial Intelligence: The Past, Present, and Future of AI*. Bloomsbury Publishing USA.
38. Fujita, A., N.P. Aoki, D. Shinmachi, M. Matsubara, S. Tsuchiya, M. Shiota, T. Ono, I. Yamada and K.F. Aoki-Kinoshita. (2021). The international glycan repository GlyTouCan version 3.0. *Nucleic Acids Research*, 49(D1): D1529–D1533.
39. German, D.M. (2003). The GNOME project: A case study of open source, global software development. *Software Process: Improvement and Practice*, 8(4): 201–215.
40. Gbormittah, F.O., J. Bones, M. Hincapie, F. Tousi, W.S. Hancock and O. Iliopoulos. (2015). Clusterin glycopeptide variant characterization reveals significant site-specific glycan changes in the plasma of clear cell renal cell carcinoma. *Journal of Proteome Research*, 14(6): 2425–2436.
41. Gomollón-Bel, F. (2022). IUPAC Top Ten Emerging Technologies in Chemistry 2022: Discover the innovations that will transform energy, health, and materials science, to tackle the most urgent societal challenges and catalyse sustainable development. *Chemistry International*, 44(4): 4–13.
42. Görlitz, O. (2015). *Distributed Query Processing for Federated RDF Data Management*. Doctoral dissertation, Universität Koblenz.
43. Grossberg, S. (2013). Recurrent neural networks. *Scholarpedia*, 8(2): 1888.
44. Gruber, T. (1993). What is Ontology? https://tomgruber.org.
45. Hadzic, M., P. Wongthongtham, T. Dillon, E. Chang, M. Hadzic, P. Wongthongtham, T. Dillon and E. Chang. (2009). Introduction to ontology. In: *Ontology-based Multi-agent Systems*, pp. 37–60. SpringerLink.

46. Harrison, P.W., M.R. Amode, O. Austine-Orimoloye, A.G. Azov, M. Barba, I. Barnes, A. Becker, R. Bennett, A. Berry, J. Bhai and S.K. Bhurji. (2024). Ensembl 2024. *Nucleic Acids Research*, 52(D1): D891–D899.
47. Hastings, J., G. Owen, A. Dekker, M. Ennis, N. Kale, V. Muthukrishnan, S. Turner, N. Swainston, P. Mendes and C. Steinbeck. (2016). ChEBI in 2016: Improved services and an expanding collection of metabolites. *Nucleic Acids Research*, 44(D1): D1214–D1219.
48. Herget, S., R. Ranzinger, K. Maass and C.W. Lieth. (2008). GlycoCT: A unifying sequence format for carbohydrates. *Carbohydrate Research*, 343(12): 2162–2171.
49. Haltiwanger, R.S. (2016). Symbol nomenclature for glycans (SNFG). *Glycobiology*, 26(3): 217–217.
50. Haukedal, H. and K.K. Freude. (2021). Implications of glycosylation in Alzheimer's disease. *Frontiers in Neuroscience*, 14: 625348.
51. Hofer, M., D. Obraczka, A. Saeedi, H. Köpcke and E. Rahm. (2024). Construction of knowledge graphs: Current state and challenges. *Information*, 15(8): 509.
52. Hu, H., K. Khatri and J. Zaia. (2017). Algorithms and design strategies towards automated glycoproteomics analysis. *Mass Spectrometry Reviews*, 36(4): 475–498.
53. Hua, H., Y. Li, T. Wang, N. Dong, W. Li and J. Cao. (2023). Edge computing with artificial intelligence: A machine learning perspective. *ACM Computing Surveys*, 55(9): 1–35.
54. Hwang, H., H.K. Jeong, H.K. Lee, G.W. Park, J.Y. Lee, S.Y. Lee, Y.M. Kang, H.J. An, J.G. Kang, J.H. Ko, J.Y. Kim and J.S. Yoo. (2020). Machine learning classifies core and outer fucosylation of N-glycoproteins using mass spectrometry. *Scientific Reports*, 10(1): 318.
55. Hsiung, S.Y., S.X. Deng, J. Li, S.Y. Huang, C.K. Liaw, S.Y. Huang, C.C. Wang and Y.S. Hsieh. (2023). Machine learning-based monosaccharide profiling for tissue-specific classification of Wolfiporia extensa samples. *Carbohydrate Polymers*, 322: 121338.
56. Jain, K. (2007). D-RDF: *Dynamic Resource Description Framework*. Iowa State University.
57. John, C., J. Sahoo, M. Madhavan and O.K. Mathew. (2023). Convolutional neural networks: A promising deep learning architecture for biological sequence analysis. *Current Bioinformatics*, 18(7): 537–558.
58. Jonquet, C., J. Graybeal, S. Bouazzouni, M. Dorf, N. Fiore, X. Kechagioglou, T. Redmond, I. Rosati, A. Skrenchuk, J.L. Vendetti, M. Musen and OntoPortal Alliance. (2023). Ontology repositories and semantic artefact catalogues with the OntoPortal technology. In: *International Semantic Web Conference*, pp. 38–58. Cham: Springer Nature Switzerland.
60. Katayama, T., M. Nakao and T. Takagi. (2010). TogoWS: integrated SOAP and REST APIs for interoperable bioinformatics Web services. *Nucleic Acids Research*, 38(suppl_2): W706–W711.
61. Kather, J.N., D. Ferber, I.C. Wiest, S. Gilbert and D. Truhn. (2024). Large language models could make natural language again the universal interface of healthcare. *Nature Medicine*, 30(10): 2708–2710.
62. Kaufmann, M. and A. Meier. (2023). *SQL and NoSQL Databases*. Switzerland: Springer. doi, 10: 978-3.
63. Kellman, B.P., J. Mariethoz, Y. Zhang, S. Shaul, M. Alteri, D. Sandoval, M. Jeffris, E. Armingol, B. Bao, F. Lisacek, D. Bojar and N.E. Lewis. (2024). Decoding glycosylation potential from protein structure across human glycoproteins with a multi-view recurrent neural network. *bioRxiv*.
64. Kim, D.N., A.D. McNaughton and N. Kumar. (2024). Leveraging artificial intelligence to expedite antibody design and enhance antibody–antigen interactions. *Bioengineering*, 11(2): 185.
65. Kok, J.N., E.J. Boers, W.A. Kosters, P. Van der Putten and M. Poel. (2009). Artificial intelligence: Definition, trends, techniques, and cases. *Artificial Intelligence*, 1(270-299): 51.

66. Kong, W.Z. and M. Fujita. 2025. GlycoMaple: Recent updates and applications in visualization and analysis of glycosylation pathways. *Analytical and Bioanalytical Chemistry*, 417(5): 885–894.
67. Kumar, S., D. Guruparan, P. Aaron, P. Telajan, K. Mahadevan, D. Davagandhi and O.X. Yue. (2023). Deep learning in computational biology: Advancements, challenges, and future outlook. *arXiv preprint arXiv*:2310.03086.
68. Lanthaler, M. (2014). *Third Generation Web Apis. Bridging the Gap between REST and Linked Data*. Doctoral dissertation. Graz.
69. Lauriola, I., A. Lavelli and F. Aiolli. (2022). An introduction to deep learning in natural language processing: Models, techniques, and tools. *Neurocomputing*, 470: 443–456.
70. Lee, S., T. Ono, S. Masaaki, A. Fujita, M. Matsubara, A. Zappa, I. Yamada and K.F. Aoki-Kinoshita. (2025). Updates implemented in version 4 of the GlyCosmos glycoscience portal. *Analytical and Bioanalytical Chemistry*, 417(5): 907–919.
71. Lemmin, T. and C. Soto. (2019). Glycosylator: A Python framework for the rapid modeling of glycans. *BMC Bioinformatics*, 20: 1–7.
72. Li, H., A.W. Chiang and N.E. Lewis. (2022). Artificial intelligence in the analysis of glycosylation data. *Biotechnology Advances*, 60: 108008.
73. Li, X., Z. Xu, X. Hong, Y. Zhang and X. Zou. (2020). Databases and bioinformatic tools for glycobiology and glycoproteomics. *International Journal of Molecular Sciences*, 21(18): 6727.
74. Liebal, U.W., A.N. Phan, M. Sudhakar, K. Raman and L.M. Blank. (2020). Machine learning applications for mass spectrometry-based metabolomics. *Metabolites*, 10(6): 243.
75. Lisacek, F., M. Tiemeyer, R. Mazumder and K.F. Aoki-Kinoshita. (2022). Worldwide glycoscience informatics infrastructure: The GlySpace Alliance. *JACS Au*, 3(1): 4–12.
76. Lnenicka, M. (2015). An in-depth analysis of open data portals as an emerging public E-service. *International Journal of Social, Education, Economics, and Management Engineering*, 9(2): 589–599.
77. Lu, J., S. Guo, Q. Liu, N. Tursumamat, S. Liu, S. Wu, H. Li and J. Wei. (2025). Recent advances in analytical methods and bioinformatic tools for quantitative glycomics. In: *Analytical and Bioanalytical Chemistry*, pp. 1–13.
78. Lu, H. and S. Uddin. (2023). Disease prediction using graph machine learning based on electronic health data: A review of approaches and trends. In: *Healthcare*, 11(7): 1031. MDPI.
79. Lundstrøm, J., E. Korhonen, F. Lisacek and D. Bojar. (2022). LectinOracle: A generalizable deep learning model for lectin–glycan binding prediction. *Advanced Science*, 9(1): 2103807.
80. Lundstrøm, J., J. Urban, L. Thomès and D. Bojar. (2023). GlycoDraw: A python implementation for generating high-quality glycan figures. *Glycobiology*, 33(11): 927–934.
81. Maeda, M., N. Fujita, Y. Suzuki, H. Sawaki, T. Shikanai and H. Narimatsu. (2015). JCGGDB: Japan consortium for glycobiology and glycotechnology database. *Glycoinformatics*, 161– 179.
82. Matsubara, M., K.F. Aoki-Kinoshita, N.P. Aoki, I. Yamada and H. Narimatsu. (2017). WURCS 2.0 update to encapsulate ambiguous carbohydrate structures. *Journal of Chemical Information and Modeling*, 57(4): 632–637.
83. Mariethoz, J., D. Alocci, A. Gastaldello, O. Horlacher, E. Gasteiger, M. Rojas-Macias, N.G. Karlsson, N.H. Packer and F. Lisacek. (2018). Glycomics@ ExPASy: Bridging the gap. *Molecular and Cellular Proteomics*, 17(11): 2164–2176.
84. Medsker, L.R. and L. Jain. (2001). Recurrent neural networks. *Design and Applications*, 5(64-67): 2.

85. Mehta, A.Y. and R.D. Cummings. (2020). GlycoGlyph: A glycan visualizing, drawing, and naming application. *Bioinformatics*, 36(11): 3613–3614.
86. Meroño-Peñuela, A., A. Ashkpour, M. Van Erp, K. Mandemakers, L. Breure, A. Scharnhorst, S. Schlobach and F. Van Harmelen. (2015). Semantic technologies for historical research: A survey. *Semantic Web*, 6(6): 539–564.
87. Mienye, I.D. and T.G. Swart. (2024). A comprehensive review of deep learning: Architectures, recent advances, and applications. *Information*, 15(12): 755.
88. Mohapatra, S., J. An and R. Gómez-Bombarelli. (2022). Chemistry-informed macromolecule graph representation for similarity computation, unsupervised and supervised learning. *Machine Learning: Science and Technology*, 3(1): 015028.
89. Montoya, G., H. Skaf-Molli, P. Molli and M.E. Vidal. (2015, October). Federated SPARQL queries processing with replicated fragments. In: *International Semantic Web Conference*, pp. 36–51. Cham: Springer International Publishing.
90. Morlandstø, M.K. (2022). *Prototyping and Evaluation of Sensor Data Integration in Cloud Platforms*. Master's thesis, The University of Bergen.
91. Navelkar, R., G. Owen, V. Mutherkrishnan, P. Thiessen, T. Cheng, E. Bolton, N. Edwards, M. Tiemeyer, M.P. Campbell, M. Martin and J. Vora. (2021). Enhancing the interoperability of glycan data flow between ChEBI, PubChem and GlyGen. *Glycobiology*, 31(11): 1510–1519.
92. Norouzi, Y. and N. Jafarifar. (2024). *Evaluating Open-Source Digital Library Systems: A Comparative Study*. Knowledge Economy Studies.
93. Okuda, S., H. Nakao and T. Kawasaki. (2017). GlycoEpitope. In: *A Practical Guide to Using Glycomics Databases*, pp. 227–245. Springer.
94. Oh, M.J., Y. Seo, N. Seo and H.J. An. (2025). MS-based glycome characterization of biotherapeutics with N-and O-Glycosylation. *Mass Spectrometry Reviews*.
95. Pauwels, P. and K. McGlinn. (eds.). (2022). *Buildings and Semantics: Data Models and Web Technologies for the Built Environment*. CRC Press.
96. Pérez, S. (2023). Computational modeling of protein–carbohydrate interactions: Current trends and future challenges. *Advances in Carbohydrate Chemistry and Biochemistry*, 83: 133–149. Academic Press.
97. Pradeep, P., H. Kang and B. Lee. (2023). Glycosylation and behavioral symptoms in neurological disorders. *Translational Psychiatry*, 13(1): 154.
98. Qin, S. and Z. Tian. (2025). Deep structure-level N-glycan identification using feature-induced structure diagnosis integrated with a deep learning model. *Analytical and Bioanalytical Chemistry*, 417(5): 1001–1014.
99. Ranzinger, R., K.F. Aoki-Kinoshita, M.P. Campbell, S. Kawano, T. Lütteke, S. Okuda, D. Shinmachi, T. Shikanai, H. Sawaki, P. Toukach and M. Matsubara. (2015). GlycoRDF: An ontology to standardize glycomics data in RDF. *Bioinformatics*, 31(6): 919–925.
100. Ranzinger, R., K.J. Kochut, J.A. Miller, M. Eavenson, T. Lütteke and W.S. York. (2017). GLYDE-II: The GLYcan data exchange format. *Perspectives in Science*, 11: 24–30.
101. Reily, C., T.J. Stewart, M.B. Renfrow and J. Novak. (2019). Glycosylation in health and disease. *Nature Reviews Nephrology*, 15(6): 346–366.
102. Reuel, N.F., B. Mu, J. Zhang, A. Hinckley and M.S. Strano. (2012). Nanoengineered glycan sensors enabling native glycoprofiling for medicinal applications: Towards profiling glycoproteins without labeling or liberation steps. *Chemical Society Reviews*, 41(17): 5744–5779.
103. Saint-Andre, P. and J. Klensin. (2017). Uniform resource names (URNs) (No. rfc8141).
104. Seber, P. and R.D. Braatz. (2024). Recurrent neural network-based prediction of O-GlcNAcylation sites in mammalian proteins. *Computers and Chemical Engineering*, 189: 108818.

105. Schloen, S.R. and M.C. Prosser. (2023). The case for a database approach. In: *Database Computing for Scholarly Research: Case Studies Using the Online Cultural and Historical Research Environment*, pp. 25–73. Cham: Springer International Publishing.
106. Smith, T.S. (1996). *Uniform Resource Locators (URLs): Powerful Reference Tools for Librarians and Information Professionals*. US Department of Education.
107. Spiro, R.G. (2002). Protein glycosylation: Nature, distribution, enzymatic formation, and disease implications of glycopeptide bonds. *Glycobiology*, 12(4): 43R–56R.
108. Solovieva, E., T. Shikanai, N. Fujita and H. Narimatsu. (2018). GGDonto ontology as a knowledge-base for genetic diseases and disorders of glycan metabolism and their causative genes. *Journal of Biomedical Semantics*, 9: 1–14.
109. Steinbeck, C. (2025). The evolution of open science in cheminformatics: A journey from closed systems to collaborative innovation. *Journal of Cheminformatics*, 17(1): 44.
110. Suchanek, M., J. Martinkova, M. Wilkinson, D. Garijo, T. Miksa, P. Manghi, R. Dyume, M.R. Mazon, K. Moilanen, M. Heinonen and D. Broeder. (2024). OSTrails D2.1: Analysis of the OSTrails interoperability framework tools compliance. https://ostrails.eu.
111. Takahashi, Y., M. Shiota, A. Fujita, I. Yamada and K.F. Aoki-Kinoshita. (2024). GlyComb: A novel glycoconjugate data repository that bridges glycomics and proteomics. *Journal of Biological Chemistry*, 300(2).
112. Tanaka, K., K.F. Aoki-Kinoshita, M. Kotera, H. Sawaki, S. Tsuchiya, N. Fujita, T. Shikanai, M. Kato, S. Kawano, I. Yamada and H. Narimatsu. (2014). WURCS: The Web3 unique representation of carbohydrate structures. *Journal of Chemical Information and Modeling*, 54(6): 1558–1566.
113. Toukach, P.V. and K.S. Egorova. (2019). New features of Carbohydrate Structure Database notation (CSDB Linear), as compared to other carbohydrate notations. *Journal of Chemical Information and Modeling*, 60(3): 1276–1289.
114. Toukach, P.V. and A.I. Shirkovskaya. (2022). Carbohydrate structure database and other glycan databases as an important element of glycoinformatics. *Russian Journal of Bioorganic Chemistry*, 48(3): 457–466.
115. Tsuchiya, S., M. Matsubara, K.F. Aoki-Kinoshita and I. Yamada. (2021). SugarDrawer: A web-based database search tool with editing glycan structures. *Molecules*, 26(23): 7149.
116. Thomer, A.K., J.R. Starks, A. Rayburn and M.C. Lenard. (2022). Maintaining repositories, databases, and digital collections in memory institutions: An integrative review. In: *Proceedings of the Association for Information Science and Technology*, 59(1): 310–323.
117. Thomès, L., R. Burkholz and D. Bojar. (2021). Glycowork: A Python package for glycan data science and machine learning. *Glycobiology*, 31(10): 1240–1244.
118. Tiemeyer, M., K. Aoki, J. Paulson, R.D. Cummings, W.S. York, N.G. Karlsson, F. Lisacek, N.H. Packer, M.P. Campbell, N.P. Aoki and A. Fujita. (2017). GlyTouCan: An accessible glycan structure repository. *Glycobiology*, 27(10): 915–919.
119. Trbojević-Akmačić, I., G.S. Lageveen-Kammeijer, B. Heijs, T. Petrovic, H. Deris, M. Wuhrer and G. Lauc. (2022). High-throughput glycomic methods. *Chemical Reviews*, 122(20): 15865–15913.
120. UniProt: The universal protein knowledgebase in 2021. (2021). *Nucleic Acids Research*, 49(D1): D480–D489.
121. Vidal, M.E., S. Castillo, M. Acosta, G. Montoya and G. Palma. (2016). On the selection of SPARQL endpoints to efficiently execute federated SPARQL queries. *Transactions on Large-scale Data-and Knowledge-centered Systems XXV*, 109–149.
122. von der Lieth, C.W. (2009). Glycobiology, Glycomics, and (Bio) Informatics. In: *Bioinformatics for Glycobiology and Glycomics: An Introduction*, pp. 1– 20. Wiley.
123. von Horsten, H.H. (2024). Artificial Intelligence applications for producing glycosylated biopharmaceutical drug modalities. In: *Innovation in Life Sciences: The Digital Revolution*, pp. 45–59. Cham: Springer Nature Switzerland.

124. Walsh, I., R. O'Flaherty and P.M. Rudd. (2017). Bioinformatics applications to aid high-throughput glycan profiling. *Perspectives in Science*, 11: 31–39.
125. Wang, Y., H. Wang, M. Hou, Y. Wang, D. Bu, C. Zhang, C. Huang and S. Sun. (2021). Glycan immunogenicity prediction based on graph neural network. In: *2021 IEEE International Conference on Bioinformatics and Biomedicine* (*BIBM*), pp. 348–353. IEEE.
126. Watanabe, Y., K.F. Aoki-Kinoshita, Y. Ishihama and S. Okuda. (2021). GlycoPOST realizes FAIR principles for glycomics mass spectrometry data. *Nucleic Acids Research*, 49(D1): D1523–D1528.
127. Xu, M., Y. Geng, Y. Zhang, L. Yang, J. Tang and W. Zhang. (2024). Glycanml: A multi-task and multi-structure benchmark for glycan machine learning. *arXiv preprint arXiv*:2405.16206.
128. Yamada, I., M.P. Campbell, N. Edwards, L.J. Castro, F. Lisacek, J. Mariethoz, T. Ono, R. Ranzinger, D. Shinmachi and K.F. Aoki-Kinoshita. (2021). The glycoconjugate ontology (GlycoCoO) for standardizing the annotation of glycoconjugate data and its application. *Glycobiology*, 31(7): 741–750.
129. Yamada, I., M. Shiota, D. Shinmachi, T. Ono, S. Tsuchiya, M. Hosoda, A. Fujita, N.P. Aoki, Y. Watanabe, N. Fujita, K. Angata, H. Kaji, H. Narimatsu, S. Okuda and K.F. Aoki-Kinoshita. (2020). The GlyCosmos Portal: A unified and comprehensive web resource for the glycosciences. *Nature Methods*, 17(7): 649–650.
130. York, W.S., R. Mazumder, R. Ranzinger, N. Edwards, R., Kahsay, K.F. Aoki-Kinoshita, M.P. Campbell, R.D. Cummings, T. Feizi, M. Martin and D.A. Natale. (2020). GlyGen: Computational and informatics resources for glycoscience. *Glycobiology*, 30(2): 72–73.
131. Zafar, I., S. Anwar, W. Yousaf, F.U. Nisa, T. Kausar, Q. ul Ain, A. Unar, M.A. Kamal, S. Rashid, K.A. Khan and R. Sharma. (2023). Reviewing methods of deep learning for intelligent healthcare systems in genomics and biomedicine. *Biomedical Signal Processing and Control*, 86: 105263.
132. Zhang, F., Q. Lu, Z. Du, X. Chen and C. Cao. (2021). A comprehensive overview of RDF for spatial and spatiotemporal data management. *The Knowledge Engineering Review*, 36: e10.
133. Zhang, W., M. Vesser and N. Edwards. (2025). GNOme, an ontology for glycan naming and subsumption. In: *Analytical and Bioanalytical Chemistry*, pp. 1–13.
134. Zheng, J., B. Hu, X. Zhang, Q. Ge, Y. Yan, J. Akresi, V. Piyush, L. Huang and Y. Yin. 2023. dbCAN-seq update: CAZyme gene clusters and substrates in microbiomes. *Nucleic Acids Research*, 51(D1): D557–D563.
135. Zhou, J., G. Cui, S. Hu, Z. Zhang, C. Yang, Z. Liu, L. Wang, C. Li and M. Sun. (2020). Graph neural networks: A review of methods and applications. *AI Open*, 1: 57–81.

CHAPTER 10

Review on the Foresight Approaches of Glycomics in Cancer Studies

Ganesh, N.,[1,*] *Prakash, P.,*[1] *Babu, S.,*[1] *Balaji, S.,*[1] *Sathish, P.,*[2] *Ambikapathy, V.,*[1] *Gomathi, S.*[1] and *Panneerselvam, A.*[1]

Glycomics, the comprehensive study of glycans has emerged as a pivotal frontier in cancer research, offering novel insights into tumor biology, progression, and diagnostics. This review explores cutting-edge quantitative glycomics techniques including Lectin Microarray-based assays, MALDI-MS, and mass spectrometry (MS) coupled with separation technologies, for their precision in profiling glycan structures in cancerous tissues. The application of these tools has uncovered a range of potential N- and O-glycobiomarkers across multiple cancer types such as breast, liver, ovarian, lung, pancreatic, and other malignancies. Emphasis is placed on the role of isomeric N-glycans which demonstrate remarkable specificity and sensitivity as cancer biomarkers. The dynamic interplay between glycan

[1] P.G. and Research Department of Botany, A.V.V.M Sri Pushpam College (Autonomous), (Affiliated to Bharathidasan University Tiruchirappalli – 24), Poondi-613 503, Thanjavur (Dt).

[2] P.G. and Research Department of Physics, Periyar Maniammai Institute of Science & Technology, Vallam, Thanjavur.

* Corresponding author: 7gnn.ganesh1374@gmail.com

DOI: 10.1201/9781003635901-10

expression and cancer pathophysiology underscores the promise of glycomics in improving early detection, prognostic stratification, and personalized therapy. This is a foresight-driven approach of integrated glycan-based diagnostics into clinical oncology, advancing precision medicine, and patient outcomes.

1. Introduction

Glycomics was defined as the comprehensive study of glycomes. They were referred to as carbohydrates; the entire set was called glycans in the biological system. Glycosylation was a crucial post-translational modification of proteins with overall half of the human proteins being glycoproteins [1–3]. This modification oligosaccharides was present in nearly all membrane proteins, secreted proteins, and intracellular proteins [4, 5]. It plays a significant contribution in cellular processes such as cell–cell interactions, communication between cells, and the extracellular environment and cellular metabolism [6, 7]. Glycomics was defined as the comprehensive study of glycomes. The glycans associated with proteins significantly influence their structure and function. Glycans can have structural and modulatory roles, facilitate recognition processes (both intrinsic and extrinsic), and mimic host glycans in molecular interactions [8]. In analogy to a switchboard, glycans act as 'on' and 'off' switches for regulating glycoproteins [9]. Glycomics is akin to other '-omics' fields such as genomics, proteomics, and metabolomics, but with a specific focus on the intricate carbohydrates that contribute to the function of cells and organisms.

Though over 700 monosaccharides that survive in nature [10], human glycans primarily consist of eight monosaccharides: glucose (Glc), galactose (Gal), mannose (Man), N-acetylglucosamine (GlcNAc), N-acetylgalactosamine (GalNAc), L-fucose (Fuc), sialic acid (Sia, Neu5Ac), and glucuronic acid (GlcA). The structure of glycans is much more complex than that of proteins or nucleic acids due to the template-free biosynthesis process which is controlled by enzymes like glycosyltransferases, exoglycosidases, and other glycan-modifying enzymes [11, 12]. Unlike the linear sequences of amino acids in proteins or nucleotides in nucleic acids, monosaccharides in glycans are connected by different configurations (α or β) and in a variety of patterns leading to a wide range of structural diversity [12–14]. Glycans are mainly classified into N-linked and O-linked categories based on whether they are attached to asparagine or serine/threonine residues in proteins, respectively [15].

The initial research linking glycosylation changes to cancer dates back to 1969, when it was first reported that there were notable differences in the membrane glycoproteins of healthy versus transformed mouse fibroblasts [16]. Since then, numerous studies have demonstrated variations in glycan expression including overexpression and underexpression as well as changes in size, branching and core structure of N-linked and O-linked glycans [17, 18]. Additional cancer-related glycans and their roles in transformation, metastasis, and tumor progression have been investigated. Today, most FDA-approved disease biomarkers are glycoproteins. While glycan-targeted therapies are increasingly emerging, they have not yet been fully explored. The identification of glycan biomarkers could enhance our understanding of disease mechanisms and lead to more precise treatments and diagnoses. However, for this to become a reality, reliable analytical techniques are essential.

Advancements in mass spectrometry (MS) and separation technologies have greatly improved the identification and quantitative analysis of glycans in biological samples. Electrospray ionization (ESI)-MS and matrix-assisted laser desorption ionization (MALDI)-MS are the primary ionization techniques used for glycan analysis. However, MS alone is insufficient for characterizing glycan isomers. As a result, separation methods like liquid chromatography (LC) and capillary electrophoresis (CE) are often combined with MS to enable high-throughput and detailed glycomic analysis.

Glycoscience encompasses a variety of areas, such as glycogenes, glycoproteins, glycopeptides, released glycans, glycolipids, and free oligosaccharides. This review will focus on the discovery of clinical glycan biomarkers through glycomic studies, which specifically analyze released glycans. We will explore MS techniques, both with and without separation, used for glycome profiling in the identification and validation of potential glyco-biomarkers linked to various diseases. The review will also summarize and discuss the potential glycobiomarkers associated with different cancers and other diseases.

2. Quantitative Glycomics Techniques

Absolute and relative quantitation: Absolute quantitation involves measuring the precise amount of analytics using a calibration curve made from known standards. In contrast, relative quantitation is typically achieved by comparing the abundance of individual glycans to the total glycome or to the most abundant glycan. Glycan standards are challenging, absolute quantitation is often not feasible in clinical glycomics and most studies focus on relative quantitation to show changes in glycan levels between healthy and diseased groups. This section will outline the techniques currently used in clinical glycomic research [19].

2.1 *Lectin Microarray-based Techniques*

Lectin binding has been extensively used to study glycan composition in biological systems. Specific lectins have been employed for clinical diagnoses or prognoses especially in cancer detection. For example, Lens culinaris agglutinin is used to assess α1,6 core fucosylation levels of α-fetoprotein (AFP) for diagnosing hepatocellular carcinoma [20]. By immobilizing different lectins onto a solid surface lectin microarrays enable high-throughput analysis of glycosylation patterns on target molecules including intact glycoproteins or whole cells [21]. These microarrays require minimal sample preparation and the glycan-binding patterns can distinguish between different glycan profiles [22]. The glycan-lectin interactions can be detected either by fluorescently labeling the analytes before applying them to the microarray or by using fluorescence-labeled antibodies targeting specific analytes. Although lectin microarrays are useful for identifying glycan patterns, they cannot provide detailed structural information about the intact glycans, a capability that MS-based methods offer [22, 23].

2.2 *MALDI-MS-based Techniques*

MALDI-MS is the most widely used MS technique in clinical glycomics due to its rapid analysis, straightforward sample preparation, and ability to provide detailed glycan structural information [24]. Extensive research has focused on optimizing MALDI matrices to enhance ionization efficiency and reduce background noise. Among these, 2,5-dihydroxybenzoic acid (DHB) remains the most commonly used matrix for glycan analysis. However, MALDI-MS tends to cause fragmentation of labile glycans such as sialic acid, leading to potential data loss [25]. To mitigate this, derivatization methods like methylamidation, esterification, or permethylation are frequently employed, not only to preserve sialic acids but also to facilitate linkage-specific identification [26].

Since MALDI-MS signal intensity does not directly correlate with glycan abundance, normalization strategies such as adjusting to the total ion intensity or the most abundant glycan are commonly used for quantitative assessments [27]. Despite this, direct peak intensity comparisons have been successfully applied in multiplexed sample analysis to track structural variations in individual glycans [28, 29]. Additionally, MALDI imaging enables spatial mapping of glycan distributions within tissue sections, offering valuable insights into glycomic alterations. Recent advancements in this area have been reviewed [30].

One major limitation of MALDI-based glycomics is the absence of prior separation, which complicates spectral interpretation when analyzing complex biological samples. Furthermore, competitive ionization among analytes reduces sensitivity, necessitating the integration of effective separation techniques with mass spectrometry to improve analytical performance.

2.3 *MS Coupled with Separation Techniques*

Separation of glycans before MS analysis is a widely used approach to minimize ion suppression during ESI. LC-MS is commonly employed in clinical glycomic profiling due to its high separation resolution and compatibility with MS [24]. Various LC techniques have been utilized to analyze glycosylation patterns including hydrophilic interaction liquid chromatography (HILIC), reverse-phase liquid chromatography (RPLC), high-pH anion exchange chromatography (HPAEC), and liquid chromatography with porous graphitic carbon (PGC). These methods have been extensively studied and reviewed in previous literature.

Additionally, different derivatization techniques have been applied to improve glycan stability and ionization efficiency. Fluorescence labeling via reductive amination using reagents such as 2-aminobenzamide (2-AB) and 2-aminobenzoic acid (2-AA) allows for simultaneous spectroscopic quantification and identification of glycans. Permethylation enhances MS signals and stabilizes labile glycan residues including sialic acids and fucoses, thereby facilitating structural characterization through tandem MS. Esterification selectively modifies sialic acids, forming lactones and esters that distinguish α2,3-linked from α2,6-linked sialic acid isomers.

CE is another widely used glycan separation technique that effectively resolves isomeric structures. Glycans are typically derivatized with charged chromophores or fluorophores for improved separation in CE. While the low flow rate of CE can cause instability in ESI, this issue is often addressed using a sheath liquid. CE-MS enables glycomic analysis with high-resolution separation, making it particularly suitable for small sample amounts.

In MS-based glycomics, quantification is generally achieved by analyzing the peak areas of extracted ion chromatograms (EICs). Relative quantification is performed by comparing individual glycan peak areas to the total EIC peak area of the entire glycome or the most intense glycan peak. Fluorescence response can also be used for quantification in cases where glycans are labeled at the reducing end. Another widely adopted quantification method is multiple reaction monitoring (MRM) tandem MS, which measures the intensities of selected transitions. Compared to conventional MS-based quantification, MRM is less affected by coeluting contaminants, providing higher sensitivity and a broader dynamic range. The use of MRM for glycomics analysis has been extensively reviewed by Lebrilla and colleagues [31].

By leveraging these quantitative glycomic techniques, numerous promising glycan biomarkers have been identified across various diseases. Table 1 summarize potential glycan biomarkers, including N-glycobiomarkers, O-glycobiomarkers, and isomeric glycobiomarkers, respectively.

Table 1: Cancer types and its marker.

Cancer Type	Glycobiomarkers	Clinical Application	References
Colorectal	CEA (Carcinoembryonic Antigen)	Diagnosis, recurrence monitoring	48
Pancreatic	CA19-9 (Sialyl Lewis A antigen)	Monitoring treatment, disease progression	80
Prostate	PSA (Prostate-Specific Antigen)	Screening, diagnosis, monitoring	81
Breast	MUC1 (under-glycosylated)	Prognosis, therapeutic target (e.g., CA15-3)	78
Liver	(HCC) AFP-L3 (Lens culinaris-reactive AFP)	Early detection and progression monitoring	79
Lung	AFP-L3 (Lens culinaris-reactive AFP)	Early detection and progression monitoring	77
Ovarian	CA125 (MUC16)	Early detection (with HE4), monitoring	47
Bladder	Lewis X antigen	Urine biomarker research	75
Melanoma	Glycosylation changes in integrins	Tumor invasiveness	76

3. Glycans as Potential Biomarkers in Cancer

Cancer remains the leading cause of death in developed nations and poses an even greater challenge in less developed regions. The rising incidence of cancer is largely attributed to population growth and aging. With the advancement of glycomics technologies, the field has become increasingly crucial in cancer research [32]. Glycosylation plays a significant role in cancer progression, contributing to processes such as tumor adhesion, malignant transformation, and metastasis. Cutting-edge glycomics approaches have enhanced the identification of glycobiomarkers in cancer, aiming to precisely detect unique N- and O-glycans, as well as specific glycan patterns in cancerous tissues compared to healthy ones. This progress supports the potential application of glycans as biomarkers for improved cancer diagnosis, prognosis, and therapeutic strategies [33].

3.1 Potential O-Glycobiomarkers in Cancer

The exploration of O-glycans as biomarkers is less developed compared to N-glycans, primarily due to the challenges in efficiently releasing O-glycans. Traditional chemical methods for cleaving O-glycans are often time-consuming and can lead to glycan degradation. However, O-glycans are still considered promising biomarkers for cancer for two key reasons. First, mucin-type O-glycans (typically GalNAc-type) are often found in transmembrane and secreted glycoproteins and these structures are frequently altered during cancer development and progression. Second, one of the most common alterations in cancer is the truncation of O-glycans, which occurs due to changes in the tumor microenvironment and in the expression of glycosyltransferases [64]. Although fewer studies have been conducted on O-glycans, they hold significant potential as biomarkers for various cancers, as summarized in recent literature.

3.1.1 Breast Cancer

Mucin 1, a well-established biomarker for breast cancer diagnosis, is extensively O-glycosylated. Alterations in mucin O-glycans are known to occur in the mammary gland during malignancy [65]. As such, developing comprehensive O-glycan profiling techniques offers a promising approach for discovering new biomarkers for breast cancer [66]. In a proof-of-concept study, Miyamoto and colleagues used MALDI-FTICR-MS to analyze O-glycans extracted from breast cancer cell line supernatants and serum samples. The study observed similar glycan patterns across three breast cancer cell lines (MDA-MB-361, MDA-MB-468, and MDA-MB-453), which were also comparable to those found in ovarian cancer serum [56]. However, certain glycans present in these MDA-MB cell lines were not detected in the less malignant BT-474 cell line or the normal MCF10A cell line. Further investigations on serum O-glycans from a mouse model of metastatic breast cancer, as well as a small cohort of patient and healthy control samples, revealed distinct differences. Mice were transplanted with mammary tumors from the PyMT breast cancer model, and over the course of 4–8 weeks, 21 glycan structures were detected in the serum at all-time points, with 11 additional structures appearing only after tumor implantation [67]. The study also demonstrated that the O-glycan profiles could distinguish breast cancer patients from healthy individuals based on principal component analysis (PCA), although these findings were preliminary due to the limited sample size.

3.1.2 Ovarian Cancer

In ovarian cancer, Lebrilla and colleagues analyzed the O-glycome of shed glycoproteins from four ovarian cancer cell lines (OVCAR-3, Caov-3, SK-OV-3, and ES-2) using MALDI-FTICR. While many glycan structures were common across all four cell lines, each exhibited unique glycan patterns. They then analyzed O-glycans from cancerous (n = 5) and normal serum samples (n = 5) to identify potential markers for ovarian cancer. The study found at least 15 unique glycans that were present in all ovarian cancer patients, but absent in healthy controls.

Vitiazeva and colleagues also profiled the O-glycans in ovarian cyst fluid and tissue samples from patients with serous and mucinous ovarian tumors [68]. Using mass spectrometry average composition (MSAC) to characterize glycan profiles, they observed that fucosylation, sialylation, and sulfation were key features differentiating the types of ovarian tumors. The MSAC method involved reducing all identified glycans to their monosaccharide compositions (hexose, N-acetylhexosamine, fucose, sialic acid, and sulfate groups) and calculating the ratio between the intensity of each monosaccharide residue and the total intensity of all oligosaccharides in the sample. This approach enabled more detailed profiling of the glycan structures which could potentially be useful for differentiating ovarian tumor types.

3.1.3 Other Cancers

Efforts to identify O-glycan biomarkers have been extended to a variety of cancers [69] conducted an analysis of released and 1-phenyl-3-methyl-5-pyrazolone (PMP)-labeled O-glycans from serum samples of 157 gastric cancer patients and 144 healthy controls using a LC-MS technique. The mass spectrometry data (average retention time between 19.14 and 20.88 minutes) of O-glycans from the serum of gastric cancer patients and healthy controls are presented [70]. To assess alterations in O-glycans between the two groups, partial least squares-discriminant analysis (PLS-DA) was used. Additionally, unpaired Student's t-tests and receiver operating characteristic (ROC) analysis were performed on the significantly altered O-glycans [71]. The ROC analysis revealed an impressive area under the curve (AUC) of 0.97. Several O-glycan structures, including core 1, core 2, ST antigen, and core 2 complex O-glycans, were found to be upregulated in the serum of gastric cancer patients compared to healthy controls, while the expression of m/z 529.75 (H1N2F2, a novel O-glycan extending from core 1) and diST-antigen were significantly reduced in cancer patients' sera [72].

3.2 Potential N-Glycobiomarkers of Different Cancers

3.2.1 Breast Cancer

Breast cancer is the most prevalent cancer among females, accounting for approximately 25% of all cancer cases worldwide. It is also the leading cause of cancer-related deaths in women. Additionally, breast cancer ranks as the second most common metastatic cancer after lung cancer [34]. Studies estimate that 10–16% of breast cancer patients develop malignant brain metastases. The incidence of breast cancer brain metastases has been rising due to improved diagnostic techniques and extended survival rates among primary breast cancer patients [35].

Study analyzing glycome expression in the sera of breast cancer patients compared to healthy individuals [36]. Using 2D gel electrophoresis, N-glycans were enzymatically released via in-gel PNGase F digestion and labeled with 2-AB. These labeled glycans were then identified and quantified through weak anion exchange high-performance liquid chromatography (WAX-HPLC), followed by amide-normal phase liquid chromatography coupled with tandem mass spectrometry (NP-LC-MS/MS). The findings revealed elevated levels of tri- and tetra-sialylated glycans in breast cancer sera, along with increased branch fucosylation. To further investigate glycan linkages an array of exoglycosidase digestions was performed.

Notably, the N5H4F1 glycan (where N represents N-acetylhexosamine, H denotes hexose, F indicates fucose, and S stands for sialic acid) exhibited a significant increase, over twofold, in patients with advanced-stage breast cancer. A longitudinal study involving 10 breast cancer patients tracked the levels of N5H4F1 from early-stage to late metastatic breast cancer. The results demonstrated a consistent increase in N5H4F1 levels in most patients at later stages, suggesting that this glycan structure serves as a more reliable biomarker for breast cancer progression and metastasis compared to commonly used markers such as cancer antigen (CA) 15-3 and carcinoembryonic antigen (CEA).

3.2.2 Liver Cancer

Liver cancer is not only one of the most prevalent cancers but also ranks second in cancer-related mortality rates [38]. Hepatocellular carcinoma (HCC) accounts for over 80% of liver malignancies, with a dismal survival rate of just 2–7%, primarily due to the absence of effective early diagnostic methods. Since most serum glycoproteins originate from the liver, glycomic research plays a crucial role in understanding liver diseases [40].

In the early stages of liver disease, elevated levels of high-mannose glycans have been uniquely associated with HCC. A study conducted by [39] analyzed 202 serum samples, including 73 HCC cases, 77 cancer-free individuals, and 52 patients with chronic liver disease, using MALDI-TOF. Researchers identified 83 distinct N-glycans and observed significant alterations in 57 of them among HCC cases. Notably, a combination of three specific glycans (N5H6S2, N6H7S3, and bisecting N5H4S1) demonstrated a 90% accuracy rate in predicting HCC [40]. Additionally, studies on salivary glycomics in liver diseases revealed distinct glycan expression changes. While various glycans exhibited alterations across different liver conditions, an increased total fucosylation level and a decreased total sialylation level emerged as the most promising biomarkers for HCC detection.

3.2.3 Ovarian Cancer

Ovarian cancer is a significant cause of mortality in women, ranking as the fifth leading cause of death in females [41]. Unfortunately, it is often diagnosed at an advanced stage after metastasis has occurred which severely limits treatment options and reduces survival rates [42]. While certain prognostic markers such as CA125 have been identified for early-stage ovarian cancer [43], there is an ongoing need to discover additional promising biomarkers to improve early detection and treatment strategies.

Recent research has focused on alterations in the glycome of ovarian cancer, with some glycan expression changes being considered as potential biomarkers. Among these, core fucosylation has been found to increase in ovarian cancer side population cells [44]. Additionally, both core fucosylation and the presence of agalactosyl biantennary glycans, along with sialyl Lewis x (SLex), have been elevated in patient sera [45]. A similar increase in agalactosyl biantennary glycan levels was observed in a study analyzing the glycome of 30 ovarian cancer patient sera [46]. Moreover, there were noted decreases in bisecting structures, while tri- and tetra-antennary glycans showed an increase with specific glycans such as N6H7S4 and N6H7F1S4 being significantly elevated in ovarian cancer patients [46].

A study by Biskup et al. [47] compared serum glycome profiles from 63 preoperative epithelial ovarian cancer (EOC) patients with those of 33 healthy individuals. Using MALDI-TOF/TOF to analyze released and permethylated N-glycans, a total of 47 glycans were identified and quantified. Among these, 11 glycans were found to be significantly up- or down-regulated. The GLYCOV value, calculated as the ratio of the average abundance of up-regulated glycans to down-regulated glycans, was used for ROC analysis. This approach demonstrated a sensitivity of 97% and a specificity of 98.4%, outperforming the widely used CA125 biomarker. Miyamoto et al. [42] also studied glycan expression in ascites fluid from ovarian cancer patients and in normal serum samples. Their findings showed significant over-expression of N-glycans such as N5H6, *F1N5H6S1, *F2N5H6S3, and *F1N6H7S1 in ovarian cancer samples, although these results were based on a single patient's ascites fluid. These findings suggest that the glycome may offer valuable insights for identifying novel biomarkers for ovarian cancer particularly in early detection and prognosis. However, further research with larger sample sizes is required to validate these potential biomarkers.

3.2.4 Lung Cancer

Lung cancer is one of the most prevalent cancers globally and remains the leading cause of cancer-related mortality [34, 41]. It is classified into two main types: non-small cell lung cancer (NSCLC), which makes up approximately 80% of cases and small cell lung cancer (SCLC), which accounts for about 20% [49]. Lung cancer is also notably associated with a high rate of brain metastasis [50]. Currently, imaging remains the primary method for diagnosing lung cancer, though there are some biomarkers like CEA and CA125. However, these biomarkers suffer from limitations in both sensitivity and specificity, making them unsuitable for widespread use and highlighting the need for improved biomarkers [51].

In early research, SLex (sialyl Lewis X) was identified as a potential biomarker, detectable in both serum and tissue samples from lung cancer patients [52, 53]. While SLex has shown potential as a prognostic marker particularly in predicting lymph node metastasis (with a sensitivity of 71%), it is not a reliable diagnostic tool for NSCLC with a sensitivity of only 24% [53]. This suggests that while SLex could serve as an indicator of lung cancer progression, it is not effective for early diagnosis. This underscores the necessity of identifying additional glycan-based markers for better diagnosis and monitoring of lung cancer.

3.2.5 Pancreatic Cancer

The survival rates for most cancers are steadily improving but pancreatic cancer continues to have one of the lowest 5-year relative survival rates at only 8% [41]. This poor prognosis is largely due to the fact that over 50% of pancreatic cancer cases are diagnosed at an advanced stage, where the survival rate drops to just 3% [54]. The challenges in managing pancreatic cancer stem from its rapid progression and the fact that many individuals remain asymptomatic during the early stages of the disease [55]. As a result, there is a critical need for reliable biomarkers that can provide early detection and accurate prognosis. Currently, carbohydrate antigen 19-9 (CA 19-9) is the most commonly used biomarker in clinical settings for diagnosing pancreatic cancer, as well as monitoring treatment and recurrence [40, 57]. However, CA 19-9 has limitations, including insufficient sensitivity and specificity, which undermine its effectiveness in some cases [57].

3.2.6 Other Cancers

Quantitative N-glycomics utilizing a variety of techniques has proven to be a valuable tool for identifying improved cancer biomarkers. Its efficiency has been demonstrated across a wide range of cancers as previously mentioned. In addition to the cancers already discussed, glycomics-based biomarker discovery has also been applied to other cancer types such as kidney cancer [58], esophageal cancer [59, 60], gastric cancer [61] and brain cancer [62, 63]. In kidney cancer, notable increases in high-mannose and highly sialylated structures were detected [58]. For esophageal cancer, the reduction of two core fucosylated glycans, F1N4H5 and F1N4H4, was identified as potential biomarkers with impressive sensitivity (>90%) and specificity (>85%) [59]. In gastric cancer, the decline in core fucosylation proved to be an effective early detection biomarker, with an area under the curve (AUC) of 0.923 [61]. Furthermore, neuroblastoma cell lines exhibited an increase in larger, sialylated glycan structures [62].

3.3 Isomeric N-glycans as Potential Cancer Biomarkers

In a separate study, Miyamoto and colleagues [82] identified 14 sulfated O-glycans as potential biomarkers by comparing the O-glycan profiles of serum samples from 11 gastric and 9 pancreatic adenocarcinoma patients with those from 20 disease-free controls. Moreover, in another study [72], they employed selected reaction monitoring (SRM) analysis to highlight the potential of core 1 sialyl Lewis A as a diagnostic marker for cancer, noting that its correlation with the cancer antigen CA19-9 varied from weak to poor depending on the cancer type.

More recently, Packer and colleagues [73] analyzed the N- and O-glycans in tissues from patients with serous ovarian and peritoneal cancers, identifying three O-glycans among several glycosylation changes that could distinguish tumor sites. In another study [74], Packer's team compared the glycome profiles of colon cancer cell cultures and tumor tissues, revealing significant differences in the N- and O-glycans between various cancer cell lines and the epithelial cells from tumor tissues. These discrepancies underscore the importance of carefully selecting appropriate cell line models for glycan biomarker discovery.

4. Conclusion

The landscape of cancer diagnostics is undergoing a transformative shift with the integration of glycomics, particularly through advanced quantitative and analytical methodologies. Techniques such as lectin microarrays, MALDI-MS, and MS-separation hybrids have proven instrumental in delineating glycan alterations associated with tumor genesis. Wide arrays of N- and O-glycans have emerged as potential biomarkers across various cancer types, reinforcing their utility in early detection and therapeutic monitoring. Furthermore, the exploration of isomeric N-glycans offers a new dimension of biomarker specificity, pushing the boundaries of current diagnostic capabilities. As the field continues to evolve, the foresight approach in glycomics presents a powerful strategy to bridge laboratory findings with clinical applications, fostering a new era of glycan-based precision oncology.

References

1. Kristic, J. and G. Lauc. (2017). Ubiquitous importance of protein glycosylation. *Methods Mol. Biol.*, 1503: 1–12.
2. Clerc, F., K.R. Reiding, B.C. Jansen, G.S. Kammeijer, A. Bondt and M. Wuhrer. (2016). Human plasma protein N-glycosylation. *Glycoconj. J.*, 33(3): 309–343.
3. Kailemia, M.J., D. Park and C.B. Lebrilla. (2017). Glycans and glycoproteins as specific biomarkers for cancer. *Anal. Bioanal. Chem.*, 409(2): 395–410.
4. Moremen, K.W., M. Tiemeyer and A.V. Nairn. (2012). Vertebrate protein glycosylation: Diversity, synthesis, and function. *Nat. Rev. Mol. Cell Biol.*, 13(7): 448–462.
5. Christiansen, M.N., J. Chik, L. Lee, M. Anugraham, J.L. Abrahams and N.H. Packer. (2014). Cell surface protein glycosylation in cancer. *Proteomics*, 14(4-5): 525–546.
6. Kailemia, M.J, G. Xu, M. Wong, Q. Li, E. Goonatilleke, F. Leon and C.B. Lebrilla. (2018). Recent advances in the mass spectrometry methods for glycomics and cancer. *Anal. Chem.*, 90(1): 208–224.
7. Wooding, K.M., W. Peng and Y. Mechref. (2016). Characterization of pharmaceutical IgG and biosimilars using miniaturized platforms and LC-MS/MS. *Curr. Pharm. Biotechnol.*, 17(9): 788–801.
8. Varki, A. (2017). Biological roles of glycans. *Glycobiology*, 27(1): 3–49.
9. Council, N.R. (2012). Transforming glycoscience: A roadmap for the future. National Academies Press, Washington, D.C.
10. Dube, D.H. and C.R. Bertozzi. (2005). Glycans in cancer and inflammation—Potential for therapeutics and diagnostics. *Nat. Rev. Drug Discov.*, 4(6): 477–488.
11. Schnaar, R.L. (2016). Glycobiology simplified: Diverse roles of glycan recognition in inflammation. *J. Leukoc. Biol.*, 99(6): 825–838.
12. Schmaltz, R.M., S.R. Hanson and C.H. Wong. (2011). Enzymes in the synthesis of glycoconjugates. *Chem. Rev.*, 111(7): 4259–4307.
13. Zhou, S., X. Dong, L. Veillon, Y. Huang and Y. Mechref. (2017). LC-MS/MS analysis of permethylated N-glycans facilitating isomeric characterization. *Anal. Bioanal. Chem.*, 409(2): 453–466.
14. Zhou, S., Y. Huang, X. Dong, W. Peng, L. Veillon, D.A.S. Kitagawa, A.J.A. Aquino and Y. Mechref. (2017). Isomeric separation of permethylated glycans by porous graphitic carbon (PGC)-LC-MS/MS at high temperatures. *Anal. Chem.*, 89(12): 6590–6597.

15. Mechref, Y., Y. Hu, A. Garcia, S. Zhou, J.L. Desantos-Garcia and A. Hussein. (2012). Defining putative glycan cancer biomarkers by MS. *Bioanalysis*, 4(20): 2457–2469.
16. Meezan, E., H.C. Wu, P.H. Black, and P.W. Robbins. (1969). Comparative studies on the carbohydrate-containing membrane components of normal and virus-transformed mouse fibroblasts. II. Separation of glycoproteins and glycopeptides by Sephadex chromatography. *Biochemistry*, 8(6): 2518–2524.
17. Dennis, J.W., S. Laferte, C. Waghorne, M.L. Breitman and R.S. Kerbel. (1987). Beta 1-6 branching of Asn-linked oligosaccharides is directly associated with metastasis. *Science*, 236(4801): 582–585.
18. Kim, Y.J. and A. Varki. (1997). Perspectives on the significance of altered glycosylation of glycoproteins in cancer. *Glycoconj. J.*, 14(5): 569–576.
19. Moh, E.S., M. Thaysen-Andersen and N.H. Packer. (2015). Relative versus absolute quantitation in disease glycomics. *Proteomics Clin. Appl.*, 9(3-4): 368–382.
20. Taketa, K., Y. Endo, C. Sekiya, K. Tanikawa, T. Koji, H. Taga, S. Satomura, S. Matsuura, T. Kawai and H. Hirai. (1993). A collaborative study for the evaluation of lectin-reactive alpha-fetoproteins in early detection of hepatocellular carcinoma. *Cancer Res.*, 53(22): 5419–5423.
21. Zheng, T., D. Peelen and L.M. Smith. (2005). Lectin arrays for profiling cell surface carbohydrate expression. *J. Am. Chem. Soc.*, 127(28): 9982–9983.
22. Syed, P., K. Gidwani, H. Kekki, J. Leivo, K. Pettersson and U. Lamminmäki. (2016). Role of lectin microarrays in cancer diagnosis. *Proteomics*, 16(8): 1257–1265.
23. Hirabayashi, J., M. Yamada, A. Kuno and H. Tateno. (2013). Lectin microarrays: Concept, principle, and applications. *Chem. Soc. Rev.*, 42(10): 4443–4458.
24. Everest-Dass, A.V., E.S.X. Moh, C. Ashwood, A.M.M. Shathili and N.H. Packer. (2018). Human disease glycomics: Technology advances enabling protein glycosylation analysis (Part 1). *Expert Rev. Proteomics*, 15(2): 165–182.
25. O'Connor, P.B., E. Mirgorodskaya and C.E. Costello. (2002). High pressure matrix-assisted laser desorption/ionization Fourier transform mass spectrometry for minimization of ganglioside fragmentation. *J. Am. Soc. Mass Spectrom.*, 13(4): 402–407.
26. Reiding, K.R, D. Blank, D.M. Kuijper, A.M. Deelder and M. Wuhrer. (2014). High-throughput profiling of protein N-glycosylation by MALDI-TOF-MS employing linkage-specific sialic acid esterification. *Anal. Chem.*, 86(12): 5784–5793.
27. Alvarez-Manilla, G., N.L. Warren, T. Abney, J. Atwood 3rd, P. Azadi, W.S. York, M. Pierce and R. Orlando. (2007) Tools for glycomics: Relative quantitation of glycans by isotopic permethylation using 13CH3I. *Glycobiology*, 17(7): 677–687.
28. Toghi Eshghi, S., S. Yang, X. Wang, P. Shah, X. Li and H. Zhang. (2014). Imaging of N-linked glycans from formalin-fixed paraffin-embedded tissue sections using MALDI mass spectrometry. *ACS Chem. Biol.*, 9(9): 2149–2156.
29. Powers, T.W., E.E. Jones, L.R. Betesh, P.R. Romano, P. Gao, J.A. Copland, A.S. Mehta and R.R. Drake. (2013). Matrix assisted laser desorption ionization imaging mass spectrometry workflow for spatial profiling analysis of N-linked glycan expression in tissues. *Anal. Chem.*, 85(20): 9799–9806.
30. Drake, R.R., T.W. Powers, E.E. Jones, E. Bruner, A.S. Mehta and P.M. Angel. (2017). MALDI mass spectrometry imaging of N-Linked Glycans in cancer tissues. *Adv. Cancer Res.*, 134: 85–116.
31. Pang, P.C, P.C. Chiu, C.L. Lee, L.Y. Chang, M. Panico, H.R. Morris, S.M. Haslam, K.H. Khoo, G.F. Clark, W.S. Yeung and A. Dell. (2011). Human sperm binding is mediated by the sialyl-Lewis(x) oligosaccharide on the zona pellucida. *Science*, 333(6050): 1761–1764.
32. Torre, L.A., F. Bray, R.L. Siegel, J. Ferlay, J. Lortet-Tieulent and A. Jemal. (2015). Global cancer statistics, 2012. *CA Cancer J. Clin.*, 65(2): 87–108.

33. Bos, P.D., X.H. Zhang, C. Nadal, W. Shu, R.R. Gomis, D.X. Nguyen, A.J. Minn, M.J. van de Vijver, W.L. Gerald, J.A. Foekens and J. Massagué. (2009). Genes that mediate breast cancer metastasis to the brain. *Nature*, 459(7249): 1005–1009.
34. Forman, D. and J. Ferlay. (2014). The global and regional burden of cancer. pp. 16–53. *In*: B.W. Stewart and C.P. Wi. (eds.). *World Cancer Report 2014, International Agency for Research on Cancer, 150 cours. Albert Thomas, 69372 Lyon Cedex 08, France.*
35. Rostami, R., S. Mittal, P. Rostami, F. Tavassoli and B. Jabbari. (2016). Brain metastasis in breast cancer: A comprehensive literature review. *J. Neurooncol.*, 127(3): 407–414.
36. Abd Hamid, U.M., L. Royle, R. Saldova, C.M. Radcliffe, D.J. Harvey, S.J. Storr, M. Pardo, R. Antrobus, C.J. Chapman, N. Zitzmann, J.F. Robertson, R.A. Dwek and P.M. Rudd. (2008). A strategy to reveal potential glycan markers from serum glycoproteins associated with breast cancer progression. *Glycobiology*, 18(12): 1105–1118.
37. Kyselova, Z., Y. Mechref, P. Kang, J.A. Goetz, L.E. Dobrolecki, G.W. Sledge, L. Schnaper, R.J. Hickey, L.H. Malkas and M.V. Novotny. (2008). Breast cancer diagnosis and prognosis through quantitative measurements of serum glycan profiles. *Clin. Chem.*, 54(7): 1166–1175.
38. Mehta, A., H. Herrera and T. Block. (2015). Glycosylation and liver cancer. *Adv. Cancer Res.*, 126: 257–279.
39. Ringelhan, M., D. Pfister, T. O'Connor, E. Pikarsky and M. Heikenwalder. (2018). The immunology of hepatocellular carcinoma. *Nat. Immunol.*, 19(3): 222–232.
40. Adamczyk, B., T. Tharmalingam and P.M. Rudd. (2012). Glycans as cancer biomarkers. *Biochim. Biophys. Acta*, 1820(9): 1347–1353.
41. Siegel, R.L., K.D. Miller and A. Jemal. (2018). Cancer statistics, 2018. *CA Cancer J. Clin.*, 68(1): 7–30.
42. Miyamoto, S., L.R. Ruhaak, C. Stroble, M.R. Salemi, B. Phinney, C.B. Lebrilla and G.S. Leiserowitz. (2016). Glycoproteomic analysis of malignant ovarian cancer ascites fluid identifies unusual glycopeptides. *J. Proteome Res.*, 15(9): 3358–3376.
43. Felder, M., A. Kapur, J. Gonzalez-Bosquet, S. Horibata, J. Heintz, R. Albrecht, L. Fass, J. Kaur, K. Hu, H. Shojaei, R.J. Whelan and M.S. Patankar. (2014). MUC16 (CA125): Tumor biomarker to cancer therapy, a work in progress. *Mol. Cancer*, 13: 129.
44. Zhao, R., X. Liu, Y. Wang, X. Jie, R. Qin, W. Qin, M. Zhang, H. Tai, C. Yang, L. Li, P. Peng, M. Shao, X. Zhang, H. Wu, Y. Ruan, C. Xu, S. Ren and J. Gu. (2016). Integrated glycomic analysis of ovarian cancer side population cells. *Clin. Proteomics*, 13: 32.
45. Saldova, R., L. Royle, C.M. Radcliffe, U.M. Abd Hamid, R. Evans, J.N. Arnold, R.E. Banks, R. Hutson, D.J. Harvey, R. Antrobus, S.M. Petrescu, R.A. Dwek and P.M. Rudd. (2007). Ovarian cancer is associated with changes in glycosylation in both acute-phase proteins and IgG. *Glycobiology*, 17(12): 1344–1356.
46. Alley, W.R., Jr., J.A. Vasseur, J.A. Goetz, M. Svoboda, B.F. Mann, D.E. Matei, N. Menning, A. Hussein, Y. Mechref and M.V. Novotny. (2012). N-linked glycan structures and their expressions change in the blood sera of ovarian cancer patients. *J. Proteome Res.*, 11(4): 2282–2300.
47. Biskup, K., E.I. Braicu, J. Sehouli, C. Fotopoulou, R. Tauber, M. Berger and V. Blanchard. (2013). Serum glycome profiling: A biomarker for diagnosis of ovarian cancer. *J. Proteome Res.*, 12(9): 4056–4063.
48. Doherty, M., E. Theodoratou, I. Walsh, B. Adamczyk, H. Stöckmann, F. Agakov, M. Timofeeva, I. Trbojević-Akmačić, F. Vučković, F. Duffy, C.A. McManus, S.M. Farrington, M.G. Dunlop, M. Perola, G. Lauc, H. Campbell and P.M. Rudd. (2018). Plasma N-glycans in colorectal cancer risk. *Sci. Rep.*, 8(1): 8655.
49. Zhang, Y., Q. Yang and S. Wang. (2014). MicroRNAs: A new key in lung cancer. *Cancer Chemother. Pharmacol.*, 74(6): 1105–1111.
50. Franchino, F., R. Ruda and R. Soffietti. (2018). Mechanisms and therapy for cancer metastasis to the brain. *Front. Oncol.*, 8: 161.

51. Lerner, L., R. Winn and A. Hulbert. (2018). Lung cancer early detection and health disparities: the intersection of epigenetics and ethnicity. *J. Thorac. Dis.*, 10(4): 2498–2507.
52. Ogawa, J., T. Tsurumi, S. Yamada, S. Koide and A. Shohtsu. (1994). Blood vessel invasion and expression of sialyl Lewisx and proliferating cell nuclear antigen in stage I non-small cell lung cancer. Relation to postoperative recurrence. *Cancer*, 73(4): 1177–1183.
53. Mizuguchi, S., K. Inoue, T. Iwata, T. Nishida, N. Izumi, T. Tsukioka, N. Nishiyama, T. Uenishi and S. Suehiro. (2006). High serum concentrations of Sialyl Lewisx predict multilevel N2 disease in non-small-cell lung cancer. *Ann. Surg. Oncol.*, 13(7): 1010–1018.
54. Siegel, R.L., K.D. Miller and A. Jemal. (2017). Cancer Statistics, 2017. *CA Cancer J. Clin.*, 67(1): 7–30.
55. Kenner, B.J., V.L.W. Go, S.T. Chari, A.E. Goldberg and L.J. Rothschild. (2017). Early detection of pancreatic cancer the role of industry in the development of biomarkers. *Pancreas*, 46(10): 1238–1241.
56. Orntoft, T.F. and E.M. Vestergaard. (1999). Clinical aspects of altered glycosylation of glycoproteins in cancer. *Electrophoresis*, 20(2): 362–371.
57. Rosty, C. and M. Goggins. (2002). Early detection of pancreatic carcinoma. *Hematology-Oncology Clinics of North America*, 16(1): 37–52.
58. Gbormittah, F.O., L.Y. Lee, K. Taylor, W.S. Hancock and O. Iliopoulos. (2014). Comparative studies of the proteome, glycoproteome, and N-glycome of clear cell renal cell carcinoma plasma before and after curative nephrectomy. *J. Proteome Res.*, 13(11): 4889–4900.
59. Mechref, Y., A. Hussein, S. Bekesova, V. Pungpapong, M. Zhang, L.E. Dobrolecki, R.J. Hickey, Z.T. Hammoud and M.V. Novotny. (2009). Quantitative serum glycomics of esophageal adenocarcinoma and other esophageal disease onsets. *J. Proteome Res.*, 8(6): 2656–2666.
60. Mitra, I., Z. Zhuang, Y. Zhang, C.Y. Yu, Z.T. Hammoud, H. Tang, Y. Mechref and S.C. Jacobson. (2012). N-glycan profiling by microchip electrophoresis to differentiate disease states related to esophageal adenocarcinoma. *Anal. Chem.*, 84(8): 3621–3627.
61. Qin, R., J. Zhao, W. Qin, Z. Zhang, R. Zhao, J. Han, Y. Yang, L. Li, X. Wang, S. Ren, Y. Sun and J. Gu. (2017). Discovery of non-invasive glycan biomarkers for detection and surveillance of gastric cancer. *J. Cancer*, 8(10): 1908–1916.
62. Hu, Y., A. Mayampurath, S. Khan, J.K. Cohen, Y. Mechref and S.L. Volchenboum. (2015). N-linked glycan profiling in neuroblastoma cell lines. *J. Proteome Res.*, 14(5): 2074–2081.
63. Wildburger, N.C., S. Zhou, L.G. Zacharias, R.A. Kroes, J.R. Moskal, M. Schmidt, P. Mirzaei, J. Gumin, F.F. Lang, Y. Mechref and C.L. Nilsson. (2015). Integrated transcriptomic and glycomic profiling of glioma stem cell xenografts. *J. Proteome Res.*, 14(9): 3932–3939.
64. Pinho, S.S. and C.A. Reis. (2015). Glycosylation in cancer: Mechanisms and clinical implications. *Nat. Rev. Cancer*, 15(9): 540–555.
65. Stowell, S.R., T. Ju and R.D. Cummings. (2015). Protein glycosylation in cancer. *Annu. Rev. Pathol.*, 10: 473–510.
66. Burchell, J.M., A. Mungul and J. Taylor-Papadimitriou. (2001). O-linked glycosylation in the mammary gland: Changes that occur during malignancy. *J. Mammary Gland Biol. Neoplasia*, 6(3): 355–364.
67. Kirmiz, C., B. Li, H.J. An, B.H. Clowers, H.K. Chew, K.S. Lam, A. Ferrige, R. Alecio, A.D. Borowsky, S. Sulaimon, C.B. Lebrilla and S. Miyamoto. (2007). A serum glycomics approach to breast cancer biomarkers. *Mol. Cell Proteomics*, 6(1): 43–55.
68. Vitiazeva, V., J.J. Kattla, S.A. Flowers, S.K. Lindén, P. Premaratne, B. Weijdegård, K. Sundfeldt and N.G. Karlsson. (2015). The O-linked glycome and blood group antigens ABO on mucin-type glycoproteins in mucinous and serous epithelial ovarian tumors. *PLoS One*, 10(6): e0130197.

69. He, Y., Q. Xie, Y. Wang, Y. Liang, X. Xu, Y. Li, J. Miao, Z. Chen and Y. Li. (2016). Liquid chromatography mass spectrometry-based Oglycomics to evaluate glycosylation alterations in gastric cancer. *Proteomics Clin. Appl.*, 10(2): 206–215.
70. Tanaka-Okamoto, M., M. Mukai, H. Takahashi, Y. Fujiwara, M. Ohue and Y. Miyamoto. (2017). Various sulfated carbohydrate tumor marker candidates identified by focused glycomic analyses. *Glycobiology*, 27(5): 400–415.
71. Tanaka-Okamoto, M., M. Yabu, M. Mukai, H. Takahashi, Y. Fujiwara, M. Ohue, Y. Kamada, E. Miyoshi and Y. Miyamoto. (2016). Elevation of CA19-9-related novel marker, Core 1 Sialyl Lewis A, in sera of adenocarcinoma patients verified by a SRM-based method. *J. Proteome Res.*, 15(1): 152–165.
72. Anugraham, M., F. Jacob, A.V. Everest-Dass, A. Schoetzau, S. Nixdorf, N.F. Hacker, D. Fink, V. Heinzelmann-Schwarz and N.H. Packer. (2017). Tissue glycomics distinguish tumour sites in women with advanced serous adenocarcinoma. *Mol. Oncol.*, 11(11): 1595–1615.
73. Hofmann, J. and K. Pagel. (2017). Glycan analysis by ion mobility–mass spectrometry. *Angew. Chem. Int. Ed. Engl.*, 56(29): 8342–8349.
74. Chik, J.H, J. Zhou, E.S. Moh, R. Christopherson, S.J. Clarke, M.P. Molloy and N.H. Packer. (2014). Comprehensive glycomics comparison between colon cancer cell cultures and tumours: Implications for biomarker studies. *J. Proteomics*, 108: 146–162.
75. Wilczak, M., M. Surman and M. Przybyło. (2023). Altered glycosylation in progression and management of bladder cancer. *Molecules*, 28(8): 3436.
76. Grzesik, K., M. Janik and D. Hoja-Łukowicz. (2023). The hidden potential of glycomarkers: Glycosylation studies in the service of cancer diagnosis and treatment. *Biochimica et Biophysica Acta (BBA)-Reviews on Cancer*, 1878(3): 188889.
77. Arnold, J.N., R. Saldova, M.C. Galligan, T.B. Murphy, Y. Mimura-Kimura, J.E. Telford, A.K. Godwin and P.M. Rudd. (2011). Novel glycan biomarkers for the detection of lung cancer. *J. Proteome Res.*, 10(4): 1755–1764.
78. Leone, J.P. and B.A. Leone. (2015). Breast cancer brain metastases: The last frontier. *Exp. Hematol. Oncol.*, 4(33).
79. Kamiyama, T., H. Yokoo, J. Furukawa, M. Kurogochi, T. Togashi, N. Miura, K. Nakanishi, H. Kamachi, T. Kakisaka, Y. Tsuruga, M. Fujiyoshi, A. Taketomi, S. Nishimura and S. Todo. (2013). Identification of novel serum biomarkers of hepatocellular carcinoma using glycomic analysis. *Hepatology*, 57(6): 2314–2325.
80. Zhao, J., W. Qiu, D.M. Simeone and D.M. Lubman. (2007). N-linked glycosylation profiling of pancreatic cancer serum using capillary liquid phase separation coupled with mass spectrometric analysis. *J. Proteome Res.*, 6(3): 1126–1138.
81. Kyselova, Z., Y. Mechref, M.M. Al Bataineh, L.E. Dobrolecki, R.J. Hickey, J. Vinson, C.J. Sweeney and M.V. Novotny. (2007). Alterations in the serum glycome due to metastatic prostate cancer. *J. Proteome Res.*, 6(5): 1822–1832.
82. He, Y., Q. Xie, Y. Wang, Y. Liang, X. Xu, Y. Li, J. Miao, Z. Chen and Y. Li. (2016). Liquid chromatography mass spectrometry-based Oglycomics to evaluate glycosylation alterations in gastric cancer. *Proteomics Clin. Appl.*, 10(2): 206–215.

CHAPTER 11

Functional Glycomics and Homeostatic Disruption

Molecular R Mechanisms and Therapeutic Pathways

Puja Nayak,[1] *Shikha Pandey*,[1] *Rishi Paliwal*[2] and *Shivani Rai Paliwal*[1,*]

Glycans, complex carbohydrates integral to proteins and lipids, constitute a structurally diverse and dynamic molecular repertoire known as the glycome. As essential mediators of cellular communication, immune regulation, and homeostatic signaling, glycans are increasingly recognized for their critical roles in both physiological function and disease progression. This chapter provides a comprehensive overview of functional glycomics, highlighting the structural complexity, biosynthetic pathways, and analytical strategies employed to investigate glycan biology. Glycosylation is regulated by a complex network of glycosyltransferases and glycosidases within the endoplasmic reticulum and golgi apparatus, generates a variety of glycan structures that modulate key processes such as protein folding, stability, receptor-ligand interactions, and immune surveillance.

[1] Department of Pharmacy, Guru Ghasidas Vishwavidyalaya (A Central University), Bilaspur, CG, India, 495009.

[2] Department of Pharmacy, Indira Gandhi National Tribal University, Amarkantak, MP, India, 484886.

* Corresponding author: srai2k@gmail.com

DOI: 10.1201/9781003635901-11

Aberrations in glycosylation patterns—driven by genetic mutations, epigenetic alterations, or environmental factors—can disrupt homeostasis and contribute to a spectrum of diseases. Inflammation is exacerbated by abnormal fucosylation and sialylation, while in cancer, altered N- and O-glycosylation patterns facilitate tumor proliferation, metastasis, and immune evasion. In metabolic disorders such as Type 2 Diabetes, hyper-O-GlcNAcylation and disrupted glycosylation of insulin signaling proteins play pivotal roles in insulin resistance. Neurological and cardiovascular diseases also exhibit glycomic imbalances that affect cell signaling and tissue integrity. Advanced analytical techniques, including mass spectrometry, lectin microarrays, and nuclear magnetic resonance, have revolutionized glycan profiling and biomarker discovery. Furthermore, the therapeutic potential of glycomics is emerging, with glycan-based diagnostics and targeted interventions showing promise for precision medicine. This chapter discussed glycosylation as a key regulator of physiological balance and a critical focus for translational research in disease treatment and prevention.

1. Introduction

The fundamental processes of life, from unicellular organisms to complex multicellular systems, are governed by the coordinated actions of various biomolecules. Although the central dogma of molecular biology, which outlines the flow of genetic information from DNA to RNA to protein, has traditionally been the foundation of biological knowledge, it does not encompass the full spectrum of molecular complexity within living systems. A critical, yet historically underrecognized, class of macromolecules is glycans, or complex carbohydrates. These carbohydrate structures are present across all domains of life and perform functions that go far beyond traditional roles in energy storage and structural integrity. Glycans are highly diverse in structure and composition, and they participate in a wide range of biological processes through their involvement in molecular recognition, cell signaling, and immune modulation. Their structural complexity encodes a form of biochemical information often referred to as the 'glycocode', which regulates cellular communication and interactions. As such, glycans represent a pivotal component of biological systems, contributing to both normal physiology and disease pathology, and have increasingly become a focus of modern biomedical research [1]. Covalently attached to proteins and lipids as glycoconjugates, glycans populate cell surfaces, pervade the extracellular matrix, reside within organelles, and exist as free entities. Collectively, they shape the dynamic molecular landscape known as the glycome [2]. This extensive and diverse display of saccharide structures orchestrates a multitude of biological functions: cell-cell recognition, cellular communication, host-pathogen interactions, immune modulation, and the maintenance of tissue and organ integrity.

The remarkable structural complexity of glycans underpins their vast functional repertoire. Unlike nucleic acids or proteins, which are composed of linear sequences of a limited set of monomers, glycans are constructed from a diverse palette of monosaccharides—such as glucose, galactose, mannose, fucose, N-acetylglucosamine (GlcNAc), N-acetylgalactosamine (GalNAc), and sialic acids. These units can be linked in various orientations (α or β), through multiple glycosidic positions (e.g., 1-2, 1-3, 1-4, 1-6), and often form highly branched architectures. This combinatorial complexity enables an exponential increase in information-carrying capacity, vastly surpassing that of linear biopolymers [3]. While this complexity presents significant analytical challenges, it also provides the molecular specificity necessary for their diverse roles in biological recognition and regulation [4].

The field of glycomics—the comprehensive study of the glycome—aims to decode this complexity by identifying, quantifying, and functionally characterizing glycans within biological systems. A specialized subfield, functional glycomics, focuses on elucidating how specific glycan structures interact with glycan-binding proteins (e.g., lectins) and other biomolecules to mediate physiological and pathological processes.

At the heart of glycan functionality lies their ability to engage in highly specific, non-covalent interactions with cognate binding partners. These glycan-protein interactions govern a vast array of cellular processes. For example, selectins on endothelial cells recognize sialylated and fucosylated glycans on leukocytes, enabling leukocyte rolling and extravasation during inflammation. Glycans on receptors such as receptor tyrosine kinases can influence receptor dimerization, ligand binding, and signal transduction, thus modulating cellular outcomes like proliferation, differentiation, and survival. In addition, glycans affect protein folding, trafficking, stability, and subcellular localization, making them essential determinants of protein functionality [5].

Maintaining cellular and organismal homeostasis is essential for health, and glycans play a pivotal—yet frequently overlooked—role in this equilibrium. Homeostasis involves the intrinsic capacity of biological systems to stabilize internal conditions in response to external fluctuations. It is governed by self-regulating mechanisms that control key physiological parameters such as temperature, pH, nutrient levels, and immune responses. Glycans function as homeostatic modulators, acting as molecular rheostats in signaling networks, regulators of immune tolerance, mediators of cell adhesion, and gatekeepers in protein quality control [3]. For instance, proper glycosylation of immunoglobulins is vital for their anti-inflammatory function, while dysregulated glycosylation promotes pro-inflammatory signaling Likewise, epithelial glycans maintain mucosal integrity and prevent pathogen invasion [6].

However, this finely tuned glycan-mediated homeostasis is vulnerable to disruption. Aberrant glycosylation—marked by alterations in glycan structure, abundance, or spatial distribution—is now recognized not merely as a byproduct but as a causative factor in various diseases. These aberrations can result from mutations in glycosylation enzymes, environmental insults, or metabolic imbalances. The consequences are profound and wide-ranging. In cancer, altered glycosylation patterns promote tumor growth, metastasis, and immune evasion [7]. In inflammatory diseases, abnormal glycosylation of immune proteins perpetuates immune activation and dysregulation. In metabolic disorders, impaired glycosylation of insulin receptors and related proteins hampers glucose homeostasis. Aberrant glycans also contribute to neurodegenerative diseases by modulating protein aggregation and synaptic function, and to cardiovascular diseases, where altered glycosylation of endothelial and plasma proteins exacerbates atherosclerosis and thrombosis [8].

This chapter aims to discuss the critical role of functional glycomics in both the maintenance and disruption of physiological homeostasis. It begins by exploring the foundational principles of glycan biology, including the structural diversity of glycans, the enzymatic pathways involved in their biosynthesis, and the cutting-edge analytical techniques enabling their characterization. The chapter then highlights the essential role of glycans as regulators of homeostasis, followed by a detailed examination of the mechanistic underpinnings of glycan dysregulation across major human diseases—including inflammation, cancer, metabolic syndromes, neurodegenerative conditions, and cardiovascular disorders. Particular focus is placed on the molecular mechanisms driving these disruptions, such as altered expression and activity of glycosyltransferases and glycosidases, changes in glycan branching and sialylation, aberrant glycan-protein interactions, and their impact on key signaling networks. Finally, the chapter discusses the therapeutic potential of glycomics: from glycan-based biomarkers enabling early disease detection and precision medicine, to innovative glycan-targeted therapies and promising strategies aimed at modulating glycosylation pathways for clinical benefit. By integrating insights from recent advances in glycoscience, this review underscores the indispensable role of functional glycomics in elucidating disease pathogenesis and accelerating the development of transformative diagnostic and therapeutic strategies.

2. Understanding Functional Glycomics

Functional glycomics explores the roles of glycans in biological processes, shedding light on how glycan structures, biosynthetic pathways, and molecular interactions regulate cell behavior, intercellular communication, and disease mechanisms. Unlike genomics and proteomics, glycomics encompasses a highly complex and dynamic regulatory layer due to the non-template-driven nature of glycan biosynthesis. Recent advances in analytical technologies have significantly enhanced our capacity to study the structure-function relationships of glycans, leading to deeper insights into their physiological and pathological significance [9].

2.1 *Glycan Structure and Diversity*

Glycans are intricate carbohydrate molecules composed of repeating monosaccharide units connected via glycosidic bonds, forming either free oligosaccharides or covalently linked structures known as glycoconjugates. These glycoconjugates include N-linked and O-linked glycoproteins, glycolipids, and proteoglycans, each serving distinct biological roles. The remarkable diversity in glycan structures is attributed to variations in monosaccharide composition, linkage types (such as α or β orientations), the position of glycosidic bonds, degrees of branching, and the presence of specific modifications like sulfation, acetylation, methylation, or phosphorylation. This structural complexity is essential for the wide range of biological functions that glycans perform, including cell-cell communication, signal transduction, immune response modulation, and host-pathogen recognition [10, 11].

Common monosaccharides found in mammalian glycans include glucose, galactose, mannose, fucose, GlcNAc, GalNAc, and various forms of sialic acid. These units can be assembled into linear chains or complex branched architectures, significantly expanding the possible glycan structures. For instance, the presence of sialic acid at terminal positions of glycan chains influences molecular recognition by pathogens like influenza viruses, which bind to sialylated glycoconjugates to infect host cells. In another example, fucosylation of N-glycans on immunoglobulin G (IgG) modulates its affinity for Fcγ receptors, impacting immune effector functions [12].

Among glycoconjugates, N-linked glycans are attached to the amide nitrogen of asparagine residues within a specific consensus sequence.

These are typically classified into three major types: high-mannose, hybrid, and complex glycans. High-mannose structures contain multiple mannose residues, whereas complex types feature diverse sugars, including GlcNAc, galactose, and sialic acids, often forming branched patterns. In contrast, O-linked glycans are connected to the hydroxyl groups of serine or threonine residues, and unlike N-glycans, they lack a consensus sequence, contributing to their greater variability. Mucin-type O-glycosylation, characterized by the initial attachment of N-acetylgalactosamine, is abundant in epithelial tissues and plays a role in forming protective mucus barriers [13–15[.

Additionally, glycosaminoglycans (GAGs) represent another class of structurally unique glycans. These are long, unbranched polysaccharides composed of repeating disaccharide units, such as glucuronic acid and GlcNAc or iduronic acid and N-acetylgalactosamine. GAGs like heparan sulfate and chondroitin sulfate are integral components of the extracellular matrix, influencing cell adhesion, proliferation, and growth factor signaling. For example, heparan sulfate interacts with fibroblast growth factors to regulate angiogenesis and wound healing [16]. The immense structural heterogeneity of glycans not only facilitates specific and dynamic biological interactions but also presents a challenge in glycomic analysis. However, advances in analytical techniques such as mass spectrometry (MS) and glycan microarrays have significantly improved the ability to decode glycan structures and functions. As a result, understanding glycan diversity is increasingly recognized as essential for uncovering molecular mechanisms in health and disease [1].

2.2 Glycosylation Pathways

Glycosylation is a highly orchestrated enzymatic process through which carbohydrate structures, known as glycans, are covalently attached to proteins and lipids, forming complex glycoconjugates. This modification is not guided by a genetic template, unlike nucleic acid or protein synthesis. Instead, it depends on a dynamic interplay of factors such as the availability and localization of glycosyltransferases and glycosidases, the concentrations of sugar donors, and the biochemical state of the cell. Glycosylation primarily occurs within the lumen of the ER and the Golgi apparatus, where distinct sets of enzymes catalyze specific steps in glycan assembly. Due to its sensitivity to cellular conditions, glycosylation patterns can vary significantly across different tissues, developmental stages, and disease states [17].

There are several major types of glycosylation, each with distinct mechanisms and biological outcomes. N-linked glycosylation, which begins in the ER, involves the en bloc transfer of a preassembled oligosaccharide—made up of 3-glucoses, 9-mannoses, and 2-N-acetylglucosamines ($Glc_3Man_9GlcNAc_2$)—to an asparagine residue within a specific consensus sequence of a nascent polypeptide. This is followed by stepwise trimming and remodeling in the Golgi, resulting in high-mannose, hybrid, or complex N-glycan structures. O-linked glycosylation, on the other hand, is initiated in the Golgi apparatus, where monosaccharide units such as GalNAc are sequentially added to the hydroxyl groups of serine or threonine residues. Unlike N-glycosylation, there is no conserved amino acid sequence that dictates O-glycan addition, making its regulation even more context-dependent [18, 19]. In addition to these canonical types, there are specialized forms of glycosylation. C-mannosylation involves the attachment of a mannose molecule to the indole ring of tryptophan residues and has been implicated in protein folding and trafficking. Glypiation refers to the addition of a glycosylphosphatidylinositol anchor, which facilitates membrane anchoring of proteins. Another rare modification, phosphoglycosylation, entails the incorporation of a sugar-phosphate group and plays a role in specific protein trafficking pathways [20].

Functionally, glycosylation is vital for proper protein folding and quality control in the ER, influencing protein solubility, stability, localization, and interactions with other biomolecules. Deficiencies in glycosylation enzymes or disruptions in these pathways can lead to a group of rare but severe inherited conditions known as congenital disorders of glycosylation (CDGs). Beyond genetic disorders, aberrant glycosylation is increasingly recognized as a hallmark of complex diseases. In cancer, for instance, abnormal patterns of sialylation and fucosylation are frequently observed. These modifications can enhance tumor cell motility, facilitate metastasis, contribute to immune evasion by masking tumor-associated antigens, and alter receptor-mediated signaling pathways crucial for proliferation and survival [21, 22]. Such glycomic alterations are not only of diagnostic relevance but also offer promising targets for therapeutic intervention.

2.3 *Analytical Techniques in Glycomics*

The inherent structural complexity, microheterogeneity, and typically low abundance of glycans in biological systems have historically made their analysis a demanding task. However, significant advancements in analytical technologies have greatly enhanced the sensitivity, specificity, and throughput of glycan characterization. MS remains one of the most powerful tools in glycomics, offering unparalleled capability in detecting glycan structures with high accuracy. Techniques such as matrix-assisted laser desorption/ionization time-of-flight (MALDI-TOF) and electrospray ionization (ESI)-MS have been extensively applied to profile both released glycans and glycopeptides. For instance, a study by Stebbins et al., 2014 [4] demonstrated the successful application of LC-ESI-MS to distinguish structural isomers of N-glycans in serum, improving diagnostic precision for cancer biomarkers. Integration of liquid chromatography or capillary electrophoresis with MS allows for efficient separation and detection of glycan isomers and modified forms, such as sulfation and acetylation, that are often missed in conventional profiling [23].

High-performance liquid chromatography and ultra-performance liquid chromatography are widely employed for the quantification of glycans, especially when combined with fluorescent labeling agents like 2-aminobenzamide (2-AB). These techniques enable precise and reproducible quantitation of glycan species across clinical samples. Fluorescent derivatization not only increases detection sensitivity but also facilitates the comparison of glycomic profiles in different disease states, such as inflammatory bowel disease or liver fibrosis, as illustrated in comparative studies of patient serum glycomes [24]. Another vital technique is the use of lectin microarrays, which consist of immobilized glycan-binding proteins that recognize specific glycan motifs. This method allows rapid and high-throughput glycan fingerprinting in biological samples such as serum, cell membranes, or extracellular vesicles. In a study, lectin arrays enabled the discrimination of cancer cell lines based on unique glycan signatures, highlighting their potential in tumor diagnostics and therapeutic stratification [25].

Nuclear magnetic resonance (NMR) spectroscopy offers high-resolution structural insights into glycan linkages, branching, and stereochemistry. Although it requires relatively large quantities of purified glycans and high-end instrumentation, NMR remains indispensable for validating novel glycan structures and understanding conformational dynamics. It has been particularly useful in defining the three-dimensional arrangement of complex glycans on viral glycoproteins, such as the HIV-1 envelope, aiding vaccine design efforts [26].

The emerging field of glycoproteomics integrates traditional proteomics with glycan analysis, allowing for the site-specific mapping of glycosylation and characterization of intact glycopeptides. Advances in MS, in combination with enrichment strategies like hydrophilic interaction chromatography and solid-phase extraction of N-linked glycans, have substantially increased the depth of glycoproteomic coverage. A study by Ruhaak et al. [27] demonstrated the ability of these combined methods to identify and quantify site-specific glycosylation changes in plasma proteins associated with early-stage breast cancer, offering potential for early diagnosis and personalized therapy.

3. Homeostasis: A Delicate Balance

Homeostasis refers to the physiological equilibrium maintained through complex and dynamic processes that regulate internal conditions in the face of external fluctuations. This equilibrium is critical for sustaining cellular and systemic function, including thermoregulation, pH balance, nutrient availability, osmolarity, and immune surveillance. Among the molecular regulators of this balance, glycans—complex carbohydrates covalently attached to proteins and lipids—play a central yet often underappreciated role. As integral components of cell membranes, signaling pathways, and immune recognition systems, glycans are essential for maintaining biological stability across diverse systems [28].

3.1 Physiological Homeostasis

Physiological homeostasis involves the dynamic regulation of internal environments that allows organisms to function optimally despite perturbations. This stability is maintained through tightly regulated feedback mechanisms that control variables such as body temperature, pH, glucose concentration, and ion gradients across tissues and organs. Homeostatic regulation is fundamental for processes including cell viability, immune responsiveness, neuronal communication, and metabolic balance. Disruptions in homeostatic pathways often lead to pathological conditions, including chronic inflammation, metabolic syndrome, neurodegenerative disorders, and cancer [29].

The orchestration of homeostasis relies on intricate molecular signaling networks that facilitate communication among various cells and tissues. These include hormonal signals, cytokines, neurotransmitters, and metabolic cues, coordinated through systems such as the hypothalamic–pituitary–adrenal (HPA) axis, insulin-glucagon signaling, and the autonomic nervous system. Notably, the concept of a fixed homeostatic "set point" has evolved to reflect its dynamic nature; set points can shift in response to aging, stress, environmental stimuli, and disease states, underscoring the plasticity and adaptability of homeostatic control [22].

3.2 Glycans as Homeostatic Regulators

Glycans play a pivotal regulatory role in maintaining cellular and systemic homeostasis. Through the process of glycosylation—the enzymatic attachment of glycans to proteins and lipids—glycans modulate protein folding, stability, trafficking, and receptor interactions. These post-translational modifications are indispensable for biological functions such as immune surveillance, intercellular adhesion, and signal transduction

[30]. Unlike nucleic acids and proteins, glycans are synthesized through non-template-driven processes. Their biosynthesis depends on the coordinated activity of glycosyltransferases, glycosidases, sugar transporters, and nucleotide-sugar donors. Consequently, glycosylation is highly sensitive to metabolic flux, cellular stress, and inflammatory signals—allowing glycans to function as adaptive modulators of physiological balance [22]. Key roles of glycans in homeostatic regulation include:

Immune Modulation: Glycan modifications on antibodies, cytokines, and cell surface receptors influence immune responses. For instance, sialylation of IgG imparts anti-inflammatory properties, whereas aberrant glycosylation of major histocompatibility complex molecules can impair antigen presentation and T-cell activation [31].

Metabolic Regulation: Glycans on hormone receptors such as the insulin and leptin receptors modulate receptor folding and signaling. Disruption of N-glycan branching—such as loss of Mgat5—has been associated with impaired glucose uptake and insulin resistance [32].

Neuronal Function: In the central nervous system, glycosylation of ion channels and neurotransmitter receptors is essential for synaptic transmission and plasticity. Genetic mutations affecting glycosylation enzymes underlie several CDGs, which often present with severe neurological symptoms [21].

Inflammatory Control: The glycan structures of acute-phase proteins and adhesion molecules are remodeled during inflammation, aiding leukocyte recruitment and cytokine signaling. Persistent alterations—such as increased sialylation or fucosylation—are hallmarks of chronic inflammation and autoimmune disorders [33].

4. Disruption of Homeostasis: Glycomics Perspective

Homeostasis, the ability of biological systems to maintain internal stability amidst changing external conditions, is essential for survival and function. Glycomics—the comprehensive study of glycan structures, biosynthesis, and function—has revealed that glycosylation plays a critical role in preserving homeostasis at the cellular, tissue, and systemic levels. Disruptions in glycan biosynthesis or glycan-protein interactions are now recognized as pivotal contributors to various disease states, from chronic inflammation to cancer and neurodegeneration. These disruptions disturb signaling, immune responses, cellular communication, and metabolism, ultimately leading to homeostatic imbalance.

4.1 Immune Dysregulation and Inflammatory Homeostasis

Glycans play a central role in maintaining physiological homeostasis by regulating a wide array of cellular processes, including immune surveillance, cell signaling, and tissue integrity. Alterations in glycosylation patterns can disrupt these functions, contributing to immune dysregulation and inflammatory pathologies. The glycome—the entire repertoire of glycan structures within a biological system—is dynamically regulated by glycosyltransferases and glycosidases, whose expression is sensitive to both physiological and pathological cues. When glycosylation is dysregulated, it can act not only as a biomarker of disease but also as a mechanistic contributor to pathogenesis [34].

Glycans in Immune Homeostasis: In the immune system, glycans act as molecular "zip codes" that influence immune cell development, trafficking, and function. They are integral to modulating cell-cell interactions, receptor activation, and immune tolerance. One key mechanism involves sialylated glycans, which interact with sialic acid-binding immunoglobulin-like lectins (Siglecs) present on immune cells. These interactions typically transmit inhibitory signals that maintain immune quiescence and prevent overactivation. Disruption of this glycan–lectin axis—such as by hyposialylation or altered sialic acid linkages—can compromise immune tolerance and promote chronic inflammation or autoimmunity [35].

Inflammation and Pro-inflammatory Glycosylation Patterns: During inflammatory responses, there is often a shift in glycosylation patterns due to altered expression of glycosyltransferases. These changes can modify the glycan structures on cytokine receptors, adhesion molecules, and immunoglobulins. For instance, endothelial cells under inflammatory stress frequently upregulate fucosylation and sialylation, enhancing the expression of selectin ligands such as sialyl Lewis^{x} (sLe^{x}). These modified glycans promote leukocyte rolling and adhesion, a critical early step in leukocyte extravasation and tissue infiltration [36]. Such glycan modifications act as enhancers of the inflammatory cascade, facilitating immune cell recruitment and activation. These dynamic alterations represent a glycomic signature of inflammation and offer potential targets for therapeutic intervention.

Glycosylation in Autoimmune Diseases: Autoimmune diseases are among the clearest examples of how glycomic dysregulation contributes to immune dysfunction. One of the most studied phenomena is the agalactosylation of IgG in rheumatoid arthritis (RA). In healthy individuals, the Fc region of IgG contains complex biantennary N-glycans with terminal galactose and sialic acid residues. In RA, there is a shift toward the "G0" glycoform—IgG molecules lacking terminal galactose—which are associated with heightened inflammatory potential due to increased binding affinity to Fcγ receptors and enhanced complement activation [37].

Similarly, systemic lupus erythematosus (SLE) patients exhibit altered sialylation and galactosylation patterns on IgG and cell surface glycoconjugates, contributing to immune complex deposition, impaired clearance, and tissue damage. Changes in glycan expression on T and B cells in autoimmune diseases can also impair peripheral tolerance and promote autoreactivity, further fueling the disease process.

4.2 Aberrant Glycosylation and Cellular Dysfunction

The integrity of glycosylation pathways is crucial for maintaining protein folding, trafficking, receptor activation, and cell– cell adhesion. Alterations in glycan branching, sialylation, fucosylation, or glycan chain length have been linked to cellular stress and loss of homeostasis. For example, increased β1,6-GlcNAc branching catalyzed by N-acetylglucosaminyltransferase V (GnT-V) has been associated with malignant transformation by altering receptor signaling and inhibiting apoptosis [38].

Disruptions in N-glycosylation can affect ER efficient control mechanisms, leading to ER stress and activation of the unfolded protein response, which further contributes to diseases such as diabetes, liver disorders, and neurodegeneration [39]. Similarly, improper O-glycosylation of mucins has been implicated in colon cancer due to defective epithelial barrier function and increased susceptibility to inflammation [40].

4.3 Cancer and Malignant Transformation

Cancer is a complex and multifactorial disease characterized by uncontrolled cell proliferation, invasion, and metastasis. One of the near-universal molecular features of cancer cells is altered glycosylation, which contributes to nearly all hallmarks of malignancy. Importantly, these glycomic alterations are often among the earliest molecular events in tumor initiation and progression [38].

Aberrant N- and O-glycosylation is commonly observed in cancer. Tumor cells frequently display increased branching of N-glycans, particularly the formation of β1,6-branched structures, which are strongly associated with enhanced metastatic potential. In parallel, truncated O-glycans, such as the Thomsen–Friedenreich (T) antigen and

Tn antigen, are frequently expressed on malignant cells, contributing to immune evasion and increased invasiveness [41]. Increased sialylation, particularly of α2,3- and α2,6-linked sialic acids, is another hallmark of cancer glycosylation. These sialic acids can mask tumor-associated antigens, thereby shielding cancer cells from immune surveillance, while also promoting cell survival, adhesion to endothelial cells, and metastatic dissemination [42].

Furthermore, lectin–glycan interactions within the tumor microenvironment play a significant role in cancer progression. Tumor-associated glycans can bind to endogenous lectins, such as galectins, thereby modulating processes like cell proliferation, apoptosis resistance, angiogenesis, and immune suppression. For instance, galectin-3, which binds to β-galactoside-containing glycans, is frequently overexpressed in cancer and facilitates tumor growth and metastasis [43].

4.4 Metabolic Disorders

Metabolic disorders such as Type 2 diabetes mellitus (T2DM), obesity, and non-alcoholic fatty liver disease are increasingly recognized to involve significant alterations in glycosylation patterns, impacting multiple aspects of metabolic regulation. Glycans modulate critical biological functions, including insulin signaling, glucose uptake, lipid metabolism, and inflammatory responses. When glycomic homeostasis is disrupted, it contributes to the pathogenesis and progression of these metabolic diseases [44].

T2DM-Hyperglycemia, a hallmark of T2DM, promotes non-enzymatic glycation of proteins, resulting in the formation of advanced glycation end-products (AGEs). These AGEs interfere with normal cellular function, alter receptor activity, and trigger oxidative stress and chronic inflammation, thereby exacerbating diabetic complications [45]. Moreover, AGEs interact with their receptor, RAGE, on various cell types, perpetuating pro-inflammatory signaling.

Another key aspect of glycomic dysregulation in T2DM is the altered glycosylation of insulin receptors and glucose transporters, which impairs insulin sensitivity. Aberrant glycosylation of the insulin receptor can disrupt ligand binding or receptor conformation, contributing to insulin resistance [32]. In addition, O-GlcNAcylation, a dynamic form of intracellular glycosylation, plays a pivotal role in glucose sensing and insulin signaling. Elevated glucose flux through the hexosamine biosynthetic pathway increases O-GlcNAcylation levels, which can negatively affect the phosphorylation of insulin signaling proteins like IRS-1 and Akt, impairing downstream signaling and metabolic control [46].

Obesity and Insulin Resistance: Obesity is associated with systemic inflammation and metabolic dysfunction, where glycan alterations in plasma proteins and immune cell receptors contribute to pathophysiology. For instance, changes in N-glycan branching and increased core fucosylation are often observed, which can affect protein stability and receptor interactions. Such modifications may exacerbate chronic inflammation and interfere with insulin signaling, promoting insulin resistance [47].

Non-Alcoholic Fatty Liver Disease (NAFLD) and Non-Alcoholic Steatohepatitis (NASH)—The liver is a central organ in glycan biosynthesis and modification. In these conditions, glycosylation changes reflect and drive metabolic stress. Altered fucosylation and sialylation patterns in serum glycoproteins have been identified as early biomarkers for liver fibrosis and NASH [48]. These changes stem from modifications in hepatic glycosyltransferase expression under lipotoxic and inflammatory conditions.

4.5 Neurodegenerative Diseases

In the central nervous system, glycosylation is essential for synaptic plasticity, myelination, and neural communication. Glycans on neural cell adhesion molecules, such as polysialic acid, modulate neuronal migration and signaling during development and repair. Disruption of glycosylation in these proteins has been implicated in neurodevelopmental and neurodegenerative disorders [49]. Neurodegenerative diseases such as Alzheimer's disease, Parkinson's disease (PD), and multiple sclerosis (MS) are marked by progressive neuronal dysfunction and structural deterioration. Growing evidence supports the contribution of aberrant glycosylation to the initiation and progression of these disorders through mechanisms involving protein aggregation, impaired signaling, and neuroinflammation.

Alzheimer's Disease (AD): AD is characterized by the accumulation of amyloid-beta (Aβ) plaques and neurofibrillary tangles composed of hyperphosphorylated tau protein. Aberrant glycosylation is a key factor influencing both tau pathology and amyloid processing:

Tau Protein: Altered O-GlcNAcylation of tau has been linked to increased tau phosphorylation and aggregation. Normally, O-GlcNAc modification competes with phosphorylation on serine/threonine residues; reduced O-GlcNAcylation facilitates tau hyperphosphorylation, enhancing its aggregation and toxicity [50].

Amyloid Precursor Protein (APP) and Aβ Peptides: Glycosylation of APP influences its trafficking and proteolytic processing. Abnormal

N-glycosylation, including increased levels of bisecting GlcNAc structures and altered sialylation patterns, has been observed in AD brains, promoting abnormal APP cleavage and Aβ accumulation [51, 52]. These glycomic changes collectively impair neuronal signaling, foster neuroinflammation, and contribute to synaptic and neuronal loss.

Parkinson's Disease (PD): PD is primarily characterized by the aggregation of α-synuclein into Lewy bodies, leading to dopaminergic neuronal death. Though less extensively studied than in AD, emerging data suggest that glycosylation also plays a role in PD: α-Synuclein Glycosylation: Aberrant glycosylation of α-synuclein may influence its conformation, aggregation propensity, and clearance, thereby affecting the formation of toxic oligomers and fibrils [53]. Altered glycan structures might modulate interactions with glial cells or disrupt lysosomal degradation pathways, both of which are implicated in PD pathogenesis.

Multiple Sclerosis (MS): MS is an autoimmune, demyelinating disease of the CNS, in which both neuronal and immune glycosylation are critically implicated. In the context of myelin glycoproteins, altered glycosylation of structural proteins such as myelin basic protein MBP, can compromise myelin integrity, rendering it more vulnerable to autoimmune attack. In terms of immune glycosylation, changes in the glycosylation of immunoglobulins and immune cell surface receptors (e.g., T cell and B cell receptors) contribute to dysregulated immune responses and chronic inflammation. These glycan alterations can modify antigen recognition, cytokine signaling, and immune cell trafficking [54].

4.6 Cardiovascular Diseases (CVDs)

Cardiovascular homeostasis is tightly regulated by the structural and functional integrity of vascular cells, lipoproteins, and immune components. Glycosylation—the enzymatic addition of carbohydrate chains to proteins and lipids—plays a central role in modulating these components. Glycans on endothelial cells, lipoproteins, and circulating immune cells are key regulators of vascular tone, lipid metabolism, immune surveillance, and hemostasis. Dysregulation of these glycosylation pathways is increasingly recognized as a contributing factor to the pathogenesis of CVDs, including atherosclerosis, hypertension, thrombosis, and heart failure [55].

Cholesterol Homeostasis and Lipoprotein Glycosylation: Glycosylation directly influences the metabolism of lipoproteins, such as low-density lipoproteins (LDL), which are central to cholesterol transport and cardiovascular risk. In the case of LDL Receptor (LDLR), N-glycans modulate both its binding affinity and recycling efficiency. Alterations in these glycan structures can impair LDL uptake by hepatocytes, leading to elevated circulating cholesterol and increased risk for atherosclerosis [56]. Additionally, glycosylation affects the structural conformation and receptor recognition of lipoprotein particles, including both LDL and high-density lipoproteins (HDL). These modifications influence their atherogenic potential, thereby impacting cardiovascular health.

Endothelial Glycosylation and Leukocyte Recruitment: The endothelium plays a central role in vascular homeostasis and immune surveillance. Glycans on endothelial surface proteins act as molecular beacons for immune cell trafficking and inflammatory responses. One key example involves selectins, which are glycoproteins on endothelial cells that mediate the initial rolling of leukocytes along the vascular wall. This interaction depends on specific glycan ligands, such as sialyl Lewis^{X}. Disruption or overexpression of these glycan structures can lead to excessive leukocyte recruitment, endothelial activation, and vascular inflammation—hallmarks of atherosclerosis and hypertension [57]. Furthermore, aberrant glycosylation can heighten endothelial sensitivity to inflammatory stimuli, thereby increasing vascular permeability and promoting plaque formation, both of which contribute to cardiovascular disease progression.

Glycosylation in Atherosclerosis: Atherosclerosis is a chronic inflammatory condition characterized by lipid accumulation and immune cell infiltration in the arterial wall. Glycomic alterations are central to several stages of disease progression. One critical mechanism involves the glycosylation of endothelial adhesion molecules, such as E-selectin and VCAM-1. The presence of N-glycans on these proteins enhances the recruitment of monocytes and T-cells, initiating an inflammatory cascade that contributes to plaque formation and vascular injury [58]. In addition, altered glycosylation of LDL can increase their retention in the subendothelial space and promote uptake by macrophages, resulting in the formation of foam cells and the development of a necrotic core key features of advanced atherosclerotic lesions.

Thrombosis and Platelet Glycobiology: Glycans play a crucial role in the function of platelets and coagulation factors, both of which are integral to the regulation of thrombosis. In platelets, glycosylation of surface receptors notably glycoprotein Ib alpha (GPIbα), is essential for their interaction with von Willebrand factor (vWF), facilitating platelet adhesion, activation, and aggregation at sites of vascular injury. In the case of vWF, its glycosylation profile significantly influences both its stability and procoagulant activity. Abnormal glycosylation can lead to either a pro-thrombotic state due to enhanced multimerization and platelet binding or bleeding disorders, resulting from reduced stability or impaired function [59].

Heart Failure and Extracellular Matrix Remodeling (ECM): In heart failure, the structural and mechanical properties of cardiac tissue are significantly altered due to extracellular matrix (ECM) remodeling. Key ECM components, including GAGs and proteoglycans—such as heparan sulfate and chondroitin sulfate undergo glycosylation changes that affect tissue stiffness, interstitial fluid dynamics, and cell–matrix interactions [60]. These alterations contribute to ventricular remodeling, reduced myocardial compliance, impaired contractility, and the overall progression of cardiac dysfunction.

5. Molecular Mechanism of Disruption

Glycosylation—the enzymatic process of attaching glycans (complex carbohydrates) to proteins and lipids—is a fundamental post-translational modification that critically governs protein function and cellular communication. Dysregulation of this intricate process, often referred to as aberrant glycosylation, is increasingly recognized as a central mechanism in the disruption of homeostasis across a spectrum of pathological conditions, including cancer, autoimmune diseases, and neurological disorders [61]. A functional glycomics perspective seeks to elucidate how these glycan alterations impact biological processes and contribute to disease pathogenesis.

5.1 Altered Glycosyltransferase and Glycosidase Expression

The biosynthesis and modification of glycans rely on a complex and precisely regulated system of enzymes, primarily glycosyltransferases, which add sugar units, and glycosidases, which remove them. This enzymatic network governs the dynamic composition of the cellular glycome—the entire repertoire of glycans present within a cell or organism. Any disruption in the expression levels or activity of these enzymes can significantly alter glycan structures, impacting cellular function and health [7].

In many cancers, certain glycosyltransferases, such as sialyltransferases and fucosyltransferases, are often found to be upregulated. This overexpression leads to modifications in the tumor cell surface glycan patterns, enhancing malignant behaviors including increased proliferation, evasion of immune detection, and enhanced metastatic capacity. For instance, elevated sialyltransferase activity can increase sialic acid residues on cancer cells, which helps them avoid immune surveillance by interacting with inhibitory immune receptors [22, 27].

Similarly, fucosyltransferase overexpression alters fucosylated glycans that contribute to tumor progression and angiogenesis.

On the other hand, decreased expression or functional impairments of glycosyltransferases and glycosidases can result in abnormal or truncated glycan structures that are unable to perform their normal biological roles. Such deficiencies have been implicated in CDGs, a group of inherited metabolic diseases characterized by multisystemic symptoms including developmental delays and immune dysfunction. Moreover, aberrant enzyme activity has been linked to chronic inflammatory conditions and metabolic disorders, where defective glycan remodeling disrupts cell signaling and immune homeostasis [62].

Thus, the balanced regulation of glycosyltransferases and glycosidases is critical for preserving normal glycan architecture and function. Dysregulation in these enzymes not only reshapes the cellular glycome but also drives diverse pathological processes, highlighting their central role in both health and disease. Understanding these enzymatic mechanisms opens pathways for therapeutic targeting aimed at restoring proper glycosylation patterns.

5.2 *Changes in Glycan Branching and Sialylation*

Glycan structural features, particularly branching and terminal sialylation, play crucial roles in regulating cellular functions and maintaining physiological balance. Glycan branching, exemplified by β1,6-GlcNAc branching catalyzed by the enzyme N-acetylglucosaminyltransferase V (GnT-V), increases the complexity and diversity of glycan structures. This branching influences how glycans interact with proteins, thereby modulating cellular behaviors such as adhesion, signaling, and migration. In the context of cancer, increased β1,6-GlcNAc branching has been closely linked to enhanced tumor aggressiveness. For example, elevated GnT-V activity promotes the formation of highly branched N-glycans on cell surface receptors, which can facilitate detachment from the primary tumor, invasion into surrounding tissues, and metastasis by altering cell-cell and cell-matrix interactions [63].

Sialylation, the enzymatic addition of sialic acid residues at the terminal ends of glycans, serves as another critical modification that regulates immune recognition, receptor function, and cellular communication. In many cancers, abnormal hypersialylation occurs, where excess sialic acid residues decorate glycoproteins and glycolipids on tumor cells. This over-sialylation effectively cloaks tumor-associated antigens, preventing their recognition by immune cells and enabling tumors to evade immune attack. For instance, hypersialylated glycans on cancer cells interact with inhibitory immune receptors such as siglecs, dampening immune responses and promoting tumor survival [59]. Conversely, diminished or altered sialylation patterns have been associated with autoimmune diseases and neurodegenerative conditions. In autoimmune disorders, changes in sialylation can disrupt immune tolerance, contributing to inappropriate immune activation. Similarly, in neurodegenerative diseases, altered sialylation of neural glycoproteins can impair neuronal communication and promote pathological processes.

Overall, variations in glycan branching and sialylation profoundly influence the structural and functional properties of glycoconjugates. These modifications not only dictate how cells interact with their environment but also impact critical biological processes such as immune regulation, cellular signaling, and disease progression. Understanding the mechanisms and consequences of altered glycan branching and sialylation offers valuable insights for developing targeted therapies across cancer, autoimmune, and neurodegenerative diseases.

5.3 *Aberrant Glycan–Protein Interactions*

Glycans execute a wide range of biological roles primarily through their specific interactions with glycan-binding proteins (GBPs) such as lectins, selectins, and siglecs. These proteins recognize distinct glycan structures on cell surfaces or secreted proteins, mediating essential physiological processes including protein folding, cell adhesion, immune signaling, and intracellular communication. The precision of these glycan–protein interactions is crucial for maintaining cellular homeostasis. However, pathological alterations in glycosylation often lead to aberrant glycan structures that disrupt these interactions and contribute to disease development.

In cancer, altered glycan patterns known as tumor-associated carbohydrate antigens (TACAs) are frequently expressed on tumor cells. These abnormal glycans have increased affinity for galectins—a lectin family involved in regulating cell proliferation, apoptosis, and immune evasion. For example, galectin-3 binds to cancer-associated glycans, promoting tumor cell clustering, invasion, and angiogenesis by enhancing endothelial migration and new blood vessel formation, all of which facilitate tumor growth and metastasis. Additionally, selectins, which normally help leukocytes navigate blood vessels, interact abnormally with glycosylated ligands on tumor cells, enabling their adhesion to the endothelium and dissemination through the bloodstream [64]. Within the immune system, glycan modifications on receptors are vital for proper immune function. Aberrant glycosylation of key receptors such as the

T-cell receptor (TCR) or natural killer (NK) cell receptors can impair antigen recognition and signaling, thereby reducing immune responses and allowing tumors or pathogens to evade detection. Altered sialylation patterns on immune checkpoints engage siglecs (sialic acid-binding immunoglobulin-like lectins), which act as inhibitory receptors that dampen immune activation. This mechanism fosters immune tolerance in cancer and chronic infections [65].

Beyond oncology and immunity, defective glycosylation is implicated in neurodegenerative disorders. Misfolded glycoproteins caused by aberrant glycan modifications tend to aggregate in neurons, a characteristic feature of diseases like Alzheimer's and Parkinson's. For instance, altered glycosylation of APP affects its processing, contributing to the accumulation of amyloid-beta plaques in AD. Similarly, abnormal glycans on alpha-synuclein promote its aggregation in Parkinson's disease, disrupting neuronal signaling and viability. These aggregates also provoke inflammatory responses that exacerbate neurodegeneration.

5.4 Impact on Signaling Pathways

Glycans play a crucial role in regulating cellular signaling pathways by modulating the behavior of key receptors and signaling molecules. They influence receptor function through various mechanisms, including altering receptor conformation, affecting ligand binding affinity, and shaping downstream signaling cascades. When glycosylation patterns are disrupted, these changes can cause abnormal activation or suppression of signaling pathways, contributing to the development and progression of diseases [66]. For instance, receptor tyrosine kinases (RTKs), which are essential for controlling cell growth and differentiation, are highly sensitive to their glycosylation status. Alterations in the glycosylation of RTKs can impact their ability to dimerize and activate signaling, resulting in continuous proliferative signals that fuel cancer progression. Studies have demonstrated that aberrant N-glycosylation on receptors like the epidermal growth factor receptor (EGFR) enhances receptor stability and signaling, promoting oncogenesis [67].

Beyond RTKs, glycans also regulate the function of cell adhesion molecules such as integrins and cadherins. These molecules govern cell-cell and cell-extracellular matrix interactions critical for maintaining tissue structure and facilitating cell migration. Changes in glycosylation can modify adhesion molecule activity, thus influencing processes like metastasis, where cancer cells detach and invade new tissues, and tissue remodeling during wound healing or fibrosis. Moreover, free glycans released into the extracellular environment can themselves act as signaling molecules, modulating immune responses, developmental pathways, and cellular reactions to stress. For example, certain free glycans interact with lectin receptors on immune cells, shaping inflammation and tolerance. Collectively, these examples highlight how glycans serve as dynamic regulators of signaling networks, and their dysregulation can profoundly alter cellular communication, leading to pathological conditions [68].

6. Therapeutic Implication

Glycosylation, a fundamental post-translational modification, plays a pivotal role in numerous biological processes including cellular signaling, protein folding, immune recognition, and host-pathogen interactions. Aberrations in glycosylation patterns are implicated in a wide range of diseases, including cancer, autoimmune disorders, infectious diseases, and neurodegenerative conditions. Advances in glycomics have opened new avenues for novel diagnostic and therapeutic strategies, giving rise to the burgeoning field of glyco-therapy. This emerging domain offers targeted interventions by modulating glycosylation pathways, with promising clinical implications [69].

Targeting Glycosylation Pathways for Therapy: The biosynthesis of glycans is intricately governed by the coordinated actions of glycosyltransferases, glycosidases, and the availability of activated sugar donors. Dysregulation of these enzymatic processes is implicated in a broad spectrum of diseases. Therapeutic modulation of glycosylation pathways, therefore, represents a promising avenue for disease correction and precision medicine.

Monosaccharide Supplementation: In CDGs, a group of rare, often multisystemic inherited metabolic conditions—deficiencies in donor substrates or defects in enzymatic machinery can be partially corrected through targeted monosaccharide supplementation. For example, exogenous mannose administration has demonstrated efficacy in improving clinical phenotypes in both preclinical models and human subjects by restoring proper N-glycosylation of essential proteins [70]. Emerging studies have extended this concept to oncology, where mannose has been explored as a metabolic anti-cancer agent. Preclinical evidence indicates that mannose can inhibit tumor progression and enhance chemotherapeutic responses, potentially by disrupting glycolysis and glycoprotein biosynthesis pathways [71]. Similarly, N-acetylglucosamine (GlcNAc), a critical precursor in the hexosamine biosynthetic pathway (HBP), has been shown to exert anti-inflammatory effects by modulating immune signaling cascades, including those involving NF-κB and MAPKs [72].

Glycoengineering of Antibodies: Glycan modulation also plays a pivotal role in the development of next-generation therapeutic antibodies. The Fc glycosylation profile of IgG critically influences its effector functions. Notably, the removal of core fucose from the Fc region enhances antibody-dependent cellular cytotoxicity (ADCC), thereby improving clinical efficacy in therapies targeting hematologic malignancies such as T-cell leukemia [73]. In contrast, increased Fc galactosylation has been associated with enhanced CDC, offering an additional mechanism for therapeutic enhancement.

Furthermore, the glycosylation-dependent activity of intravenous immunoglobulin (IVIG), a mainstay therapy in various autoimmune diseases, has been linked to its anti-inflammatory potential. Studies have shown that α2,6-linked sialylation of IgG Fc domains enhances binding on dendritic cells, facilitating immunomodulatory effects and offering new strategies to improve IVIG formulations [74].

Glycoantigen Modification and Tumor Immunogenicity: Innovative strategies are also leveraging glycosylation to enhance tumor antigenicity. The α1-3 galactose (α-Gal) epitope, absent in humans due to evolutionary gene loss, elicits robust natural antibody responses and is considered a major obstacle in xenotransplantation. However, this immunogenicity is now being repurposed in cancer immunotherapy. For instance, Mucin (MUC) 1 tumor antigens engineered to express α-Gal glycolipids have demonstrated enhanced antigen presentation and immune-mediated tumor clearance [75]. Synthetic α-Gal-containing constructs have similarly shown potential in stimulating cytotoxic immune responses against tumors, thus broadening the utility of glyco-immunotherapy platforms.

6.1 Glycan-based Biomarkers

Glycans—complex carbohydrate structures covalently linked to proteins and lipids—play integral roles in modulating key physiological functions. Alterations in glycan structure, composition, or presentation often accompany disease progression, making glycan-based biomarkers valuable tools for diagnosis, prognosis, and therapeutic monitoring.

Cancer Glycomics: In cancer, abnormal glycosylation is well documented. Tumors often exhibit elevated levels of sialylated and fucosylated antigens such as sialyl-Lewis^X and sialyl-Lewis^a, which facilitate metastasis by enhancing selectin-mediated adhesion. The sLe^a-based CA19-9 antigen is clinically employed as a biomarker for pancreatic and colorectal cancers [76]. Additionally, α-fetoprotein-L3 (AFP-L3), a core-fucosylated variant of AFP, provides improved specificity for hepatocellular carcinoma.

Advances in detection technologies—such as glycan microarrays, lectin-affinity chromatography, and MS—have enhanced sensitivity and specificity for glycan biomarker identification [77].

Autoimmune and Inflammatory Disorders: In autoimmune conditions like RA, IgG glycosylation profiles are markedly altered. A decrease in terminal galactose and sialic acid residues on the Fc region correlates with increased inflammatory potential via Fcγ receptor activation and complement pathways. These hypogalactosylated IgGs serve as predictive and monitoring biomarkers for RA [78]. Similarly, in systemic lupus erythematosus, aberrant glycosylation of serum glycoproteins is associated with disease activity and flares [47].

Neurodegenerative Diseases: Glycomic alterations have been observed in neurodegenerative diseases such as AD. N-glycan profiling of cerebrospinal fluid (CSF) and brain tissue in AD patients reveals increased mannose -rich and bisecting GlcNAc-containing structures, which may influence amyloid-beta and tau pathology [79]. These glycan signatures offer potential as early-stage biomarkers for AD.

Infectious Diseases and Host–Pathogen Interactions: Glycan alterations also provide insight into infectious diseases. Chronic hepatitis B and C infections are associated with increased serum glycoprotein fucosylation and sialylation, correlating with fibrosis and liver cancer progression [80]. Such glycomic profiles can be employed as non-invasive biomarkers for disease staging and treatment response.

6.2 Glycan-Targeted Therapies

Targeting disease-specific glycan structures has become an innovative and promising approach in therapeutic development, especially in cancer treatment. One notable strategy involves the use of glycan-targeted nanoparticles designed to recognize and bind specific tumor-associated carbohydrate antigens, such as sialyl-Tn, which are often overexpressed on cancer cell surfaces. These nanoparticles can be engineered to deliver therapeutic agents directly to tumor cells, for example, by carrying double-stranded RNA (dsRNA) molecules aimed at silencing oncogenic pathways in gastric cancer. This targeted delivery not only enhances treatment precision but also significantly reduces systemic toxicity by minimizing drug exposure to healthy tissues, thereby improving the overall safety and efficacy profile of anticancer therapies [81].

In addition to nanoparticles, glycan-functionalized drug delivery systems, including scaffolds and nanocarriers, are being developed to improve selective targeting of malignancies. By decorating these delivery vehicles with specific glycans or glycan-binding ligands, they can preferentially home to cancer cells that express complementary glycan receptors. This approach increases drug bioavailability at the tumor site and decreases off-target effects, which are common drawbacks of conventional chemotherapy. These glycan-based delivery platforms offer a versatile and effective means to enhance the therapeutic index of anticancer drugs across various tumor types [82]. In a study, solid lipid nanoparticles decorated with fucose showed enhanced uptake of methotrexate by breast cancer cells, improved bioavailability *in vivo*, and superior tumor suppression with reduced systemic toxicity [83].

Furthermore, Targeting glycan structures represents a cutting-edge strategy in chimeric antigen receptor (CAR) T-cell therapy, enabling heightened precision in tumor targeting. A prominent development is the engineering of "sweet CARs" that recognize aberrant glycan epitopes such as the Tn antigen—a truncated O-glycan rarely found on normal cells but frequently expressed on solid tumors. For instance, Posey et al. demonstrated that CAR T-cells engineered with an scFv specific to the Tn-glycoform of MUC1 (Tn-MUC1) effectively erased tumors in preclinical models of leukemia and pancreatic cancer, without affecting normal tissues [84]. Building on this, newer "sweet CARs" designed via structure-guided engineering have extended their binding to multiple glycoproteins presenting Tn antigen, showing improved anti-tumor cytotoxicity across diverse cancer cell lines [83]. Reviews also highlight other glycan targets like sialyl-Tn, GD2, and Lewis^{Y}, with the first generation of glycan-directed CARs exhibiting promising safety and efficacy profiles in early studies. These breakthroughs in glycan-targeted CAR T-cells offer a novel solution to the challenge of identifying tumor-specific antigens in solid cancers, paving the way for more specific, effective, and versatile cell-based cancer immunotherapies [85]. Together, these glycan-targeted therapeutic approaches—ranging from nanoparticles and drug delivery systems to engineered immune cells—represent a cutting-edge frontier in oncology, harnessing the unique glycan signatures of tumors to develop safer, more effective, and highly specific cancer treatments.

6.3 Modulating Glycosylation Pathways

Modulating endogenous glycosylation pathways presents promising therapeutic opportunities by directly influencing the biochemical processes that govern cellular function and disease progression [86]. One significant approach involves the regulation of O-GlcNAcylation, a dynamic and reversible post-translational modification where a single GlcNAc molecule is added to serine or threonine residues on proteins. This modification plays a critical role in controlling cellular stress responses, metabolism, and signaling pathways. By using small molecule inhibitors that specifically target the enzymes responsible for adding or removing O-GlcNAc groups—namely, O-GlcNAc transferase and O-GlcNAcase—it is possible to precisely adjust O-GlcNAcylation levels. Such fine-tuning holds therapeutic potential for managing metabolic disorders and certain cancers by restoring normal cellular function or halting abnormal growth processes [87, 88].

Another important strategy focuses on inhibiting the HBP, which supplies Uridine Diphosphate (UDP)-GlcNAc, a fundamental substrate required for multiple glycosylation reactions. The HBP's rate-limiting enzyme, glutamine: fructose-6-phosphate amidotransferase (GFAT), catalyzes the initial step in this pathway, making it a prime target for intervention. Compounds like azaserine and 6-diazo-5-oxo-L-norleucine (DON) inhibit GFAT activity, thereby reducing the availability of UDP-GlcNAc and ultimately decreasing aberrant glycosylation often observed in tumor cells. By disrupting these abnormal glycosylation patterns, these agents help suppress tumor growth and metastasis, offering a valuable approach in cancer therapy [89, 90].

Furthermore, targeted enzymatic degradation of specific glycans on cell surfaces represents a novel immunotherapeutic avenue. Tumors frequently display immunosuppressive glycans such as sialic acids that interact with immune checkpoint receptors like Siglecs (sialic acid-binding immunoglobulin-like lectins) on immune cells, effectively dampening the immune response and allowing tumor evasion. Therapies that employ sialidase enzymes conjugated to targeting molecules can selectively remove these sialic acid residues from tumor cells, thereby disrupting the Siglec-mediated inhibitory signals. This removal enhances the immune system's ability to recognize and attack cancer cells. For example, sialidase-conjugated treatments have been shown to improve the efficacy of CAR macrophages by lifting the suppressive effects mediated through Siglec-5 and Siglec-10, thus potentiating anti-tumor immunity. This approach represents a cutting-edge strategy to overcome tumor immune escape by modifying the glycan landscape on cancer cells [64, 91,92].

7. Conclusion

Functional glycomics has revolutionized our understanding of biology by highlighting the crucial regulatory roles of glycans in maintaining cellular and systemic homeostasis. Once considered merely structural elements, glycans are now recognized as dynamic mediators of various physiological and pathological processes. This shift in perspective has been supported by technological advances in glycomics and glycoproteomics, allowing detailed analysis of glycan structures and functions. Central to this field is the concept of dynamic glycosylation—a context-specific process governed by glycosyltransferases, glycosidases, and sugar donors. These enzymes orchestrate the formation of diverse glycan patterns that influence protein folding, trafficking, and signaling.

Importantly, glycosylation patterns are not static; they change with developmental stage, cell type, and environmental stimuli, making the glycome a responsive code that integrates internal and external signals. Aberrations in this code, often due to genetic, metabolic, or inflammatory disruptions, have been implicated in numerous diseases. For example, altered sialylation and fucosylation in cancer cells facilitate immune evasion and metastasis, while in neurodegenerative diseases, abnormal glycosylation of tau and amyloid proteins is linked to pathogenic aggregation. Misregulation of glycosylation enzymes further disrupts protein stability and cell signaling, contributing to disease progression. Glycans actively modulate receptor-ligand interactions and signal transduction pathways. In cancers, altered N-glycosylation of growth factor receptors like EGFR enhances tumor growth and therapeutic resistance. In metabolic disorders, aberrant glycosylation of insulin-related proteins impairs glucose uptake. Moreover, host-pathogen interactions are shaped by glycomic dynamics, with pathogens exploiting host glycans for entry, while host cells mount glycan-mediated defenses.

Clinically, glycomics hold promise for diagnostics and therapeutics. Disease-specific glycan patterns, such as serum N-glycan profiles in cancers or IgG glycosylation in autoimmune diseases, are being explored as biomarkers. Therapeutic strategies include glycosylation inhibitors, glycan-mimetic drugs, and glycoengineered biologics—such as fucosylated monoclonal antibodies with enhanced efficacy. Despite its potential, the field faces challenges due to the structural complexity and context-specific nature of glycans. Standardized analytical methods and integrative bioinformatics are needed to interpret glycomic data effectively. The integration of glycomics with other omics, along with emerging technologies like machine learning, spatial glycomics, and CRISPR-based glyco-editing, is paving the way for deeper insights and more precise interventions.

Fundamentally, functional glycomics offers a crucial perspective for understanding how glycans regulate both health and disease. As advancements in this field progress, it is poised to play a central role in precision medicine by supporting the development of treatments that not only address the root causes of diseases but also help re-establish cellular equilibrium at its core.

Table 1: Examples of glycomics dysregulation contributing to disease progression.

S. No.	Disease	Glycan Type/ Modification	Findings	References
1.	Amyloid neurodegenerative diseases	Glyceraldehyde-3-phosphate dehydrogenase (GAPDH)	• GAPDH loses enzymatic activity after undergoing modifications, leading to impaired glucose metabolism. • Misfolded GAPDH interferes with cellular protein-folding systems, encouraging amyloid formation. • Denatured and modified GAPDH facilitates amyloid-like conversion of proteins such as amyloid-β-peptide, tau protein, α-synuclein, and prions, contributing to neurodegeneration.	[93]
2.	Traumatic Brain Injury (TBI)	Chondroitin sulfate proteoglycans (CSPGs)	• TBI leads to specific changes in CSPGs and CS-GAGs, marked by a prominent increase in 4-sulfated GAGs around the injury core. CSPGs like neurocan and NG2 are upregulated in this region, co-localizing with C4S but not C6S, with contributions from various cell types, and this response persists from 7 to 28 days post-injury.	[94]
3.	AD and other neurodegenerative diseases	Advanced glycation end products (AGEs)	• Non-enzymatic glycosylation and the AGEs are closely linked to aging and cognitive decline. The brain possesses receptor-mediated mechanisms to eliminate AGE-modified proteins, and impairment of these systems may contribute to chronic inflammation in aging-related neurodegenerative conditions.	[95]
4.	AD	protein O-GlcNAcylation and N-/ O-glycosylation	• Altered O-GlcNAcylation and N-/O-glycosylation in brain and serum distinguish AD from other neurodegenerative diseases. • Decreased O-GlcNAcylation and increased PAS staining suggest disrupted glycan homeostasis. • A negative correlation between O-GlcNAcylation and N-/O-glycosylation points to a novel intracellular regulatory mechanism.	[96]

Table 1 contd. ...

...Table 1 contd.

S. No.	Disease	Glycan Type/ Modification	Findings	References
5.	AD	O-GlcNAcylation	• AD brains show reduced O-GlcNAcylation and unique glycan patterns absent in other neurodegenerative diseases. • Glycosylation modulates Asp-2 (β-secretase) activity; complex glycans enhance function, while high-mannose forms (from insect cells) reduce it. • Precise N-glycosylation sites are crucial for Asp-2 enzymatic activity.	[97]
6.	Schizophrenia	N-glycosylation	• Schizophrenia patients show elevated agalactosylated N-glycans in CSF and serum. • Reduced sialylation indicates defects in terminal glycan processing affecting protein function and immunity. • Altered expression of bi-/triantennary N-glycans lacking galactose/sialic acid.	[98]
7.	Schizophrenia	N-glycosylation of excitatory amino acid transporters (EAATs)	• Reduced N-glycosylation of EAAT1 and EAAT2 in schizophrenia brains suggests glia-specific impairment. • Neuronal Transporter Unaffected: EAAT3 shows no N-glycosylation in both schizophrenia and controls. • Pathophysiological Insight: Glial glycosylation deficits may disrupt glutamate homeostasis, contributing to schizophrenia.	[99]
8.	Colorectal cancer (CRC)	Glycosylation of MUC1 and Carcinoembryonic Antigen (CEA)	• Colorectal cancer shows increased TF antigen and sialylation on MUC1, enhancing PNA/MGL lectin binding. • Elevated Le^X/Le^Y, high-mannose, and branched N-glycans on CEA promote DC-SIGN and Galectin-3 binding. • These tumor-specific glycan changes in MUC1 and CEA may drive immune evasion and tumor progression.	[100]

9.	Colorectal cancer (CRC)	CEA Glycosylation	• Tumor CEA shows increased fucosylation, GalNAc, and branching, with decreased bisected structures and T antigen. • A 9-lectin panel distinguished tumors with 83% sensitivity and ~90% specificity. • Biomarker Potential: CEA glycan profiles offer promise for early CRC detection and staging.	[101]
10.	Gastric cancer (GC)	N-glycans	• CEA (CEACAM5) identified as the main carrier of sialyl-Lewis X (sLe^x) in gastric cancer cells overexpressing ST3Gal IV. • N-glycomics showed CEA bearing complex N-glycans capped with α2-3 sialic acids. • PLA-based detection of CEA–sLe^x proposed as a prognostic and theranostic biomarker.	[102]
11.	Breast cancer (BC)	Trisialylated, Triantennary N-glycan	• Identified a trisialylated, triantennary N-glycan with α1,3-fucose (part of sLe^x) elevated in advanced breast cancer serum. • A monogalactosylated form of this glycan was 2× higher in patients and correlated with disease progression. • Serum glycoforms as sensitive biomarkers for breast cancer progression.	[103]
12.	Lung cancer	α-1-antitrypsin (A1AT)	• Altered glycoforms of A1AT identified as potential lung cancer biomarkers. • Galactosylated A1AT distinguished NSCLC from benign lung disease. • Indicates A1AT glycosylation profiling as a promising diagnostic approach for lung cancer.	[104]
13.	Malignant melanoma (MM)	–	• HILIC-MS/MS identified 102 N-glycan isomers from AGP in serum. • Melanoma patients showed increased fucosylation, increased α-2,6-sialylation, and decreased nonfucosylated glycans. • Differentiation of sialic acid linkages and fucose positions was achieved. • Suggests strong potential of AGP glycan isomers as melanoma biomarkers.	[105]

Table 1 contd. ...

...Table 1 contd.

S. No.	Disease	Glycan Type/ Modification	Findings	References
14.	Hepatocellular Carcinoma (HCC)	N-glycopeptides	• They identified 93 (HCC), 87 (cirrhosis), and 68 (control) unique N-glycopeptides. • Increased branching, fucosylation, and sialylation in HCC and cirrhosis. • Supports glycopeptide-based biomarkers for early HCC diagnosis and disease staging.	[106, 107]
15.	Pancreatic ductal adenocarcinoma (PDAC)	–	• CEA is identified as the primary rBC2LCN-binding glycoprotein in PDAC via LC-MS and IHC. • Suggests potential for rBC2LCN-CEA–based diagnostics/therapeutics in PDAC.	[108]
16.	Inflammatory diseases	O-GlcNAc transferase dependent O-GlcNAcylation	• OGT drives inflammation by O-GlcNAcylating NF-κB p65, enhancing its nuclear activity. • OGT Inhibition: ST045849 blocks LPS-induced NF-κB activation, confirming glycosylation-dependent signaling. • Chitosan oligosaccharides inhibit OGT, reduce NF-κB O-GlcNAcylation, MAPK activation, and ROS, suppressing inflammation.	[109]
17.	Neurodegenerative diseases Acute respiratory distress CVDs	O-GlcNAcylation	• Aging/Cardiac Stress, alters genes in oxidative phosphorylation, impairing mitochondria. • O-GlcNAcylation, modified in hypertrophy, disrupting energy metabolism. • O-GlcNAc signaling upregulates NOX, increasing ROS. • Excess NOX-driven ROS promotes dysfunction; targeting NOX offers therapeutic potential.	[110]
18.	Ageing-associated disease such as cardiovascular and mental diseases	N- and O-glycosylation	• In male mice, aging leads to an increase in high-mannose N-glycans and a decrease in complex N-glycans in the cardiac glycoproteome. • These glycosylation changes are likely contributors to cardiovascular functional decline with age.	[111]

References

1. Cummings, R.D. and J.M. Pierce. (2014). The challenge and promise of glycomics. *Chemistry and Biology*, 21: 1–5.
2. Turnbull, J.E. and R.A. Field. (2007). Emerging glycomics technologies. *Nature Chemical Biology*, 3: 74–77.
3. Varki, A. (2017). Biological roles of glycans. *Glycobiology*, 27: 3–49.
4. Stebbins, N.W. and R. Sasisekharan. (2014). Functional glycomics analysis: Challenges and methodologies. *Comprehensive Analytical Chemistry*, 63: 255–280.
5. Helenius, A. and A. Aebi. (2001). Intracellular functions of N-linked glycans. *Science*, 291: 2364–2369.
6. Nimmerjahn, F. and J.V. Ravetch. (2008). Anti-inflammatory actions of intravenous immunoglobulin. *Annual Review of Immunology*, 26: 513–533.
7. Munkley, J. and D.J. Elliott. (2016). Hallmarks of glycosylation in cancer. *Oncotarget*, 7: 35478–35478.
8. Groux-Degroote, S., S. Cavdarli, K. Uchimura, F. Allain and P. Delannoy. (2020). Glycosylation changes in inflammatory diseases. *Advances in Protein Chemistry and Structural Biology*, 119: 111–156.
9. Miyoshi, E., Y. Kamada and T. Suzuki. (2020). Functional glycomics: Application to medical science and hepatology. *Hepatology Research*, 50: 153–164.
10. Bertozzi, C.R. and D. Rabuka. (2009). Structural basis of glycan diversity. In: *Essentials of Glycobiology* (2nd Edn.).
11. Mulloy, B., G.W. Hart and P. Stanley. (2009). Structural analysis of glycans. In: *Essentials of Glycobiology* (2nd Edn.).
12. Cambay, F., C. Forest-Nault, L. Dumoulin, A. Seguin, O. Henry, Y. Durocher and G. De Crescenzo. (2020). Glycosylation of Fcγ receptors influences their interaction with various IgG1 glycoforms. *Molecular Immunology*, 121: 144–158.
13. Brockhausen, I., H. Schachter and P. Stanley. (2009). O-GalNAc glycans. *Essentials of Glycobiology* (2nd Edn.).
14. Chen, Y.Z., Y.R. Tang, Z.Y. Sheng and Z. Zhang. (2008). Prediction of mucin-type O-glycosylation sites in mammalian proteins using the composition of k-spaced amino acid pairs. *BMC Bioinformatics*, 9: 1–2.
15. Tian, E. and K.G. Ten Hagen. (2009). Recent insights into the biological roles of mucin-type O-glycosylation. *Glycoconjugate Journal*, 26: 325–334.
16. Orlińska, K., K. Komosińska-Vassev, K. Olczyk, A. Kowalczyk and P. Olczyk. (2023). Glycosaminoglycans–types, structure, functions, and the role in wound healing processes *Annales Academiae Medicae Silesiensis*, 77: 204–216.
17. He, M., X. Zhou and X. Wang. (2024). Glycosylation: Mechanisms, biological functions, and clinical implications. *Signal Transduction and Targeted Therapy*, 9: 194.
18. Kuribara, T. and K. Totani. (2022). Oligomannose-type glycan processing in the endoplasmic reticulum and its importance in misfolding diseases. *Biology*, 11: 199.
19. Lis, H. and N. Sharon. (1993). Protein glycosylation: Structural and functional aspects. *European Journal of Biochemistry*, 218: 1–27.
20. Schjoldager, K.T., Y. Narimatsu, H.J. Joshi and H. Clausen. (2020). Global view of human protein glycosylation pathways and functions. *Nature Reviews Molecular Cell Biology*, 21: 729–749.
21. Freeze, H.H., J.X. Chong, M.J. Bamshad and B.G. Ng. (2014). Solving glycosylation disorders: Fundamental approaches reveal complicated pathways. *The American Journal of Human Genetics*, 94: 161–175.
22. Pinho, S.S. and C.A. Reis. (2015). Glycosylation in cancer: Mechanisms and clinical implications. *Nature Reviews Cancer*, 15: 540–555.

23. Wang, Y., Y. Liu, S. Liu, L. Cheng and X. Liu. (2024). Recent advances in N-glycan biomarker discovery among human diseases: N-glycan biomarkers in diseases. *Acta Biochimica et Biophysica Sinica*, 56: 1156.
24. Bones, J., S. Mittermayr, N. O'Donoghue, A. Guttman and P.M. Rudd. (2010). Ultra performance liquid chromatographic profiling of serum N-glycans for fast and efficient identification of cancer associated alterations in glycosylation. *Analytical Chemistry*, 82: 10208–10215.
25. Hirabayashi, J., M. Yamada, A. Kuno and H. Tateno. (2013). Lectin microarrays: Concept, principle, and applications. *Chemical Society Reviews*, 42: 4443–4458.
26. Brisson, J.R., E. Vinogradov, D.J. McNally, N.H. Khieu, I.C. Schoenhofen, S.M. Logan and H. Jarrell. (2010). The application of NMR spectroscopy to functional glycomics. In: *Functional Glycomics: Methods and Protocols*, pp. 155–173. MDPI.
27. Ruhaak, L.R., C. Huhn, W.J. Waterreus, A.R. De Boer, C. Neusüss, C.H. Hokke, A.M. Deelder and M. Wuhrer. (2008). Hydrophilic interaction chromatography-based high-throughput sample preparation method for N-glycan analysis from total human plasma glycoproteins. *Analytical Chemistry*, 80: 6119–6126.
28. Ohland, C.L. and C. Jobin. (2015). Microbial activities and intestinal homeostasis: A delicate balance between health and disease *Cellular and Molecular Gastroenterology and Hepatology*, 1: 28–40.
29. McEwen, B.S. and J.C. Wingfield. (2003). The concept of allostasis in biology and biomedicine. *Hormones and Behavior*, 43: 2–15.
30. Ohtsubo, K. and J.D. Marth. (2006). Glycosylation in cellular mechanisms of health and disease. *Cell*, 126: 855–867.
31. Maverakis, E., K. Kim, M. Shimoda, M.E. Gershwin, F. Patel, R. Wilken, S. Raychaudhuri, L.R. Ruhaak and C.B. Lebrilla. (2015). Glycans in the immune system and the altered glycan theory of autoimmunity: A critical review. *Journal of Autoimmunity*, 57: 1–3.
32. Dennis, J.W., I.R. Nabi and M. Demetriou. (2009). Metabolism, cell surface organization, and disease. *Cell*, 139: 1229–1241.
33. Radovani, B. and I. Gudelj. (2022). N-glycosylation and inflammation; the not-so-sweet relation. *Frontiers in Immunology*, 13: 893365.
34. Brestoff, J.R. and D. Artis. (2015). Immune regulation of metabolic homeostasis in health and disease. *Cell*, 161: 146–160.
35. Macauley, M.S., P.R. Crocker and J.C. Paulson. (2014). Siglec-mediated regulation of immune cell function in disease. *Nature Reviews Immunology*, 14: 653–666.
36. Jin, F. and F. Wang. (2020). The physiological and pathological roles and applications of sialyl Lewis x, a common carbohydrate ligand of the three selectins. *Glycoconjugate Journal*, 37(2): 277–291.
37. Arnold, J.N., M.R. Wormald, R.B. Sim, P.M. Rudd and R.A. Dwek. (2007). The impact of glycosylation on the biological function and structure of human immunoglobulins. *Annual Review of Immunology*, 25: 21–50.
38. Taniguchi, N. and Y. Kizuka. (2015). Glycans and cancer: Role of N-glycans in cancer biomarker, progression and metastasis, and therapeutics. *Advances in Cancer Research*, 126: 11–51.
39. Cherepanova, N., S. Shrimal and R. Gilmore. (2016). N-linked glycosylation and homeostasis of the endoplasmic reticulum. *Current Opinion in Cell Biology*, 41: 57–65.
40. Duarte, H.O., D. Freitas, C. Gomes, J. Gomes, A. Magalhães and C.A. Reis. (2016). Mucin-type O-glycosylation in gastric carcinogenesis. *Biomolecules*, 6(3): 33.
41. Stuchlová Horynová, M., M. Raška, H. Clausen and J. Novak. (2013). Aberrant O-glycosylation and anti-glycan antibodies in an autoimmune disease IgA nephropathy and breast adenocarcinoma. *Cellular and Molecular Life Sciences*, 70: 829–839.

42. Dobie, C. and D. Skropeta. (2021). Insights into the role of sialylation in cancer progression and metastasis. *British Journal of Cancer*, 124: 76–90.
43. Rabinovich, G.A. and D.O. Croci. (2012). Regulatory circuits mediated by lectin-glycan interactions in autoimmunity and cancer. *Immunity*, 36: 322–335.
44. Kaddour, N., F. Benyettou, K. Moulai, A. Mebarki, K. Allal-Taouli, R. Ghemrawi, J. Whelan, H. Merzouk, A. Trabolsi and N.A. Mokhtari-Soulimane. (2024). Effects of subcutaneous vs. oral nanoparticle-mediated insulin delivery on hemostasis disorders in Type 1 diabetes: A rat model study. *Heliyon*, 10: 9.
45. Ramasamy, R., S.J. Vannucci, S.S. Yan, K. Herold, S.F. Yan and A.M. Schmidt. (2005). Advanced glycation end products and RAGE: A common thread in aging, diabetes, neurodegeneration, and inflammation. *Glycobiology*, 15: 16R–28R.
46. Hart, G.W., C. Slawson, G. Ramirez-Correa and O. Lagerlöf. (2011). Cross talk between O-GlcNAcylation and phosphorylation: Roles in signaling, transcription, and chronic disease. *Annual Review of Biochemistry*, 80: 825–858.
47. Lauc, G., J.E. Huffman, M. Pučić, L. Zgaga, B. Adamczyk, A. Mužinić, M. Novokmet, O. Polašek, O. Gornik, J. Krištić and T. Keser. (2013). Loci associated with N-glycosylation of human immunoglobulin G show pleiotropy with autoimmune diseases and haematological cancers. *PLoS Genetics*, 9(1): e1003225.
48. Tacke, F. and R. Weiskirchen. (2021). Non-alcoholic fatty liver disease (NAFLD)/ non-alcoholic steatohepatitis (NASH)-related liver fibrosis: Mechanisms, treatment, and prevention. *Annals of Translational Medicine*, 9(8): 729.
49. Schnaar, R.L., R. Gerardy-Schahn and H. Hildebrandt. (2014). Sialic acids in the brain: Gangliosides and polysialic acid in nervous system development, stability, disease, and regeneration. *Physiological Reviews*, 94: 461–518.
50. Gong, C.X., F. Liu and K. Iqbal. (2016). O-GlcNAcylation: A regulator of tau pathology and neurodegeneration. *Alzheimer's and Dementia*, 12(10): 1078–1089.
51. Zhao, J. and M. Lang. (2023). New insight into protein glycosylation in the development of Alzheimer's disease. *Cell Death Discovery*, 9: 314.
52. Song, X.J., H.Y. Zhou, Y.Y. Sun and H.C. Huang. (2021). Phosphorylation and glycosylation of amyloid-β protein precursor: The relationship to trafficking and cleavage in Alzheimer's disease. *Journal of Alzheimer's Disease*, 84(3): 937–957.
53. Yang, J., H. Li and Y. Zhao. (2023). Dessert or poison? The roles of glycosylation in Alzheimer's, Parkinson's, Huntington's disease, and amyotrophic lateral sclerosis. *ChemBioChem*, 24: e202300017.
54. Martinsen, V. and P. Kursula. (2022). Multiple sclerosis and myelin basic protein: Insights into protein disorder and disease. *Amino Acids*, 54: 99–109.
55. Loaeza-Reyes, K.J., E. Zenteno, A. Moreno-Rodríguez, R. Torres-Rosas, L. Argueta-Figueroa, R. Salinas-Marín, L.M. Castillo-Real, S. Pina-Canseco and Y.P. Cervera. (2021). An overview of glycosylation and its impact on cardiovascular health and disease. *Frontiers in Molecular Biosciences*, 8: 751637.
56. Donahue, T.C., C. Ou, Q. Yang, R. Flinko, X. Zhang, G. Zong, G.K. Lewis and L.X. Wang. (2023). Synthetic site-specific antibody–ligand conjugates promote asialoglycoprotein receptor-mediated degradation of extracellular human PCSK9. *ACS Chemical Biology*, 18: 1611–1623.
57. McEver, R.P. (2005). Selectin-dependent leukocyte interactions with vascular surfaces. In: *Organ Microcirculation: A Gateway to Diagnostic and Therapeutic Interventions*. Tokyo: Springer, pp. 203–211.
58. Pu, Q. and C. Yu. (2014). Glycosyltransferases, glycosylation, and atherosclerosis. *Glycoconjugate Journal*, 31: 605–611.

59. Ward, S., J.M. O'Sullivan and J.S. O'Donnell. (2019). von Willebrand factor sialylation: A critical regulator of biological function. *Journal of Thrombosis and Haemostasis*, 17: 1018–1029.
60. Christensen, G., K.M. Herum and I.G. Lunde. (2019). Sweet, yet underappreciated: Proteoglycans and extracellular matrix remodeling in heart disease. *Matrix Biology*, 75: 286–299.
61. Yan, A. and W.J. Lennarz. (2005). Unraveling the mechanism of protein N-glycosylation. *Journal of Biological Chemistry*, 280: 3121–3124.
62. Marquardt, T. and J. Denecke. (2003). Congenital disorders of glycosylation: Review of their molecular bases, clinical presentations, and specific therapies. *European Journal of Pediatrics*, 162(6): 359–379.
63. Munkley, J. (2019). The glycosylation landscape of pancreatic cancer. *Oncology Letters*, 17: 2569–2575.
64. Gray, M.A., M.A. Stanczak, N.R. Mantuano, H. Xiao, J.F. Pijnenborg, S.A. Malaker, C.L. Miller, P.A. Weidenbacher, J.T. Tanzo, G. Ahn and E.C. Woods. (2020). Targeted glycan degradation potentiates the anticancer immune response *in vivo*. *Nature Chemical Biology*, 16(12): 1376–1384.
65. Pereira, M.S., I. Alves, M. Vicente, A. Campar, M.C. Silva, N.A. Padrão, V. Pinto, Â. Fernandes, A.M. Dias and S.S. Pinho. (2018). Glycans as key checkpoints of T cell activity and function. *Frontiers in Immunology*, 9: 2754.
66. Zhao, Y.Y., M. Takahashi, J.G. Gu, E. Miyoshi, A. Matsumoto, S. Kitazume and N. Taniguchi. (2008). Functional roles of N-glycans in cell signaling and cell adhesion in cancer. *Cancer Science*, 99: 1304–1310.
67. Butti, R., S. Das, V.P. Gunasekaran, A.S. Yadav, D. Kumar and G.C. Kundu. (2018). Receptor tyrosine kinases (RTKs) in breast cancer: Signaling, therapeutic implications, and challenges. *Molecular Cancer*, 17: 112.
68. Dewald, J.H., F. Colomb, M. Bobowski-Gerard, S. Groux-Degroote and P. Delannoy. (2016). Role of cytokine-induced glycosylation changes in regulating cell interactions and cell signaling in inflammatory diseases and cancer. *Cells*, 5: 43.
69. Reily, C., T.J. Stewart, M.B. Renfrow and J. Novak. (2019). Glycosylation in health and disease. *Nature Reviews Nephrology*, 15: 346–366.
70. Chen, X., L. Zhang, H. He, Y. Sun, Q. Shen and L. Shi. (2020). Increased O-GlcNAcylation induces myocardial hypertrophy. *In Vitro Cellular and Developmental Biology-Animal*, 56: 735–743.
71. Gonzalez, P.S., J. O'Prey, S. Cardaci, V.J. Barthet, J.I. Sakamaki, F. Beaumatin, A. Roseweir, D.M. Gay, G. Mackay, G. Malviya and E. Kania. (2018). Mannose impairs tumour growth and enhances chemotherapy. *Nature*, 563(7733): 719–723.
72. Paneque, A., H. Fortus, J. Zheng, G. Werlen and E. Jacinto. (2023). The hexosamine biosynthesis pathway: Regulation and function. *Genes*, 14(4): 933.
73. Shinkawa, T., K. Nakamura, N. Yamane, E. Shoji-Hosaka, Y. Kanda, M. Sakurada, K. Uchida, H. Anazawa, M. Satoh, M. Yamasaki and N. Hanai. (2003). The absence of fucose but not the presence of galactose or bisecting N-acetylglucosamine of human IgG1 complex-type oligosaccharides shows the critical role of enhancing antibody-dependent cellular cytotoxicity. *Journal of Biological Chemistry*, 278(5): 3466–3473.
74. Brückner, C., C. Lehmann, D. Dudziak and F. Nimmerjahn. (2017). Sweet SIGNs: IgG glycosylation leads the way in IVIG-mediated resolution of inflammation. *International Immunology*, 29: 499–509.
75. Galili, U., K. Wigglesworth and U.M. Abdel-Motal. (2007). Intratumoral injection of α-gal glycolipids induces xenograft-like destruction and conversion of lesions into endogenous vaccines. *The Journal of Immunology*, 178: 4676–4687.
76. Dube, D.H. and C.R. Bertozzi. (2005). Glycans in cancer and inflammation: Potential for therapeutics and diagnostics. *Nature Reviews Drug Discovery*, 4(6): 477–488.

77. Miyoshi, E., Y. Kamada, H. Kawakami, K. Hirata and N. Taniguchi. (2012). Glycoproteomics and clinical applications in hepatocellular carcinoma. *Molecular and Cellular Proteomics*, 11(4): O111.010132.
78. Ercan, A., J. Cui, M.M. Hazen, F. Batliwalla, L. Royle, P.M. Rudd, J.S. Coblyn, N. Shadick, M.E. Weinblatt, P. Gregersen and D.M. Lee. (2012). Hypogalactosylation of serum N-glycans fails to predict clinical response to methotrexate and TNF inhibition in rheumatoid arthritis. *Arthritis Research and Therapy*, 14: 1–8.
79. Schedin-Weiss, S., S. Gaunitz, P. Sui, Q. Chen, S.M. Haslam, K. Blennow, B. Winblad, A. Dell and L.O. Tjernberg. (2020). Glycan biomarkers for Alzheimer disease correlate with T-tau and P-tau in cerebrospinal fluid in subjective cognitive impairment. *The FEBS Journal*, 287(15): 3221–3234.
80. Higashi, M., T. Yoshimura, N. Usui, Y. Kano, A. Deguchi, K. Tanabe, Y. Uchimura, S. Kuriyama, Y. Suzuki, T. Masaki and K. Ikenaka. (2020). A potential serum N-glycan biomarker for hepatitis C virus-related early-stage hepatocellular carcinoma with liver cirrhosis. *International Journal of Molecular Sciences*, 21(23): 8913.
81. dos Santos, S.N., D.S. Junior, J.P. Pereira, N.M. Iadocicco, A.H. Silva, T. do Nascimento, L.A. Dias, F.R. de Oliveira Silva, E. Ricci-Junior, R. Santos-Oliveira and E.S. Bernardes. (2023). Development of glycan-targeted nanoparticles as a novel therapeutic opportunity for gastric cancer treatment. *Cancer Nanotechnology*, 14(1): 26.
82. Qin, H., Y. Teng, R. Dai, A. Wang and J. Liu. (2024). Glycan-based scaffolds and nanoparticles as drug delivery system in cancer therapy. *Frontiers in Immunology*, 15: 1395187.
83. Garg, N.K., B. Singh, A. Jain, P. Nirbhavane, R. Sharma, R.K. Tyagi, V. Kushwah, S. Jain and O.P. Katare. (2016). Fucose decorated solid-lipid nanocarriers mediate efficient delivery of methotrexate in breast cancer therapeutics. *Colloids and Surfaces B: Biointerfaces*, 146: 114–126.
84. Posey, A.D., R.D. Schwab, A.C. Boesteanu, C. Steentoft, U. Mandel, B. Engels, J.D. Stone, T.D. Madsen, K. Schreiber, K.M. Haines and A.P. Cogdill. (2016). Engineered CAR T cells targeting the cancer-associated Tn-glycoform of the membrane mucin MUC1 control adenocarcinoma. *Immunity*, 44(6): 1444–1454.
85. Anderluh, M., F. Berti, A. Bzducha-Wróbel, F. Chiodo, C. Colombo, F. Compostella, K. Durlik, X. Ferhati, R. Holmdahl, D. Jovanovic and W. Kaca. (2021). Emerging glyco-based strategies to steer immune responses. *The FEBS Journal*, 288(16): 4746–4772.
86. Seyrek, K., N.V. Ivanisenko, C. König and I.N. Lavrik. (2024). Modulation of extrinsic apoptotic pathway by intracellular glycosylation. *Trends in Cell Biology*, 34: 728–741.
87. Lu, Q., X. Zhang, T. Liang and X. Bai. (2022). O-GlcNAcylation: An important post-translational modification and a potential therapeutic target for cancer therapy. *Molecular Medicine*, 28(1): 115.
88. Ma, B., K.S. Khan, T. Xu, J. Xeque Amada, Z. Guo, Y. Huang, Y. Yan, H. Lam, A.S. Cheng and B.W. Ng. (2024). Targeted protein O-GlcNAcylation using bifunctional small molecules. *Journal of the American Chemical Society*, 146(14): 9779–9789.
89. Grigorian, A., S.U. Lee, W. Tian, I.J. Chen, G. Gao, R. Mendelsohn, J.W. Dennis and M. Demetriou. (2007). Control of T cell-mediated autoimmunity by metabolite flux to N-glycan biosynthesis. *Journal of Biological Chemistry*, 282(27): 20027–20035.
90. Vasconcelos-dos-Santos, A., I.A. Oliveira, M.C. Lucena, N.R. Mantuano, S.A. Whelan, W.B. Dias and A.R. Todeschini. (2015). Biosynthetic machinery involved in aberrant glycosylation: Promising targets for developing of drugs against cancer. *Frontiers in Oncology*, 5: 138.
91. RodrÍguez, E., S.T. Schetters and Y. van Kooyk. (2018). The tumour glyco-code as a novel immune checkpoint for immunotherapy *Nature Reviews Immunology*, 18(3): 204–211.

92. Yu, Y. and W. Peng. (2023). Recent progress in targeting the sialylated glycan-SIGLEC axis in cancer immunotherapy. *Cancer Biology and Medicine*, 20(5): 369–384.
93. Muronetz, V.I., K.V. Barinova, Y.Y. Stroylova, P.I. Semenyuk and E.V. Schmalhausen. (2017). Glyceraldehyde-3-phosphate dehydrogenase: Aggregation mechanisms and impact on amyloid neurodegenerative diseases. *International Journal of Biological Macromolecules*, 100: 55–66.
94. Yi, J.H., Y. Katagiri, B. Susarla, D. Figge, A.J. Symes and H.M. Geller. (2012). Alterations in sulfated chondroitin glycosaminoglycans following controlled cortical impact injury in mice. *Journal of Comparative Neurology*, 520: 3295–3313.
95. Li, J.J., D. Dickson, P.R. Hof and H. Vlassara. (1998). Receptors for advanced glycosylation endproducts in human brain: Role in brain homeostasis. *Molecular Medicine*, 4: 46–60.
96. Frenkel-Pinter, M., M.D. Shmueli, C. Raz, M. Yanku, S. Zilberzwige, E. Gazit and D. Segal. (2017). Interplay between protein glycosylation pathways in Alzheimer's disease. *Science Advances*, 3: e1601576.
97. Charlwood, J., C. Dingwall, R. Matico, I. Hussain, K. Johanson, S. Moore, D.J. Powell, J.M. Skehel, S. Ratcliffe, B. Clarke and J. Trill. (2001). Characterization of the glycosylation profiles of Alzheimer's β-secretase protein Asp-2 expressed in a variety of cell lines. *Journal of Biological Chemistry*, 276: 16739–16748.
98. Stanta, J.L., R. Saldova, W.B. Struwe, J.C. Byrne, F.M. Leweke, M. Rothermund, H. Rahmoune, Y. Levin, P.C. Guest, S. Bahn and P.M. Rudd. (2010). Identification of N-glycosylation changes in the CSF and serum in patients with schizophrenia. *Journal of Proteome Research*, 9: 4476–4489.
99. Bauer, D., V. Haroutunian, J.H. Meador-Woodruff and R.E. McCullumsmith. (2010). Abnormal glycosylation of EAAT1 and EAAT2 in prefrontal cortex of elderly patients with schizophrenia. *Schizophrenia Research*, 117: 92–98.
100. Saeland, E., A.I. Belo, S. Mongera, I. van Die, G.A. Meijer and Y. van Kooyk. (2012). Differential glycosylation of MUC1 and CEACAM5 between normal mucosa and tumour tissue of colon cancer patients. *International Journal of Cancer*, 131: 117–128.
101. Zhao, Q., T. Zhan, Z. Deng, Q. Li, Y. Liu, S. Yang, D. Ji and Y. Li. (2018). Glycan analysis of colorectal cancer samples reveals stage-dependent changes in CEA glycosylation patterns. *Clinical Proteomics*, 15: 1.
102. Gomes, C., A. Almeida, A. Barreira, J. Calheiros, F. Pinto, R. Abrantes, A. Costa, A. Polonia, D. Campos, H. Osório and H. Sousa. (2019). Carcinoembryonic antigen carrying SLeX as a new biomarker of more aggressive gastric carcinomas. *Theranostics*, 9: 7431–7443.
103. Abd Hamid, U.M., L. Royle, R. Saldova, C.M. Radcliffe, D.J. Harvey, S.J. Storr, M. Pardo, R. Antrobus, C.J. Chapman, N. Zitzmann and J.F. Robertson. (2008). A strategy to reveal potential glycan markers from serum glycoproteins associated with breast cancer progression. *Glycobiology*, 18: 1105–1118.
104. Liang, Y., T. Ma, A. Thakur, H. Yu, L. Gao, P. Shi, X. Li, H. Ren, L. Jia, S. Zhang and Z. Li. (2015). Differentially expressed glycosylated patterns of α-1-antitrypsin as serum biomarkers for the diagnosis of lung cancer. *Glycobiology*, 25: 331–340.
105. Virág, D., T. Kremmer, K. Lőrincz, N. Kiss, A. Jobbágy, S. Bozsányi, L. Gulyás, N. Wikonkál, G. Schlosser, A. Borbély and Z. Huba. (2021). Altered glycosylation of human alpha-1-acid glycoprotein as a biomarker for malignant melanoma. *Molecules*, 26: 6003.
106. Zhu, J., Z. Chen, J. Zhang, M. An, J. Wu, Q. Yu, S.J. Skilton, M. Bern, K.I. Sen, L. Li and D.M. Lubman. (2018). Differential quantitative determination of site-specific intact N-glycopeptides in serum haptoglobin between hepatocellular carcinoma and cirrhosis using LC-EThcD-MS/MS. *Journal of Proteome Research*, 18: 359–371.
107. Comunale, M.A., L. Rodemich-Betesh, J. Hafner, M. Wang, P. Norton, A.M. di Bisceglie, T. Block and A. Mehta. (2010). Linkage specific fucosylation of alpha-1-antitrypsin

in liver cirrhosis and cancer patients: Implications for a biomarker of hepatocellular carcinoma. *PLoS One*, 5: e12419.

108. Furuta, T., T. Oda, K. Kiyoi, Y. Otsubo, S. Kimura, K. Kurimori, Y. Miyazaki, Y. Yu, K. Furuya, Y. Akashi and O. Shimomura. (2021). Carcinoembryonic antigen as a specific glycoprotein ligand of rBC2LCN lectin on pancreatic ductal adenocarcinoma cells. *Cancer Science*, 112: 3722–3731.
109. Li, Y., H. Liu, Q.S. Xu, Y.G. Du and J. Xu. (2014). Chitosan oligosaccharides block LPS-induced O-GlcNAcylation of NF-κB and endothelial inflammatory response. *Carbohydrate Polymers*, 99: 568–578.
110. Begum, R., S. Thota, A. Abdulkadir, G. Kaur, P. Bagam and S. Batra. (2022). NADPH oxidase family proteins: Signaling dynamics to disease management. *Cellular and Molecular Immunology*, 19(6): 660–686.
111. Paton, B., M. Suarez, P. Herrero and N. Canela. (2021). Glycosylation biomarkers associated with age-related diseases and current methods for glycan analysis. *International Journal of Molecular Sciences*, 22(11): 5788.

CHAPTER 12

Role of Glynomics in Progression of Infections

Shikha Pandey,[1] *Puja Nayak*,[1] *Rishi Paliwal*[2] and *Shivani Rai Paliwal*[1,*]

Glycomics, the comprehensive study of glycans, has emerged as a vital field in understanding infectious disease progression. Glycans are involved in key biological processes including cell signaling, immunity, and host-pathogen interactions. Pathogens such as bacteria, viruses, and fungi exploit host glycan structures for adhesion, invasion, and immune evasion, while infections can dynamically alter host glycosylation profiles. These changes offer potential biomarkers for diagnosis, disease staging, and therapeutic monitoring. This chapter provides a detailed overview of glycan-mediated mechanisms in infection progression, supported by examples of glycan-targeted diagnostics, vaccines, glycomimetics, and inhibitors. It also explores the therapeutic value of targeting microbial and host glycosylation pathways. While challenges remain due to glycan complexity and analytical limitations, advances in glycomics are paving the way for innovative diagnostic tools and next-generation therapeutics. Overall, glycomics holds promise in transforming infectious disease research

[1] Department of Pharmacy, Guru Ghasidas Vishwavidyalaya (A Central University), Bilaspur, CG, India, 495009.
[2] Department of Pharmacy, Indira Gandhi National Tribal University, Amarkantak, MP, India, 484886.
Emails: shikha.dnp@gmail.com; poojanayak2303@gmail.com; rishipaliwal@gmail.com
* Corresponding author: srai2k@gmail.com

DOI: 10.1201/9781003635901-12

and improving clinical outcomes through precision glycan-based strategies.

1. Introduction

In the rapidly evolving field of molecular biology, glycomics has emerged as a transformative discipline that expands our understanding of cellular functions and disease mechanisms. Glycomics refers to the comprehensive study of glycans—complex carbohydrates that are covalently attached to proteins and lipids—and encompasses their structures, biosynthesis, and roles in biological systems. Unlike the linear and template-based synthesis of nucleic acids and proteins, glycan biosynthesis is non-template driven, occurring through a complex and dynamic interplay of glycosyltransferases and glycosidases within cellular organelles such as the endoplasmic reticulum and Golgi apparatus. This inherent structural and functional diversity renders glycans highly sensitive to changes in cellular states, such as differentiation, malignancy, or infection [1].

Glycans are not mere structural components; they are integral to cell–cell communication, immune responses, pathogen recognition, and molecular trafficking. In particular, glycan-mediated interactions on cell surfaces govern critical physiological and pathological events, including immune activation, microbial colonization, and tissue repair. The glycocalyx—a dense, carbohydrate-rich layer on the surface of cells—is often the first point of contact between a host and an invading pathogen. As such, glycans serve as a biochemical interface between the host and external stimuli, making them both targets and modulators in the progression of infections [2].

The significance of glycans is perhaps most striking in the context of host-pathogen interactions. Many pathogens—including viruses, bacteria, fungi, and parasites—have evolved sophisticated mechanisms to exploit host glycosylation pathways. They recognize and bind to host glycans through surface proteins such as adhesins, lectins, and glycan-binding toxins, enabling adherence, invasion, and colonization of host tissues [3]. For example, the hemagglutinin protein of influenza viruses binds specifically to sialic acid residues on epithelial cells, facilitating viral entry and replication [4]. Similarly, Helicobacter pylori, the causative agent of gastric ulcers and cancer, targets Lewis blood group antigens on gastric mucosa using its BabA adhesin [5].

Pathogens not only recognize host glycans but also actively mimic or alter them to evade immune detection. This phenomenon, known as molecular mimicry, allows microbial invaders to camouflage themselves with host-like glycans. Group B *Streptococcus*, for instance, expresses sialylated capsular polysaccharides that closely resemble those found on host cells, thereby inhibiting complement deposition and phagocytosis [6].

In other cases, pathogens decorate their own surfaces with glycan shields, as seen in HIV-1, where the envelope glycoprotein gp120 is extensively glycosylated to obscure antigenic epitopes from neutralizing antibodies [7]. These strategies exemplify how glycan interactions are not passive but are central to pathogen survival and virulence.

Equally important is the plasticity of host glycosylation during infection. Upon exposure to pathogens, host cells often undergo extensive glycan remodeling, influenced by inflammatory cytokines and pathogen-derived effectors. This dynamic modulation can either enhance or impair immune responses. For example, certain viruses like cytomegalovirus can downregulate host sialyltransferases, reducing NK cell activation and promoting viral persistence [8]. Conversely, host immune activation can alter fucosylation and sialylation patterns, shaping the recruitment of leukocytes and the secretion of antimicrobial peptides. Understanding these bidirectional modifications is crucial to deciphering the full scope of infection biology.

The rise of glycomics technologies has empowered researchers to systematically analyze glycan structures and their biological functions. Mass spectrometry, lectin microarrays, and glycan sequencing now allow for high-throughput, site-specific glycan profiling. These advancements have led to the discovery of disease-specific glycan signatures that serve as biomarkers for infection, inflammation, and cancer. For example, increased core fucosylation and sialylation of serum glycoproteins have been linked to chronic liver diseases and viral hepatitis [9]. In infectious diseases, such glycomic changes can offer valuable prognostic and diagnostic insights, potentially preceding traditional clinical symptoms.

Additionally, glycomics has spurred innovation in diagnostics and therapeutics. Glycan-based diagnostics harness unique glycan patterns for pathogen identification, while glycomimetic drugs and antiglycan antibodies are being developed to inhibit pathogen adhesion or block glycan-mediated immune suppression. Glycoconjugate vaccines—such as those for *Haemophilus influenzae* and *Streptococcus pneumoniae*—have demonstrated the power of targeting microbial glycans to elicit protective immunity [10]. New efforts are underway to apply similar strategies to viruses like SARS-CoV-2 and fungi like *Candida albicans*.

Yet, despite its promise, glycomics still face formidable challenges. The structural complexity of glycans, including branching, linkage variability, and context-specific expression, complicates their analysis and annotation. Unlike genomics or proteomics, glycomics lack a universal coding system, making computational modeling and integration more difficult. Moreover, the context-dependency of glycan expression—sensitive to cell type, environment, and developmental stage—demands high-resolution, systems-level approaches to understand functional outcomes.

This chapter explores the multifaceted roles of glycomics in infection biology, focusing on how glycan-mediated mechanisms contribute to pathogen adherence, immune modulation, and disease progression. We begin with an overview of glycan-host pathogen interactions, detailing both general mechanisms and organism-specific strategies employed by bacteria, viruses, fungi, and protozoa. This is followed by a discussion of how glycosylation patterns shift during infection and inflammation, influencing disease severity and persistence. We then examine the translational implications of these insights, highlighting emerging glycan-based diagnostics and therapeutics. Finally, we address current challenges and future directions, including the need for standardized analytical methods, integrative omics platforms, and targeted modulation of glycosylation pathways.

In understanding how glycans orchestrate the complex molecular dialogue between hosts and pathogens, glycomics offer not only a deeper appreciation of infectious disease pathogenesis but also a promising frontier for novel interventions. As we face growing threats from antimicrobial resistance and emerging pathogens, glycan-centered strategies could prove vital in developing next-generation diagnostics, broad-spectrum antivirals, and precision vaccines.

2. Glycan—Host Pathogen Interaction

Glycans—complex sugar molecules found extensively on the surface of cells and within the extracellular matrix that play a very important role in mediating interaction playing a pivotal role in mediating interactions between a host organism and invading microbes. These intricate sugar structures are far more than mere decorations; they function as crucial docking sites, sophisticated communication signals, and even as camouflage. Upon encountering a host, pathogens (including viruses, bacteria, and fungi) face the immediate challenge of adhering to and subsequently entering host cells. To overcome this, many pathogens strategically exploit host glycans, utilizing them like specific keys to unlock cellular entry points.

Host glycans are typically attached to proteins (forming glycoproteins) or lipids (forming glycolipids) on cell surfaces. Pathogens possess specialized proteins, commonly known as adhesins or lectins, that precisely recognize and bind to distinct glycan patterns. This initial recognition and binding process is fundamentally critical for establishing an infection; without this essential first contact, the microbe cannot effectively colonize, invade, or exert its pathogenic effects on host cells.

Conversely, pathogens are not passive invaders. Many have undergone evolutionary adaptations to employ sophisticated strategies, such as mimicking host glycans, creating protective sugar coats, or even subtly manipulating the host's own glycosylation machinery. These tactics are primarily aimed at evading detection and neutralization by the host immune system. A deeper understanding of these multifaceted, glycan-mediated interactions is therefore crucial for elucidating the mechanisms of infection initiation and explaining why certain microbes are exceptionally adept at immune evasion and successful colonization.

2.1 Basic Types of Glycan Interactions in Infections

Pathogens rely on diverse strategies to infect host organisms, with many of these pivotal strategies intimately involving glycans. Glycans found on the host cell surface are not randomly distributed; instead, they carry essential biological information that microbes have evolved to recognize and exploit for their survival and proliferation. This section explores three fundamental but crucial ways pathogens interact with host glycans: by leveraging them as specific docking points (receptors), by employing molecular mimicry to evade immune surveillance, and by modifying their own proteins with glycans to enhance their infectious potential.

2.1.1 Glycan Receptors on Host Cells

Glycans situated on the surfaces of host cells serve as crucial "molecular signposts" that guide pathogens to their initial points of attachment. These sugar structures are commonly presented as components of glycoproteins or glycolipids. Numerous pathogens have evolved specialized proteins, termed adhesins or lectins, which are specifically adapted to bind with high affinity to these host glycans. This initial interaction frequently represents the first and often indispensable step in establishing an infection.

Once a pathogen successfully binds to a specific glycan receptor, it can firmly anchor itself to the host cell, initiate colonization of tissues, and in many cases, gain direct entry into the cell itself.

The influenza virus initiates infection by attaching to epithelial cells in the respiratory tract through its hemagglutinin (HA) protein, which specifically recognizes and binds to sialic acid residues present on host glycoproteins. This sialic acid-mediated interaction is crucial for viral entry and infectivity [4]. Similarly, *Helicobacter pylori*, a bacterium linked to gastric ulcers and cancer, employs its outer membrane adhesin BabA to bind selectively to Lewis b antigens—fucosylated glycan structures found on the surface of gastric epithelial cells. This strong glycan-specific adhesion allows *H. pylori* to persist in the acidic environment of the stomach [5]. In the case of norovirus, the virus targets histo-blood group antigens (HBGAs) expressed on intestinal epithelial cells. These specific glycan interactions not only determine viral attachment and host susceptibility, but also explain why individuals lacking certain HBGA structures may exhibit natural resistance to some norovirus strains, highlighting the importance of host glycan diversity in infection outcomes. These examples vividly illustrate how pathogens exploit host glycans as critical entry points, frequently determining their specific tissue tropism and host species specificity.

2.1.2 Glycan-Mediated Immune Evasion

The host immune system continuously monitors the body for signs of foreign invaders. However, many pathogens have developed sophisticated mechanisms to elude this surveillance. One of the most effective strategies involves using glycans to "blend in" with host cells, a tactic known as molecular mimicry. In this strategy, microbes coat their surfaces with glycan structures that closely resemble those naturally abundant in the host.

By adopting these host-like glycan disguises, pathogens become less recognizable to immune cells, thereby making it more challenging for the host to mount an effective defensive response. Furthermore, some pathogens utilize dense glycan coatings as a protective shield, effectively masking more immunogenic (easily recognized) molecular components on their surface.

Group B *Streptococcus* (GBS) utilizes a capsular polysaccharide densely coated with sialic acid, a sugar also commonly present on human cell surfaces. This glycan mimicry enables GBS to avoid recognition by the complement system and reduces its uptake by phagocytic immune cells, facilitating immune evasion and persistence in the host [11]. Similarly, *Neisseria gonorrhoeae*, the pathogen responsible for gonorrhea, decorates its surface lipooligosaccharide (LOS) with sialic acid. This strategic glycosylation masks the bacterial surface from immune detection, enhancing resistance to complement-mediated lysis and immune clearance [12]. In the case of *Trypanosoma cruzi*, the protozoan parasite causing Chagas disease cloaks itself in glycan structures that closely resemble those of its host. This mimicry interferes with normal host immune signaling and delays immune recognition, allowing the parasite to survive and replicate within host tissues for extended periods.

These sophisticated glycan-based tactics allow pathogens to persist within the host for extended periods, significantly increasing the likelihood of dissemination and contributing to the development of chronic infections.

2.1.3 Pathogen Glycosylation and Virulence

Glycosylation is not exclusively a host cell process; many pathogens also extensively modify their own proteins and lipids with sugars. These pathogen-derived glycans serve multiple critical functions that contribute to microbial survival and virulence. They can facilitate adhesion to host cells, actively suppress the host immune system, or ensure the proper folding and function of essential pathogen proteins. This intrinsic self-glycosylation by pathogens often directly enhances their virulence, which is their ability to cause disease.

In viruses, glycosylation frequently forms a dense, protective "sugar coating" around surface proteins, effectively hiding crucial parts of the virus from neutralizing antibodies and other immune recognition mechanisms. In bacteria and parasites, specific glycosylation patterns can play roles in host cell recognition, active immune suppression, and even motility or biofilm formation.

HIV employs extensive glycosylation of its envelope glycoprotein gp120 to form a dense "glycan shield", which obscures critical viral epitopes and prevents effective recognition by neutralizing antibodies. This structural masking significantly hinders the immune system's ability to mount a lasting response, posing a major obstacle to the development of effective vaccines [7]. *Campylobacter jejuni*, a leading cause of bacterial gastroenteritis, strategically glycosylates its flagellin protein—essential for motility—to avoid detection by host innate immune sensors. This immune evasion strategy not only enhances bacterial survival, but also promotes efficient colonization and invasion of the intestinal epithelium. Similarly, *Plasmodium falciparum*, the causative agent of severe malaria, expresses glycosylated proteins on the surface of infected erythrocytes. These glycoconjugates mediate strong adhesion to vascular endothelium, allowing the parasite to evade splenic clearance and contributing to complications such as cerebral malaria. Understanding precisely how pathogens glycosylate their own proteins provides invaluable insights into their survival mechanisms and how they successfully thrive within the complex environment of the host.

2.2 Specific Interactions between Pathogens and Host Glycans

Microbial pathogens—encompassing bacteria, viruses, and fungi—are highly dependent on interacting with host cell surface glycans at virtually every critical step of the infection process. These interactions are far from random; instead, pathogens have evolved remarkably specific and diverse mechanisms to bind to, manipulate, or mimic host glycans to their distinct advantage.

The following subsections delineate how different classes of pathogens strategically employ glycans for key steps in infection, including initial adherence to host cells, subsequent entry, intracellular replication, evasion of immune detection, and ultimately, the induction of disease and long-term damage.

2.2.1 *Bacterial Infections and Glycan-Mediated Mechanisms*

Bacteria leverage host glycans as precise docking sites, complex communication signals, or as effective camouflage to establish and sustain infections. These multifaceted glycan interactions are fundamentally critical for initiating colonization and successfully evading the host's immune destruction.

Adhesion and Invasion: Adhesion represents the indispensable first step in bacterial infection. Bacteria must firmly attach to host surfaces to resist being mechanically cleared by bodily fluids such as urine, mucus, or blood flow. Many host cell surfaces are extensively coated in glycoproteins and glycolipids that prominently display various glycan structures—these serve as molecular ' handles' for bacteria to specifically grab onto.

Uropathogenic *Escherichia coli* (UPEC), a predominant causative agent of urinary tract infections, initiates infection by specifically binding to mannose-containing glycans present on the surface of bladder epithelial cells. This adhesion is mediated through fimbriae, particularly Type 1 pili, which recognize and attach to these glycans, facilitating colonization and infection [13]. Similarly, *H. pylori,* a bacterium closely associated with gastritis and peptic ulcers, utilizes its outer membrane protein BabA to bind specifically to Lewis b antigens, a type of fucosylated glycan expressed on gastric epithelial cells. This targeted interaction allows *H. pylori* to establish a strong foothold in the harsh acidic environment of the stomach, thereby promoting persistent infection and inflammation [5]. Following successful adhesion, some bacteria are capable of invading host cells. They often achieve this by hijacking natural endocytosis pathways, which are frequently regulated by glycan-receptor interactions, thereby allowing the bacteria to enter the cell stealthily without triggering immediate host alarm signals.

Toxin Binding: Many clinically significant bacteria release potent protein toxins that directly mediate disease pathology. For these toxins to exert their devastating effects, they must first bind with high specificity to particular glycans on the surface of target host cells, facilitating their entry or activation.

Cholera toxin, secreted by *Vibrio cholerae,* exhibits a high specificity for GM1 ganglioside, a glycolipid abundantly present on the surface of intestinal epithelial cells. This binding is essential for the internalization of the toxin, which subsequently disrupts ion transport and water homeostasis in the gut, resulting in the hallmark profuse diarrhea characteristic of cholera infection [14]. In a similar manner, Shiga toxin, produced by certain pathogenic strains of UPEC, targets the glycosphingolipid receptor Gb3, which is particularly enriched in kidney endothelial cells. This selective interaction not only facilitates toxin uptake but also underpins the pathogenesis of severe complications such as hemolytic uremic syndrome, leading to acute kidney injury [15]. These examples underscore the critical role of specific glycan–toxin interactions in mediating host tissue targeting and disease severity. This glycan-dependent entry mechanism often dictates the specific tissue tropism of bacterial toxins, explaining why certain organs are more vulnerable to their damaging effects.

Immune Evasion: Beyond initial entry, bacteria must also survive within the hostile host environment. To evade destruction by the immune system, many bacteria have evolved sophisticated strategies, including covering themselves in glycan structures that intricately mimic those naturally found on human cells, thereby flying under the immune radar.

Group B *Streptococcus* employs a clever immune evasion strategy by incorporating sialic acids into its polysaccharide capsule, effectively mimicking host cell surfaces. This molecular camouflage prevents activation of the complement system, allowing the pathogen to evade early immune attack [11]. Similarly, *Mycobacterium tuberculosis* modifies its surface lipoglycans, such as lipoarabinomannan, to interfere with host immune recognition and signaling, thereby enabling prolonged intracellular survival within macrophages [16]. *C. jejuni*, another prominent pathogen, evades immune detection by altering the glycosylation of its flagellin protein. Since flagellin is a major target of innate immune sensors like Toll-like receptor 5 (TLR5), these modifications reduce immune recognition and enhance bacterial persistence [17]. Collectively, these glycan-mediated strategies illustrate how pathogens exploit glycosylation to subvert host immunity and establish infection.

2.2.2 Viral Infections and Glycan Interactions

Viruses, as obligate intracellular parasites, are entirely dependent on host cells for their replication. They exploit host glycans extensively at nearly every stage of their life cycle—from initial attachment to immune escape.

Attachment and Entry: For a virus to initiate infection, it must first successfully attach to the host cell surface. Most viruses achieve this by utilizing specific surface proteins that bind to particular glycans displayed on the host cell membrane. This initial glycan binding helps to firmly anchor the virus, allowing it sufficient time to locate and engage with a secondary protein receptor that ultimately mediates viral entry into the cell.

SARS-CoV-2, the causative agent of COVID-19, initiates infection by binding to heparan sulfate glycans present on the surface of lung epithelial cells, a crucial interaction that occurs alongside engagement with its primary receptor, angiotensin-converting enzyme 2 (ACE2). This dual binding mechanism enhances viral attachment and facilitates entry into the host cells [18]. Similarly, influenza viruses exploit glycan-mediated interactions to establish infection, using their HA surface proteins to bind to terminal sialic acid residues on host cell glycoproteins. The type of sialic acid linkage—α2,3 or α2,6—plays a pivotal role in determining host specificity, with α2,3-linked sialic acids commonly found in avian species and α2,6-linked forms predominating in the human respiratory tract. This glycan recognition pattern influences both host range and transmission efficiency [4]. Glycan binding is frequently the first and most critical step in viral infection, and subtle mutations that alter this binding specificity can profoundly influence which species or tissues a virus can infect (host tropism).

Replication and Assembly: Once inside a host cell, the virus must replicate its genetic material and synthesize new viral components, which are then assembled into new infectious particles. Many viral proteins, especially those forming the outer envelope of enveloped viruses, undergo extensive glycosylation (decoration with glycans) during this intracellular replication and assembly process.

The glycosylation of viral envelope proteins plays a vital role in the life cycle and infectivity of many viruses. For example, the dengue virus envelope protein undergoes essential N-glycosylation within host cells, a modification that is crucial for proper folding and assembly of the protein. This glycosylation directly impacts the formation of infectious viral particles, ensuring their stability and ability to infect new cells [19]. Similarly, in the case of Hepatitis C virus, glycosylation of its envelope proteins is indispensable for correct folding and efficient assembly of new virions. These glycan modifications not only facilitate viral maturation but also influence viral infectivity and immune evasion [20]. Without proper glycosylation, many viruses are unable to replicate efficiently, produce properly structured progeny, or generate infectious particles, underscoring the vital role of host glycosylation machinery for viral success.

Immune Evasion: Many viruses employ dense layers of glycans on their surfaces as a strategic mechanism to protect themselves from recognition and neutralization by the host immune system.

HIV famously utilizes a "glycan shield" on its gp120 envelope protein, where a dense layer of carbohydrates physically masks vulnerable regions of the viral protein from neutralizing antibodies. This glycan shield presents a major obstacle to effective vaccine development, as it prevents the immune system from easily recognizing and targeting the virus [7]. Similarly, the Ebola virus coats its surface glycoproteins with heavy glycosylation, which effectively conceals it from immune detection and facilitates infection of host cells. This glycan-based camouflage is crucial for the virus's ability to evade immune responses and establish infection [21].

These extensive glycan shields make it exceedingly challenging for the host's immune system—and even vaccine-induced responses—to generate broadly effective antibodies against these highly glycosylated viral pathogens.

2.2.3 *Fungal Infections and Glycan Roles*

Fungal pathogens interact with host glycans through mechanisms that often differ significantly from those employed by bacteria or viruses. Fungi are characterized by complex cell walls rich in various polysaccharides, predominantly glucans and mannans. These cell wall glycans serve dual functions: providing structural integrity and protection for the fungus, while also acting as crucial signals that are recognized by the host immune system.

Adhesion and Invasion: These are critical early steps in fungal pathogenesis, much like in bacterial infections. Fungi initiate infection by adhering to host tissues, a process largely mediated by glycosylated proteins on their surfaces that recognize and bind specific host glycans. *Candida albicans*, a common opportunistic fungal pathogen, exemplifies this mechanism through the expression of mannoproteins on its cell surface. These mannoproteins facilitate strong adhesion to epithelial cells, allowing the fungus to colonize mucosal surfaces efficiently. Furthermore, this interaction promotes the development of biofilms—structured microbial communities that are exceptionally resilient against both antifungal therapies and host immune defenses. The biofilm matrix, enriched with extracellular glycans, not only anchors the fungal cells but also acts as a barrier, significantly contributing to the persistence and chronicity of *C. albicans* infections [22]. Once firmly attached, some fungi are capable of invading host tissues, a process often aided by the secretion of enzymes that degrade host extracellular matrix components, allowing deeper penetration.

Virulence Factors: The polysaccharide components of fungal cell walls are not merely structural elements; they also function as critical virulence factors that aid fungal survival and pathogenesis within the host.

Cryptococcus neoformans, a life-threatening fungal pathogen, is distinguished by its thick polysaccharide capsule primarily composed of glucuronoxylomannan (GXM). This capsule acts as a critical virulence factor by effectively protecting the fungus from phagocytosis by the host immune cells, thereby enabling it to persist and proliferate within the host [23]. In addition to this defensive barrier, the glycan structures within the fungal cell wall play vital roles in survival and pathogenicity. Even subtle modifications in glycan composition or branching patterns can dramatically alter the fungus's resistance to environmental stressors, antifungal agents, and immune-mediated attacks. These structural changes can enhance fungal robustness, making infections more persistent and difficult to treat.

Immune Recognition: The host immune system employs specialized **pattern recognition receptors (PRRs)**, primarily located on immune cells, to specifically detect conserved glycan patterns present on fungal cell walls. These PRRs are crucial for initiating an appropriate antifungal immune response.

Dectin-1, a prominent PRR expressed on macrophages and dendritic cells, plays a critical role in the immune system's defense against fungal pathogens. It specifically binds to β-glucans, which are integral components of fungal cell walls, initiating a cascade of intracellular signaling that promotes phagocytosis, cytokine production, and other immune effector mechanisms aimed at pathogen clearance [24]. This glycan-specific recognition by Dectin-1 does not function in isolation; rather, it operates synergistically with other PRRs, including Toll-like receptors (TLRs), to orchestrate a coordinated and potent antifungal immune response. The interplay between these receptor systems ensures precise recognition and effective elimination of fungal invaders [25]. Despite the immune system's sophisticated recognition mechanisms, some fungi have evolved strategies to subtly hide or modify their surface glycans to delay or dampen immune detection, thereby prolonging their persistence within the host. Different types of Pathogen-Glycan interactions and their role in infection and disease progression are presented in Table 1.

3. Role of Glycomics in Infection Progression

The trajectory of an infection—from the initial encounter between a pathogen and a host cell to the eventual development of symptoms or, ideally, disease resolution—is intricately shaped by a dynamic and highly interactive molecular landscape. Among the most pivotal players in this complex biological drama are glycans, the sophisticated sugar structures that intricately decorate the surface of virtually all living cells. Far from being mere passive components, these sugar structures actively orchestrate a wide array of processes, including how pathogens adhere to host cells, gain entry, skilfully evade the host's immune system, and strategically manipulate the host environment to their own advantage [1].

Glycomics, the large-scale and systematic study of glycan structures and their diverse biological functions, stand as a cornerstone for unraveling these intricate host-pathogen interactions. Throughout the course of an infection, both the invading pathogens and their host organisms undergo significant and often characteristic changes in their glycan expression profiles. These changes are not arbitrary; rather, they are a direct reflection

Table 1: Pathogen-Glycan interactions and their role in infection and disease progression.

Pathogen Type	Glycan Interaction	Mechanism of Action	Example Pathogens	Impact on Infection	References
Bacteria	Adhesion to host glycan receptors	Fimbriae/lectins bind to host glycans for colonization	*E. coli, H. pylori, P. aeruginosa*	Promotes bacterial attachment and persistence	[26, 27]
	Toxin binding to glycan structures	Bacterial toxins recognize glycosylated receptors	*V. cholerae, C. difficile*	Facilitates toxin entry and cytotoxicity	[28]
	Immune evasion through glycan mimicry	Bacterial surface glycans mimic host glycans to evade detection	*Campylobacter jejuni, N. meningitidis*	Reduces immune recognition and response	[29]
Viruses	Attachment to sialic acid or other glycan motifs	Viral envelope proteins bind specific glycans to initiate cell entry	Influenza virus, SARS-CoV-2, HIV	Enables host cell invasion and determines tissue tropism	[30, 31]
	Glycan shielding of viral antigens	Dense glycosylation hides viral epitopes from immune system	HIV, Lassa virus	Facilitates immune evasion and viral persistence	[32]
	Altered host glycosylation during infection	Host protein glycosylation changes in response to viral load	HBV, HCV, Dengue	Serves as biomarker for disease stage and prognosis	[9, 33]
Fungi	Glycan-mediated adhesion to host tissue	Surface mannoproteins interact with host epithelial glycans	*Candida albicans, Aspergillus fumigatus*	Facilitates fungal colonization and invasion	[34]
	Immune evasion via glycan camouflage	Fungal cell wall glycans modulate immune detection	*Cryptococcus neoformans*	Dampens immune response and enhances virulence	[35]
	Induction of host glycan remodeling	Infection triggers altered mucin glycosylation or glycosyltransferase expression	*Candida, Aspergillus, Pneumocystis*	Affects host-pathogen interaction and immune signaling	[36]

of an ongoing evolutionary "arms race"—a relentless struggle for survival between microbial invaders striving to establish themselves and host systems relentlessly attempting to defend themselves.

Gaining a deep understanding of how glycan structures evolve and transform during the various stages of an infection provides researchers with an exceptionally powerful lens through which to examine and comprehend the nuanced progression of disease. In this section, we embark on an exploration of how these multifaceted sugar molecules contribute fundamentally to three critical phases of infection: the initial and crucial process of pathogen entry into host cells, the subsequent and continuous phenomenon of glycan remodeling, and the overall profound impact of these glycan dynamics on the ultimate severity of the disease [37].

3.1 Glycan-mediated Entry of Pathogens into Host Cells

Before a pathogen can effectively invade a host and inflict harm, it must first successfully attach itself to the surface of host cells. This is by no means a random or haphazard encounter; instead, many pathogens exhibit a remarkable specificity, recognizing and binding with high affinity to particular glycans situated on the host cell membrane. These specific glycans effectively function as molecular "docking stations", providing the pathogen with a secure foothold. These highly selective interactions are fundamental in determining the pathogen's ability to infect specific tissues or organs—a phenomenon precisely known as tissue tropism. Understanding these initial binding events is critical for developing strategies to block infection at its earliest stage [38].

A prime and widely recognized example of this glycan-mediated entry mechanism is observed with the influenza virus. This pervasive respiratory pathogen utilizes a key surface protein called HA to bind with precision to sialic acid residues. These sialic acids are abundantly found on the epithelial cells lining the respiratory tract, providing numerous potential attachment points. Interestingly, human influenza strains display a preferential recognition for α2,6-linked sialic acids, which are predominantly found in human airways. In stark contrast, avian influenza viruses exhibit a strong preference for α2,3 linkages, which are commonly abundant in birds. This distinct glycan specificity elegantly explains why avian flu viruses rarely manage to infect humans efficiently unless they undergo specific genetic mutations that alter their HA to recognize human-type sialic acids [4]. This precise molecular distinction is a fundamental barrier to zoonotic transmission.

Similarly, *H. pylori*, a bacterium strongly implicated in the development of gastric ulcers and certain forms of stomach cancer, strategically targets Lewis b antigens. These antigens are specific fucosylated glycans naturally present in the human stomach lining. A critical aspect of *H. pylori* pathogenesis is that these Lewis b antigens are often observed to increase in abundance during states of gastric inflammation, providing the bacteria with an even greater number of binding sites and thereby actively promoting their successful colonization and persistence within the gastric environment. This illustrates how host responses can inadvertently facilitate pathogen establishment [5].

Another compelling illustration comes from *C. jejuni*, a common culprit behind foodborne gastroenteritis. This bacterium effectively attaches to ganglioside-like sialylated glycans found on the surface of intestinal cells. Importantly, this attachment process is often enhanced by the presence of inflammation in the gut, which can alter the availability or conformation of these binding glycans. Such intricate glycan-pathogen interactions are crucial for the bacterium to gain a firm foothold and successfully establish an infection, leading to the characteristic symptoms of gastroenteritis [39].

Even complex parasitic protozoa, such as *Plasmodium falciparum*, the causative agent of the most severe form of malaria, cunningly exploit host glycans as part of their life cycle. During the crucial liver and red blood cell stages of infection, the parasite relies on binding to specific host glycans. For instance, in the liver stage, the parasite binds to heparan sulfate proteoglycans on hepatocytes (liver cells), a necessary step for initiating the infection in the human host. Subsequently, during the red blood cell stage, the parasite binds to receptors like glycophorin A on red blood cells. These glycan-dependent interactions are absolutely key steps in the parasite's complex life cycle, making them potential targets for antimalarial interventions. These diverse examples highlight the universal importance of glycans as gatekeepers for pathogen entry across bacterial, viral, and parasitic infections [40].

3.2 Glycan Modification or Alteration during Infection

Once an infection has successfully commenced, the glycan landscape within the host continues to undergo significant and dynamic shifts. Both the host organism and the invading pathogen actively modify their respective glycan structures as a direct and continuous response to the ongoing molecular battle. These modifications are not arbitrary; rather, they serve distinct purposes, which can include promoting the pathogen's survival, subtly altering the host's immune responses, or contributing to the establishment and persistence of chronic infection. This glycan remodeling is a crucial aspect of the host-pathogen arms race that dictates the outcome of the infection [41].

Some viruses, notably human cytomegalovirus (HCMV), display a remarkable ability to directly interfere with host glycosylation pathways. A key mechanism involves HCMV downregulating the expression or activity of sialyltransferases—a family of enzymes responsible for adding sialic acid residues to host glycoproteins. By reducing host sialylation, HCMV effectively prevents the optimal activation of certain immune cells, such as Natural Killer (NK) cells. NK cells often rely on detecting specific, often abnormal, glycosylation patterns on the surface of infected cells to initiate their cytotoxic killing functions. Thus, HCMV's glycan manipulation provides a cunning escape mechanism from innate immune surveillance [42].

Pathogens can also employ an ingenious strategy of molecular mimicry to evade host immune surveillance. A classic example is *Group B Streptococcus* (GBS), a bacterium notorious for causing severe infections in newborns. GBS effectively cloaks itself in sialylated polysaccharides that bear a striking resemblance to human glycans. This molecular disguise allows GBS to effectively reduce complement activation (a crucial part of the innate immune system that marks pathogens for destruction) and successfully evade phagocytosis by host immune cells, thereby enhancing its ability to persist and cause disease [6].

On the host side, the presence of infection or inflammation often triggers a significant and widespread glycan remodeling process. These changes in host glycosylation can profoundly influence immune signaling pathways. For instance, chronic Hepatitis B and C infections, which often lead to persistent liver inflammation, cause liver cells to produce an increased quantity of fucosylated glycoproteins, particularly those exhibiting core fucosylation. These specific fucosylation patterns have been strongly linked to the progression of liver fibrosis and ultimately, cirrhosis, a severe scarring of the liver. Crucially, these specific glycan changes are detectable in the blood, making them promising candidates for development as potential biomarkers for monitoring disease progression and assessing liver damage in chronic viral hepatitis [28].

In the context of HIV infection, the virus itself actively manipulates glycan structures on its critical envelope protein, gp120. This manipulation results in the formation of a dense "glycan shield" that effectively protects the virus from recognition and neutralization by the host's antibodies [7]. This glycan shield is primarily composed of high-mannose and complex N-linked glycans. Not only does this shield aid in immune evasion, but it also subtly affects how the virus engages with host cell receptors, optimizing its entry and replication. This intricate interplay between viral and host glycosylation highlights the sophisticated strategies pathogens employ to survive and thrive within the host [43].

3.3 *Dynamic Interplay of Glycans and Disease Severity*

As infections progress, the continuous and evolving interplay between the host's glycome (the sum total of its glycans) and the pathogen's glycome profoundly determines the ultimate severity of the disease and whether the infection will resolve spontaneously or, conversely, become chronic. This intricate interaction is highly dynamic and is influenced by a multitude of factors, including the host's immune status, the specific tissue microenvironment, and the pathogen's adaptive capabilities.

In tuberculosis, caused by *Mycobacterium tuberculosis*, the bacterium exhibits a remarkable ability to modify its outer surface glycolipids, particularly mannosylated lipoarabinomannan (ManLAM). This modification allows the bacterium to actively suppress host immune detection and promote its intracellular survival, particularly within host macrophages. This adaptation is absolutely critical for the pathogen's notorious ability to persist within these immune cells, often leading to prolonged latent infections or chronic disease, making tuberculosis a formidable challenge for global health [44].

The host immune system, in turn, responds to infection by adjusting its own glycosylation patterns. For instance, activated T-cells and dendritic cells, crucial players in initiating and regulating immune responses, can upregulate specific patterns of sialylation and fucosylation on their surfaces. These changes in surface glycans directly affect how these immune cells interact with pathogens and with other immune cells. Depending on the precise context of the infection, these glycan modifications may either enhance the efficiency of pathogen clearance or, paradoxically, contribute to immune tolerance, potentially allowing the pathogen to persist. This delicate balance underscores the complexity of glycan-mediated immune regulation [45].

In fungal infections, such as those caused by *C. albicans*, the host immune system typically recognizes specific fungal glycans, particularly β-glucans, through specialized pattern recognition receptors like Dectin-1. This recognition normally triggers robust antifungal immune responses. However, *C. albicans* has evolved a clever strategy to counteract this. It can effectively mask these highly immunogenic β-glucans under a protective layer of mannans. This masking modulates the fungal pathogen's immune visibility, allowing it to evade efficient immune clearance and consequently increasing its pathogenic potential [46].

In some specific cases, such as with the Zika virus, glycan-mediated changes have been unequivocally linked to more severe clinical outcomes. The Zika virus utilizes its glycosylated envelope proteins to bind to host lectins, such as DC-SIGN, which are found on the surface of certain immune cells. This binding interaction enhances the infection of these immune cells, potentially aiding viral dissemination. In pregnant women, these glycan-dependent mechanisms are thought to contribute significantly to the virus's ability to cross the placental barrier and infect the fetus, leading to severe congenital abnormalities, most notably microcephaly. This highlights how glycan interactions can directly impact disease severity and long-term consequences [47].

3.4 Host Glycome Plasticity: Influence of Age, Diet, and Microbiota

The human glycome—the complete and intricate set of glycans produced throughout the body—is by no means a static entity. It is, in fact, remarkably dynamic and highly susceptible to continuous modifications influenced by a myriad of factors, including an individual's age, dietary habits, environmental exposures, and the unique composition of their gut microbiota. These multifactorial influences exert a profound impact on an individual's susceptibility to infections, the nature and strength of their immune response, and even the behavior and virulence of microbial pathogens. Understanding this plasticity is crucial for a holistic view of infection [48].

Age plays a demonstrably critical role in shaping the glycan landscape. Neonates and infants possess an immature glycosylation machinery, which significantly affects the types of glycans expressed on their epithelial surfaces and on secreted glycoproteins like immunoglobulins (antibodies). For example, infants often exhibit lower levels of sialylation and fucosylation on their mucosal glycans compared to adults. This developmental immaturity can render them inherently more susceptible to certain bacterial pathogens that specifically target these underdeveloped glycan structures for attachment and invasion [49]. Conversely, human milk oligosaccharides (HMOs), which are abundant and diverse glycans found in maternal milk, serve as ingenious "decoy receptors". These HMOs prevent pathogen binding to the infant's own cells, providing a crucial form of passive glycan-mediated immunity in the vulnerable neonatal period [50].

As individuals advance through the lifespan, their glycosylation patterns continue to evolve. Elderly individuals, for instance, tend to exhibit increased levels of agalactosylated IgG glycans. This specific glycan modification is strongly associated with heightened inflammation and a state of immune dysregulation, often referred to as 'inflammaging'. This inflammaging phenotype may contribute significantly to the increased susceptibility to severe infections, such as pneumonia and influenza, observed in older populations [51]. These age-related glycan changes highlight the importance of age-specific preventative strategies.

Dietary habits also exert a significant and direct influence on the host glycome. Essential nutrients like glucose, galactose, and fructose serve as fundamental substrates for the complex pathways of glycan biosynthesis. Consequently, diets that are consistently high in sugar or fat can alter these crucial glycosylation pathways, potentially leading to an increased expression of certain glycan structures that opportunistic pathogens can readily exploit. For example, high-fat diets have been shown to increase the abundance of α2,3-sialylated glycans in the respiratory tract, which are preferred binding targets for highly pathogenic avian influenza viruses [52]. Conversely, specific dietary components, such as prebiotics and dietary fiber, can actively promote beneficial glycan profiles by favorably shaping the composition and metabolic activity of the gut microbiota.

The gut microbiota itself, a vast and complex community of microorganisms residing in the human gut, exerts a profound and multifaceted influence on host glycosylation. Commensal bacteria (the beneficial microbes) directly interact with host epithelial cells, modulating the expression of glycosyltransferase enzymes and thereby altering the patterns of mucosal glycans. Some microbes even possess the enzymatic machinery to directly cleave host glycans, utilizing them as a nutrient source. *Bacteroides thetaiotaomicron*, for instance, harbors a wide array of glycosidases that efficiently break down mucin glycans in the gut. This activity indirectly affects the host's susceptibility to enteric pathogens by altering the protective mucin layer [53]. Furthermore, dysbiosis—a significant imbalance in the microbial community—has been strongly associated with abnormal glycosylation patterns, which can promote chronic inflammation and increase the risk of both infections and immune-mediated diseases [54].

Therefore, a comprehensive understanding of these diverse factors that shape the human glycome is absolutely crucial for designing highly effective and personalized interventions aimed at reducing infection risk. This includes developing age- and diet-specific nutritional strategies, such as personalized nutrition plans tailored to an individual's glycomic profile, or exploring probiotic therapies specifically designed to restore and maintain healthy glycosylation patterns. By appreciating the dynamic plasticity of the glycome, we can unlock new avenues for enhancing host defense against a wide array of infectious threats.

4. Glycan-based Diagnostic Tools and Therapeutic Implications

Glycomics, the comprehensive study of glycans (sugar chains), has profoundly transformed our comprehension of infectious diseases by illuminating the intricate roles played by both host and pathogen glycans in the complex mechanisms of disease. This burgeoning field has, in turn, paved the way for the development of innovative diagnostic and therapeutic strategies. This section delves into the multifaceted ways in which glycan structures and their dynamic interactions are being leveraged for the crucial purposes of disease diagnosis, precise monitoring of disease progression, and the targeted treatment of infections.

4.1 Glycan Profiling for Disease Diagnosis

Glycan profiling, a systematic and meticulous analysis of glycan structures present on proteins, lipids, or cellular surfaces, offers a powerful tool for disease detection. These glycan structures frequently undergo distinct and measurable alterations during the course of an infection. Such modifications can be directly attributed to inflammatory processes, the host's immune response, or even direct manipulation by microbial pathogens, thereby serving as highly valuable biomarkers for the early and accurate detection of disease.

A compelling illustration of this diagnostic potential is observed in active tuberculosis (TB). Patients afflicted with active TB exhibit unique glycan patterns on their serum glycoproteins. These patterns are notably characterized by an increase in core fucosylation and alterations in sialylation. Crucially, these specific glycan signatures are significantly different from those found in individuals with latent TB or healthy subjects. Furthermore, these altered glycan patterns demonstrate a strong correlation with the bacterial burden, making them exceptionally valuable for the precise diagnosis of active TB [55].

Similarly, in cases of sepsis, a life-threatening condition resulting from the body's overwhelming response to infection, lectin microarray-based profiling has proven instrumental. This technique has consistently revealed elevated levels of mannose-rich glycans in septic patients. This discovery enables clinicians to differentiate sepsis from non-infectious systemic inflammatory response syndrome (SIRS) with significantly improved specificity [56], thus facilitating more timely and appropriate interventions.

In the context of chronic viral infections, such as Hepatitis B and C, changes in the glycosylation of serum proteins, notably haptoglobin and alpha-1-acid glycoprotein, serve as reliable indicators of the stage of liver damage. These specific glycan alterations have been unequivocally shown to differentiate between mild inflammation and more advanced stages of liver disease, including fibrosis or cirrhosis [57]. Likewise, in dengue virus infections, a major global health concern, researchers have successfully identified distinct glycomic signatures. These signatures are characterized by increased fucosylation and altered glycan branching, and they are strongly associated with disease severity and the modulation of the immune response [33].

H. pylori, a bacterium notoriously responsible for causing gastritis and gastric ulcers, induces significant and observable changes in the glycosylation within the gastric mucosa. Lectin histochemistry, a technique that uses lectins (proteins that bind to specific sugar structures) to stain tissues, has unveiled shifts in sialylated and fucosylated glycan structures. These alterations not only contribute to the pathogenesis of the disease but also offer a robust biomarker platform for non-invasive diagnosis [57]. These diverse examples collectively underscore the immense diagnostic potential of glycan profiling across an extensive array of infectious diseases, highlighting its versatility and precision in identifying unique disease-associated signatures.

4.2 *Glycan-based Biomarkers for Disease Progression*

Beyond their utility in initial diagnosis, glycan alterations also function as highly dynamic and sensitive indicators of disease progression. As infections evolve and the host-pathogen interplay shifts, both host and pathogen glycans can undergo characteristic structural changes that accurately reflect the severity of the disease or the evolving immune status of the host. This dynamic nature makes them invaluable for real-time monitoring and prognostic assessment.

In viral hepatitis, for instance, a notable increase in the core fucosylation of alpha-fetoprotein has been shown to correlate directly with progressive liver fibrosis. This crucial correlation allows clinicians to stratify patients according to the severity of their liver disease without the need for invasive liver biopsies, thereby reducing patient discomfort and risks associated with such procedures [9]. Similarly, in the context of HIV infection, modifications within the glycan shield of the viral envelope protein gp120 are intricately linked to immune escape mechanisms and the ongoing evolution of the virus. Tracking the gain or loss of specific glycans on this viral protein can provide profound insights into viral adaptation strategies and, importantly, the effectiveness of antiviral treatment responses [58].

In respiratory infections, particularly those caused by influenza viruses, changes in the host's sialic acid linkages on mucosal surfaces significantly influence viral tropism (the ability of a virus to infect specific cell types) and infectivity. During severe influenza infection, a remodeling of mucin glycosylation can serve as a clear signal of airway inflammation and can therefore be utilized as a reliable marker of disease progression [59]. The recent COVID-19 pandemic offers another compelling and widely recognized example. Patients suffering from severe COVID-19 illness consistently exhibit high levels of agalactosylated IgG glycans. These specific glycan patterns are strongly associated with enhanced inflammatory responses and, unfortunately, worse clinical outcomes. These glycan patterns not only assist in predicting prognosis but also provide critical insights into the underlying mechanisms of immune dysregulation in severe COVID-19 [60].

In the case of malaria, infection with *Plasmodium* parasites induces significant changes in the glycosylation of red blood cell surface proteins. These alterations directly impact the parasite's ability to adhere to host cells and evade the host's immune system. These glycomic alterations have been definitively linked to disease severity and the development of cerebral malaria, a particularly severe complication, thereby underscoring their profound importance in pathophysiological monitoring [61]. The ability to track these dynamic glycan changes offers unprecedented opportunities for real-time disease management and personalized therapeutic approaches.

4.3 *Glycan-based Vaccines*

Glycan-based vaccines represent a highly successful and innovative approach that leverages the immunogenic properties of microbial polysaccharides or glycan-protein conjugates to induce robust and protective immunity against various pathogens. This strategy has proven particularly effective in targeting encapsulated bacteria, which are often difficult to combat with traditional vaccine approaches due to their protective capsules.

One of the earliest and most impactful examples of a glycan-based vaccine is the pneumococcal vaccine. This vaccine utilizes capsular polysaccharides derived from *Streptococcus pneumoniae*, the bacterium responsible for a wide range of infections including pneumonia, meningitis, and otitis media. These polysaccharides are conjugated to protein carriers, enhancing their immunogenicity, especially in young children. These conjugate vaccines have achieved a dramatic reduction in the incidence of invasive pneumococcal disease, particularly among pediatric populations, significantly improving global child health [62]. A similar and equally successful strategy has been employed in the *Haemophilus influenzae* type b (Hib) vaccine, where the bacterial polysaccharide is conjugated to tetanus toxoid. This vaccine has been so effective that it has nearly eradicated Hib-related meningitis in many parts of the world, serving as a testament to the power of glycan-based vaccination [63].

Meningococcal vaccines, designed to combat *Neisseria meningitidis* (the causative agent of meningococcal meningitis), are formulated using serogroup-specific polysaccharide antigens. These vaccines have proven highly effective in preventing outbreaks of meningococcal meningitis, particularly in regions such as sub-Saharan Africa, where the disease burden is significant [64]. In the critical case of GBS, a leading cause of severe neonatal infections, advanced vaccine candidates are actively being developed. These candidates are based on capsular polysaccharides, with the aim of inducing protective maternal antibodies that can be transferred to the infant, providing crucial early protection against GBS [65].

Furthermore, recent advancements in fungal immunology have opened new avenues for the development of experimental vaccines targeting β-glucans and mannan structures found in pathogenic fungi like *C. albicans* and *Aspergillus fumigatus*. These innovative vaccines aim to boost antifungal immunity, particularly in immunocompromised individuals who are highly susceptible to severe fungal infections, offering a promising new frontier in antifungal prophylaxis and treatment [66]. The success of glycan-based vaccines highlights the critical role of carbohydrates in stimulating protective immune responses.

4.4 Glycomimetic Drugs

Glycomimetics represent a sophisticated class of synthetic molecules meticulously designed to structurally resemble natural glycans. However, unlike their natural counterparts, glycomimetics are engineered to resist enzymatic degradation, thereby exhibiting enhanced stability and bioavailability. Their primary mode of action involves blocking specific glycan-mediated interactions between pathogens and host cells, thus preventing or disrupting the infectious process.

A prime example of the therapeutic potential of glycomimetics is seen with heparan sulfate mimetics. These compounds have demonstrated efficacy in inhibiting SARS-CoV-2, the virus responsible for COVID-19, by binding to the viral spike protein. This binding action prevents the viral spike protein from attaching to host heparan sulfate and ACE2 receptors, thereby effectively blocking viral entry into host cells [67]. In the context of urinary tract infections (UTIs), mannose analogs act as competitive inhibitors. They work by binding to the fimbriae (hair-like appendages) of *E. coli*, preventing the bacteria from attaching to mannose receptors on uroepithelial cells. This competitive inhibition significantly reduces bacterial colonization and subsequent infection [68].

Similarly, fucose-based glycomimetics have been developed to specifically target the lectins (carbohydrate-binding proteins) of *Pseudomonas aeruginosa*, a notorious opportunistic pathogen. *P. aeruginosa* utilizes fucosylated receptors for adhesion and the crucial process of biofilm formation, which contributes significantly to its virulence. By blocking these interactions, fucose-based glycomimetics have shown considerable promise in reducing bacterial virulence and pathogenicity [69]. In the fight against influenza, synthetic sialic acid mimetics effectively bind to the viral hemagglutinin protein. This binding prevents the virus from engaging with host sialylated receptors, thereby significantly reducing viral infectivity [70].

H. pylori, known for its tenacious colonization of the stomach lining and its role in gastritis and ulcers, employs lectins to bind to galactose-containing glycans on mucosal surfaces. Galactose-mimicking compounds have been shown to effectively disrupt this crucial adhesion, thereby mitigating bacterial persistence and the associated inflammation [71]. These examples collectively demonstrate the versatile and potent therapeutic potential of glycomimetic drugs in interfering with critical host-pathogen interactions.

4.5 Antiglycan Antibodies and Inhibitors

Antibodies, specifically targeting unique glycan epitopes or small molecule inhibitors that interfere with glycosylation processes, represent powerful strategies to directly combat pathogens or disrupt host glycan-mediated immune evasion mechanisms. This approach offers highly specific and targeted therapeutic interventions.

In the field of oncology, for instance, anti-GD2 monoclonal antibodies have been successfully employed to target neuroblastoma cells, which often express this specific glycan. Similarly, in infectious diseases, antiglycan antibodies are being actively explored to neutralize pathogens that exhibit unique glycan signatures [71]. Antibodies directed against the capsular polysaccharides of *Streptococcus pneumoniae* are a classic example; they enhance opsonization (the process by which pathogens are marked for destruction) and phagocytosis, leading to more efficient bacterial clearance by immune cells [72].

Broadly neutralizing antibodies (bNAbs) against HIV often recognize conserved glycans on gp120, a heavily glycosylated viral envelope protein. These antibodies possess the remarkable ability to neutralize a wide range of HIV strains by targeting glycan-dependent epitopes that are crucial for viral survival and replication [73]. Concurrently, small molecule inhibitors like tunicamycin interfere with the N-linked glycosylation of viral proteins. This disruption prevents proper protein folding and assembly, thereby significantly reducing viral replication and infectivity [74].

Iminosugars constitute another important class of antiviral agents. These compounds function by inhibiting glycosidase enzymes, which are critical for processing glycoproteins in a variety of viruses, including dengue and hepatitis C viruses. This inhibition results in the production of non-infectious viral particles, effectively halting the spread of the infection [75]. The development of such targeted antibodies and inhibitors offers highly precise and effective avenues for therapeutic intervention in infectious diseases.

4.6 Targeting Glycosylation Pathways

Targeting the biosynthetic pathways responsible for the creation of glycan structures offers an indirect but highly effective strategy for combating infections. This approach works by disrupting essential processes in both pathogens and host cells, ultimately impairing the pathogen's ability to survive, replicate, or evade the host's immune system.

In bacterial pathogens such as *Streptococcus pneumoniae* and *Neisseria meningitidis*, inhibitors of glycosyltransferase enzymes have demonstrated significant promise. These inhibitors prevent the crucial formation of the bacterial capsule, rendering the bacteria more vulnerable to immune clearance mechanisms, including phagocytosis by host immune cells [74]. Host-directed therapies that modulate host glycosylation pathways can also play a vital role in limiting infection. For example, inhibiting host fucosyltransferases can reduce the expression of fucosylated receptors on host cells that are cunningly exploited by pathogens like *H. pylori* and certain viruses for attachment and entry [76].

Viruses, including notorious pathogens like HIV, influenza, and coronaviruses, are highly dependent on the host cell's glycosylation machinery for the proper folding, maturation, and assembly of their glycoproteins, which are essential for infectivity. Targeting these host enzymes with specific inhibitors can significantly impair viral infectivity and consequently reduce disease severity [77]. In the fight against tuberculosis, enzymes involved in synthesizing unique cell wall glycans, such as arabinogalactan, have been identified as crucial targets. Disrupting these enzymes directly impacts *Mycobacterium tuberculosis* survival and its ability to cause disease [16].

Fungal pathogens also heavily rely on the glycosylation of their cell wall components for structural integrity, which is vital for their survival, and for evading the host's immune system. Compounds that specifically

Table 2: Glycan-based diagnostic and therapeutic approaches in infectious diseases.

Approach	Mechanism/Target	Examples	References
Glycan Profiling for Diagnosis	Identification of infection-specific glycan alterations in host proteins and cells	Core fucosylation in TB, mannose-rich glycans in sepsis	[36, 56]
Glycan-Based Biomarkers	Monitoring glycan changes reflecting disease progression or immune modulation	Fucosylated AFP in hepatitis, agalactosylated IgG in COVID-19	[9, 60]
Glycan-Based Vaccines	Use of microbial polysaccharides or glycan-protein conjugates to elicit immunity	Pneumococcal, Hib, Meningococcal	[62–64]
Glycomimetic Drugs	Synthetic glycan analogs block pathogen–host interactions	Heparan mimetics in COVID-19, mannose analogs in UTIs	[67, 68]
Antiglycan Antibodies	Monoclonal antibodies or inhibitors targeting pathogen-specific glycan epitopes	Anti-GD2 in neuroblastoma, anti-gp120 antibodies in HIV	[79, 73]
Targeting Glycosylation Pathways	Disruption of glycosylation enzymes in pathogens or host	Inhibitors of arabinogalactan synthesis in TB, β-glucan synthesis in fungi	[16, 78]

inhibit β-glucan or mannan synthesis have been found to compromise fungal viability, offering promising therapeutic options against invasive fungal infections, which are often difficult to treat, especially in immunocompromised patients [78].

Collectively, these diverse examples unequivocally underscore the transformative potential of glycan-targeted diagnostics and therapeutics in the comprehensive management of infectious diseases (Table 2). The insights gained from glycomics are not only revolutionizing our understanding of host-pathogen interactions but also paving the way for the development of a new generation of highly specific, effective, and innovative tools to diagnose, monitor, and treat a wide spectrum of infectious diseases.

5. Challenges and Future Directions

Glycomics, the study of glycans—complex carbohydrates found on the surfaces of nearly all cells and molecules—plays an increasingly important role in infectious disease research. However, several persistent challenges continue to limit its full potential. A primary difficulty lies in the structural complexity and variability of glycans. Unlike nucleic acids or proteins, glycans are not produced by a genetic template. Instead, their biosynthesis depends on the action of various glycosyltransferase enzymes, and their structures vary according to tissue type, developmental stage, disease state, and environmental conditions. This non-templated synthesis results in microheterogeneity, where the same protein can be decorated with different glycans, complicating systematic analysis. Structurally similar glycans can also differ in function, yet remain difficult to distinguish analytically.

The tools available to study glycans—such as high-resolution mass spectrometry and lectin microarrays—are advanced, but limited by their high cost, technical demands, and need for specialized data analysis. These techniques are not yet standardized or widely accessible, particularly in clinical settings. Additionally, bioinformatics tools and databases tailored to glycomics are underdeveloped compared to those available for genomics and proteomics. This limits both the speed and depth of research, as well as the ability to integrate glycomic findings into broader biomedical applications.

Another major hurdle is the limited availability of glycan-specific reagents. The development of high-affinity antibodies and small molecules that target specific glycans is complicated by the poor immunogenicity of many glycan structures. Additionally, synthesizing complex glycans for research or drug development remains technically challenging. These limitations restrict both fundamental research and the creation of targeted therapies.

Adding to the complexity is the context-dependent nature of glycan functions in host-pathogen interactions. Glycans can facilitate or inhibit infection. For instance, some pathogens exploit host glycans to gain entry into cells, while glycan-rich structures like mucins help defend against microbial invasion. Pathogens may also modify their own glycan profiles or hijack host glycosylation to avoid immune detection. This duality complicates therapeutic development, as interventions must avoid disrupting essential host glycan functions.

Despite these challenges, the future of glycomics in infectious disease research holds great promise. One major opportunity lies in integrating glycomic data with other "omics" platforms—such as genomics, proteomics, and metabolomics. A multi-omics approach offers a more comprehensive understanding of infection mechanisms, linking gene expression with changes in glycosylation and host-pathogen interactions. This integration can reveal new pathways involved in disease progression and novel targets for treatment.

Technological advances in glycan synthesis and engineering are also opening new avenues for innovation. Chemically or enzymatically synthesized glycans can be used to develop carbohydrate-based vaccines, diagnostic probes, and glycomimetic drugs that inhibit pathogen adhesion. Gene editing tools such as CRISPR offer the possibility of altering host glycosylation patterns to improve resistance to infection or enhance immune detection. Improvements in analytical technologies are helping to make glycomics more practical in clinical settings. Miniaturized, automated platforms and point-of-care devices for glycan profiling promise faster, cheaper, and more accurate diagnostic tools. These innovations are essential for bringing glycomics from research labs into hospitals and clinics, where they can directly impact patient care.

Lastly, building global capacity in glycoscience is vital. Investing in education, training, and international collaboration will expand the reach of glycomic tools and knowledge. By supporting emerging researchers and fostering interdisciplinary cooperation, the field can advance more quickly and equitably. With continued innovation and investment, glycomics is poised to transform our understanding of infectious diseases and contribute significantly to the development of novel diagnostics and therapies.

6. Conclusion

Glycomics has become a vital field in understanding how infections develop and progress, by revealing the diverse and dynamic roles that glycans play in host-pathogen interactions. These carbohydrate structures are no longer seen as passive components but as active mediators in processes such as pathogen adherence, cellular entry, immune evasion, and tissue targeting. Pathogens often exploit host glycans to establish infection, while the host can alter glycan expression in response to infection and inflammation. The insights gained from glycomic studies have enabled the identification of glycan-based biomarkers and the development of novel diagnostics, vaccines, and therapeutic agents. Despite these advances, the field still grapples with challenges including the structural complexity of glycans, limited standardization in analytical techniques, and gaps in functional annotation. However, the integration of glycomics with other omics platforms—such as genomics and proteomics—and advancements in analytical tools and glycoengineering are helping to overcome these barriers. As these technologies evolve, glycomics holds significant potential to drive innovation in infectious disease diagnosis and treatment, ultimately improving global health outcomes through more targeted and effective interventions.

References

1. Varki, A., R.L. Schnaar and R. Schauer. (2017). Sialic acids and other nonulosonic acids. *Essentials of Glycobiology* (3rd Edn.), pp. 1–15. Col Spring Harbour.
2. Cummings, R.D. and J.M. Pierce. (2014). The challenge and promise of glycomics. *Chem. Biol.* [Internet], 21(1): 1–15. Available from: http://dx.doi.org/10.1016/j.chembiol.2013.12.010.
3. Sharon, N. and H. Lis. (2004). History of lectins: From hemagglutinins to biological recognition molecules. *Glycobiology*, 14(11): 6–8.
4. Matrosovich, M.N., T.Y. Matrosovich, T. Gray, N.A. Roberts and H.D. Klenk. (2004). Human and avian influenza viruses target different cell types in cultures of human airway epithelium. *Proc. Natl. Acad. Sci. USA*, 101(13): 4620–4624.
5. Ilver, D., A. Arnqvist, J. Ögren, I.M. Frick, D. Kersulyte, E.T. Incecik, D.E. Berg, A. Covacci, L. Engstrand and T. Borén. (1998). Helicobacter pylori adhesin binding fucosylated histo-blood group antigens revealed by retagging. *Science*, 279(5349): 373–377.
6. Carlin, A.F., A.L. Lewis, A. Varki and V. Nizet. (2007). Group B streptococcal capsular sialic acids interact with siglecs (immunoglobulin-like lectins) on human leukocytes. *J. Bacteriol.*, 189(4): 1231–1237.
7. Wyatt, R. and J. Sodroski. (1998). The HIV-1 envelope glycoproteins: Fusogens, antigens, and immunogens. *Science*, 280(5371): 1884–1888.
8. Zheng, L., H. Li, L. Fu, S. Liu, Q. Yan and S.X. Leng. (2020). Blocking cellular N-glycosylation suppresses human cytomegalovirus entry in human fibroblasts. *Microb. Pathog.*, 138(January): 1–8.
9. Comunale, M.A., M. Lowman, R.E. Long, J. Krakover, R. Philip, S. Seeholzer, A.A. Evans, H.W. Hann, T.M. Block and A.S. Mehta. (2006). Proteomic analysis of serum associated fucosylated glycoproteins in the development of primary hepatocellular carcinoma. *J. Proteome Res.*, 5(2): 308–315.
10. Avci, F.Y., X. Li, M. Tsuji and D.L. Kasper. (2011). A mechanism for glycoconjugate vaccine activation of the adaptive immune system and its implications for vaccine design. *Nat. Med.*, 17(12): 1602–1609.
11. Carlin, A.F., S. Uchiyama, Y.C. Chang, A.L. Lewis, V. Nizet and A. Varki. (2009). Molecular mimicry of host sialylated glycans allows a bacterial pathogen to engage neutrophil Siglec-9 and dampen the innate immune response. *Blood*, 113(14): 3333–3336.
12. Ketchen, D.J. and C.L. Shook. (1996). Institutional login. *Strateg. Manag.*, (December 2018): 23–25.
13. Whelan, S., B. Lucey and K. Finn. (2023). Uropathogenic *Escherichia coli* (UPEC)—Associated urinary tract infections: The molecular basis for challenges to effective treatment. *Microorganisms*, 11(9).
14. Bailey, J., A. Oliveri and E. Levin. (2013). NIH public access. *Bone*, 23(1): 1–7.
15. Lingwood, C. (2020). Verotoxin receptor-based pathology and therapies. *Front. Cell Infect. Microbiol.*, 10(March): 1–10.
16. Nigou, J., M. Gilleron and G. Puzo. (2003). Lipoarabinomannans: From structure to biosynthesis. *Biochimie*, 85(1-2): 153–166.
17. Day, C.J., E.A. Semchenko and V. Korolik. (2012). Glycoconjugates play a key role in Campylobacter jejuni Infection: Interactions between host and pathogen. *Front. Cell Infect. Microbiol.*, 2(February): 9.
18. Clausen, T.M., D.R. Sandoval, C.B. Spliid, J. Pihl, H.R. Perret, C.D. Painter, A. Narayanan, S.A. Majowicz, E.M. Kwong, R.N. McVicar and B.E. Thacker. (2020). SARS-CoV-2 infection depends on cellular heparan sulfate and ACE2. *Cell* [Internet], 183(4): 1043–1057.e15. Available from: https://doi.org/10.1016/j.cell.2020.09.033.

19. Yap, S.S.L., T. Nguyen-Khuong, P.M. Rudd and S. Alonso. (2017). Dengue virus glycosylation: What do we know? *Front. Microbiol.*, 8(Jul.): 1–16.
20. Helle, F., G. Vieyres, L. Elkrief, C.I. Popescu, C. Wychowski, V. Descamps, S. Castelain, P. Roingeard, G. Duverlie and J. Dubuisson. (2010). Role of N-linked glycans in the functions of Hepatitis c virus envelope proteins incorporated into infectious virions. *J. Virol.*, 84(22): 11905–11915.
21. Collar, A.L., E.C. Clarke, E. Anaya, D. Merrill, S. Yarborough, S.M. Anthony, J.H. Kuhn C. Merle M. Theisen and S.B. Bradfute. (2017). Comparison of N- and O-linked glycosylation patterns of ebolavirus glycoproteins. *Virology* [Internet], 502(December 2016): 39–47. Available from: http://dx.doi.org/10.1016/j.virol.2016.12.010.
22. Netea, M.G., G.D. Brown, B.J. Kullberg and N.A.R. Gow. (2008). An integrated model of the recognition of Candida albicans by the innate immune system. *Nat. Rev. Microbiol.*, 6(1): 67–78.
23. da Silva, T.A., P.J. Hauser, I. Bandey, T. Laskowski, Q. Wang, A.M. Najjar and P.R. Kumaresan. (2021). Glucuronoxylomannan in the Cryptococcus species capsule as a target for chimeric antigen receptor T-cell therapy. *Cytotherapy*, 23(2): 119–130.
24. Brown, G.D., J. Herre, D.L. Williams, J.A. Willment, A.S.J. Marshall and S. Gordon. (2003). Dectin-1 mediates the biological effects of β-glucans. *J. Exp. Med.*, 197(9): 1119–1124.
25. Reid, D.M., N.A. Gow and G.D. Brown. (2009). Pattern recognition: Recent insights from Dectin-1. *Curr. Opin. Immunol.*, 21(1): 30–37.
26. Kline, K.A., S. Fälker, S. Dahlberg, S. Normark and B. Henriques-Normark. (2009). Bacterial adhesins in host-microbe interactions. *Cell Host Microbe*, 5(6): 580–592.
27. Borén, T., P. Falk, K.A. Roth, G. Larson and S. Normark. (1993). Attachment of Helicobacter pylori to human gastric epithelium mediated by blood group antigens. *Science*, 262(5141): 1892–1895.
28. Kristi L. Stringer, Bulent Turan, Lisa McCormick, Modupeoluwa Durojaiye, Laura Nyblade, Mirjam-Colette Kempf and J.M.T. Bronwen Lichtenstein. (2017). HHS public access. *Physiol. Behav.*, 176(3): 139–148.
29. Moran, A.P., A. Gupta and L. Joshi. (2011). Sweet-talk: Role of host glycosylation in bacterial pathogenesis of the gastrointestinal tract. *Gut*, 60(10): 1412–1425.
30. Gamblin, S.J., L.F. Haire, R.J. Russell, D.J., Stevens, B. Xiao, Y. Ha, N. Vasisht, D.A. Steinhauer, R.S. Daniels, A. Elliot and D.C. Wiley. (2004). The structure and receptor binding properties of the 1918 influenza hemagglutinin. *Science*, 303(5665): 1838–1842.
31. Glycoprotein, C.S., A.C. Walls, Park Y. Jun, M.A. Tortorici, A. Wall, A.T. Mcguire and D. Veesler. (2020). Structure, function, and antigenicity of the SARS-CoV-2 spike glycoprotein. *Cell* [Internet], 181(2): 281–292.e6. Available from: http://dx.doi.org/10.1016/j.cell.2020.02.058.
32. Watanabe, Y., T.A. Bowden, I.A. Wilson and M. Crispin. (2019). Exploitation of glycosylation in enveloped virus pathobiology. *Biochim. Biophys. Acta—Gen Subj.*, 1863(10): 1480–1497.
33. Voge, N.V., R. Perera, S. Mahapatra, L. Gresh, A. Balmaseda, M.A. Loroño-Pino, A.S. Hopf-Jannasch, J.T. Belisle, E. Harris, C.D. Blair and B.J. Beaty. (2016). Metabolomics-based discovery of small molecule biomarkers in serum associated with dengue virus infections and disease outcomes. *PLoS Negl. Trop. Dis.*, 10(2): 1–27.
34. McKenzie, C.G.J., U. Koser, L.E. Lewis, J.M. Bain, H.M. Mora-Montes, R.N. Barker , N.A. Gow and L.P. Erwig. (2010). Contribution of Candida albicans cell wall components to recognition by and escape from murine macrophages. *Infect. Immun.*, 78(4): 1650–1658.
35. O'Meara, T.R. and J. Andrew Alspaugh. (2012). The Cryptococcus neoformans capsule: A sword and a shield. *Clin. Microbiol. Rev.*, 25(3): 387–408.
36. Hall, R.A. and N.A.R. Gow. (2013). Mannosylation in candida albicans: Role in cell wall function and immune recognition. *Mol. Microbiol.*, 90(6): 1147–1161.

37. Pinho, S.S. and C.A. Reis. (2015). Glycosylation in cancer: Mechanisms and clinical implications. *Nat. Rev. Cancer*, 15(9): 540–555.
38. Lee, S., S. Inzerillo, G.Y. Lee, E.M. Bosire, S.K. Mahato and J. Song. (2022). Glycan-mediated molecular interactions in bacterial pathogenesis. *Trends Microbiol.*, 30(3): 254–267.
39. Kalischuk, L.D., G.D. Inglis and A.G. Buret. (2009). Campylobacter jejuni induces transcellular translocation of commensal bacteria via lipid rafts. *Gut Pathog.*, 1(1): 1–10.
40. Miller, L.H., D.I. Baruch, K. Marsh and O.K. Doumbo. (2002). The pathogenic basis of malaria. *Nature*, 415(6872): 673–679.
41. Li, Y., D. Liu, Y. Wang, W. Su, G. Liu and W. Dong. (2021). The importance of glycans of viral and host proteins in enveloped virus infection. *Front. Immunol.*, 12(April): 1–12.
42. Holtappels, R., J.K. Büttner, K. Freitag, M.J.Reddehase and N.A. Lemmermann. (2024). Modulation of cytomegalovirus immune evasion identifies direct antigen presentation as the predominant mode of CD8 T-cell priming during immune reconstitution after hematopoietic cell transplantation. *Front. Immunol.*, 2024: 15.
43. Bailey, J., A. Oliveri and E. Levin. (2013). NIH public access. *Bone*, 23(1): 1–7.
44. Richardson, E.T., S. Shukla, D.R. Sweet, P.A. Wearsch, P.N. Tsichlis, W. Henry Boom and C.V. Harding. (2015). Toll-like receptor 2-dependent extracellular signal-regulated kinase signaling in Mycobacterium tuberculosis-infected macrophages drives anti-inflammatory responses and inhibits Th1 polarization of responding T cells. *Infect. Immun.*, 83(6): 2242–2254.
45. Zheng, L., Q. Yang, F. Li, M. Zhu, H. Yang, T. Tan, B. Wu, M. Liu, C. Xu, J. Yin and C. Cao. (2022). The glycosylation of immune checkpoints and their applications in oncology. *Pharmaceuticals*, 15(12): 1–17.
46. Wheeler, R.T., D. Kombe, S.D. Agarwala and G.R. Fink. (2008). Dynamic, morphotype-specific *Candida albicans* β-glucan exposure during infection and drug treatment. *PLoS Pathog.*, 4(12): 1–12.
47. Hamel R., O. Dejarnac, S. Wichit, P. Ekchariyawat, A. Neyret, N. Luplertlop, M. Perera-Lecoin, P. Surasombatpattana, L. Talignani, F. Thomas and V.M. Cao-Lormeau. (2015). Biology of Zika virus infection in human skin cells. J Virol, 89(17), 8880–8896.
48. Newburg, D.S. (2005). Innate immunity and human milk. *J. Nutr.*, 135(5): 1308–1312.
49. Ruhaak, L.R. and C.B. Lebrilla. (2012). Advances in analysis of human milk oligosaccharides. *Adv. Nutr.* [Internet], 3(3): 406S–414S. Available from: http://dx.doi.org/10.3945/an.112.001883.
50. Bode, L. (2012). Human milk oligosaccharides: Every baby needs a sugar mama. *Glycobiology*, 22(9): 1147–1162.
51. Franceschi, C. and J. Campisi. (2014). Chronic inflammation (Inflammaging) and its potential contribution to age-associated diseases. *Journals Gerontol.—Ser. A, Biol. Sci. Med. Sci.*, 69: S4–S9.
52. Siegers, J.Y., B. Novakovic, K.D. Hulme, R.J. Marshall, C.J. Bloxham, W.G. Thomas, M.E. Reichelt, L. Leijten, P. van Run, K. Knox and K.A. Sokolowski. (2020). A high-fat diet increases influenza, a virus-associated cardiovascular damage. *J. Infect. Dis.*, 222(5): 820–831.
53. Martens, E.C., H.C. Chiang and J.I. Gordon. (2008). Mucosal glycan foraging enhances fitness and transmission of a saccharolytic human gut bacterial symbiont. *Cell Host Microbe* [Internet], 4(5): 447–457. Available from: http://dx.doi.org/10.1016/j.chom.2008.09.007.
54. Wang, B., M. Yao, L. Lv, Z. Ling and L. Li. (2017). The human microbiota in health and disease. *Engineering* [Internet], 3(1): 71–82. Available from: http://dx.doi.org/10.1016/J.ENG.2017.01.008.

55. Kumagai, T., A. Palacios, A. Casadevall, M.J. García, C. Toro, M. Tiemeyer and R. Prados-Rosales. (2019). Serum IgM glycosylation associated with tuberculosis infection in mice. *mSphere*, 4(2): 1–11.
56. De Coux, A., Y. Tian, K.Y. Deleon-Pennell, N.T. Nguyen, L.E. De Castro Brás, E.R. Flynn, P.L. Cannon, M.E. Griswold, Y.F. Jin, M.A. Puskarich and A.E. Jones. (2015). Plasma glycoproteomics reveals sepsis outcomes linked to distinct proteins in common pathways. *Crit. Care Med.*, 43(10): 2049–2058.
57. Kopenhaver, B., R.E. Long, O. Junaidi and A.M. Bisceglie. (2015). Identification and development of fucosylated glycoproteins biomarkers of primary hepatocellular carcinoma. In: *Biomarkers for Hepatocellular Carcinoma: Diagnostic and Therapeutic Utility*, 8(2): 595–602. National Institute of Health.
58. Mouquet, H., L. Scharf, Z. Euler, Y. Liu, C. Eden, J.F. Scheid, A. Halper-Stromberg, P.N. Gnanapragasam, D.I. Spencer, M.S. Seaman and H. Schuitemaker. (2012). Complex-type N-glycan recognition by potent broadly neutralizing HIV antibodies. *Proc. Natl. Acad. Sci. USA*, 109(47).
59. Pangeni, R., T. Meng, S. Poudel, D. Sharma, H. Hutsell, J. Ma, B.K. Rubin, W. Longest, M. Hindle and Q. Xu. (2023). Airway mucus in pulmonary diseases: Muco-adhesive and muco-penetrating particles to overcome the airway mucus barriers. Int J Pharm, 634, 122661.
60. Wu, J., B. Liang, C. Chen, H. Wang, Y. Fang, S. Shen, X. Yang, B. Wang, L. Chen, Q. Chen and Y. Wu. (2021). SARS-CoV-2 infection induces sustained humoral immune responses in convalescent patients following symptomatic COVID-19. *Nat. Commun.* [Internet], 12(1): 1–9. Available from: http://dx.doi.org/10.1038/s41467-021-22034-1.
61. Bautista, J.M., P. Marín-García, A. Diez, I.G. Azcárate and A. Puyet. (2014). Malaria proteomics: Insights into the parasite-host interactions in the pathogenic space. *J. Proteomics*, 97: 107–125.
62. Plotkin, S.A. (2010). Correlates of protection induced by vaccination. *Clin. Vaccine Immunol.*, 17(7): 1055–1065.
63. Peltola, H. (2000). Worldwide Haemophilus influenzae Type B disease at the beginning of the 21st Century: Global analysis of the disease burden 25 years after the use of the polysaccharide vaccine and a decade after the advent of conjugates, 13(2): 302–317.
64. Harrison, L.H., S.I. Pelton, A. Wilder-Smith, J. Holst, M.A.P. Safadi, J.A. Vazquez, M.K. Taha, F.M. LaForce, A. von Gottberg, R. Borrow and S.A. Plotkin. (2011). The Global Meningococcal Initiative: Recommendations for reducing the global burden of meningococcal disease. *Vaccine* [Internet]. Apr. 18 [cited 2025 Jun. 9]; 29(18): 3363–3371. Available from: https://www.sciencedirect.com/science/article/abs/pii/S0264410X11002933.
65. Nuccitelli, A., C.D. Rinaudo and D. Maione. (2015). Group B Streptococcus vaccine: State of the art . *Ther. Adv. Vaccines*, 3(3): 76–90.
66. Cassone, A. and A. Casadevall. (2012). Recent progress in vaccines against fungal diseases. *Curr. Opin. Microbiol.* [Internet]. Aug. 1 [cited 2025 Jun. 9]; 15(4): 427–433. Available from: https://www.sciencedirect.com/science/article/abs/pii/S1369527412000471.
67. Zhang, Q., I. Pavlinov, Y. Ye and W. Zheng. (2024). Therapeutic development targeting host heparan sulfate proteoglycan in SARS-CoV-2 infection. *Front Med.*, 11(March): 1–11.
68. Ofek, I., D.L. Hasty and N. Sharon. (2003). Anti-adhesion therapy of bacterial diseases: Prospects and problems. *FEMS Immunol. Med. Microbiol.*, 38(3): 181–191.
69. Sommer, R., K. Rox, S. Wagner, D. Hauck, S.S. Henrikus, S. Newsad, T. Arnold T. Ryckmans, M. Brönstrup, A. Imberty and A. Varrot. (2019). Anti-biofilm agents against Pseudomonas aeruginosa: A structure-activity relationship study of C-Glycosidic LecB inhibitors. *J. Med. Chem.*, 62(20): 9201–9216.

70. Heida, R., Y.C. Bhide, M. Gasbarri, Ö. Kocabiyik, F. Stellacci, A.L.W. Huckriede, W.L. Hinrichs and H.W. Frijlink. (2021). Advances in the development of entry inhibitors for sialic-acid-targeting viruses. *Drug Discov. Today*, 26(1): 122–137.
71. Crocker, P.R., J.C. Paulson and A. Varki. (2007). Siglecs and their roles in the immune system. *Nat. Rev. Immunol.*, 7(4): 255–266.
72. van der Put, R.M.F., B. Metz and R.J. Pieters. (2023). Carriers and antigens: New developments in glycoconjugate vaccines. *Vaccines*, 11(2): 1–25.
73. Pejchal, R., K.J. Doores, L.M. Walker, R. Khayat, P.S. Huang, S.K. Wang, R.L. Stanfield, J.P. Julien, A. Ramos, M. Crispin and R. Depetris. (2011). A potent and broad neutralizing antibody recognizes and penetrates the HIV glycan shield. *Science*, 334(6059): 1097–1103.
74. Almahayni, K., M. Spiekermann, A. Fiore, G. Yu, K. Pedram and L. Möckl. (2022). Small molecule inhibitors of mammalian glycosylation. *Matrix Biol. Plus*, 16: 100108.
75. Alonzi, D.S., K.A. Scott, R.A. Dwek and N. Zitzmann. (2017). Iminosugar antivirals: The therapeutic sweet spot. *Biochem. Soc. Trans.*, 45(2): 571–582.
76. Moran, A.P. (2008). Relevance of fucosylation and Lewis antigen expression in the bacterial gastroduodenal pathogen Helicobacter pylori. *Carbohydr. Res.* [Internet]. Aug. 11 [cited 2025 Jun. 9], 343(12): 1952–1965. Available from: https://www.sciencedirect.com/science/article/abs/pii/S0008621507005472.
77. An, Y. and J.F. Cipollo. (2011). An unbiased approach for analysis of protein glycosylation and application to influenza vaccine hemagglutinin. *Anal. Biochem.* [Internet]. Aug. 1 [cited 2025 Jun 9], 415(1): 67–80. Available from: https://www.sciencedirect.com/science/article/abs/pii/S0003269711002582.
78. Gow, N.A., M.G. Netea, C.A. Munro, G. Ferwerda, S. Bates, H.M. Mora-Montes, L. Walker, T. Jansen, L. Jacobs, V. Tsoni and G.D. Brown. (2007). Immune recognition of Candida albicans β-glucan by dectin-1. *J. Infect. Dis.*, 196(10): 1565–1571.
79. Raska, M., K. Takahashi, L. Czernekova, K. Zachova, S. Hall, Z. Moldoveanu, M.C. Elliott, L. Wilson, R. Brown, D. Jancova and S. Barnes. (2010). Glycosylation patterns of HIV-1 gp120 depend on the type of expressing cells and affect antibody recognition. *J. Biol. Chem.*, 285(27): 20860–20869.

Index

2-AB labelling 54, 64, 65

A

Analytical techniques 224, 226, 228, 229
Antigen presentation 169, 170, 172–176, 179
Antimicrobial activity 123, 126
Antimicrobial vaccine 25, 32
Artificial Intelligence 183, 184, 192, 196
Autoimmune diseases 233, 234, 239, 241, 244, 248

B

Behavioral disorders 24, 47
Biomarkers 26, 34–37, 40, 43, 45, 47, 48, 95–99, 101–104, 106–110, 112–114, 169, 171, 176–180, 224, 226, 230, 233, 236, 244, 245, 248, 251, 252

C

Cell adhesion 1, 5, 8, 9, 12, 15
Cervicovaginal glycome 141
Colonization 119, 120, 130
Complex type N-glycan's 56

D

Database 183, 184, 186, 187, 189–193, 195, 197
Decidual glycans 141, 148
Disease diagnosis 26, 33, 36, 39, 47
Disease diagnostics 98
Disease pathogenesis 169, 178
Diseases 24–29, 32–39, 42, 43, 47, 48
Disruption of homeostasis 232, 239
Dissemination 130

E

Endo H or Endoglycosidase H 57
Environmental stress response 75

F

Functional glycomics 1–3, 7, 16, 17, 19, 72, 223, 225–227, 239, 247, 248

G

Glycan 1–20, 95–114, 183–194, 196–198, 260–287
Glycan biomarkers 34, 48, 260, 262, 275, 280, 285, 287
Glycan dysfunction 1, 3, 12, 14, 15
Glycan signaling 70, 71
Glycan structure 223, 225–230, 232, 233, 237–241, 244, 245, 246, 248
Glycan-based diagnostics 262, 263
Glycan-based vaccines 281, 282, 285
Glycoinformatics 183–185, 189, 192, 194, 197, 198
Glycomic-based therapies 120, 130, 132
Glycomics 24–29, 32, 33, 36, 45, 47, 48, 54, 59, 67, 95–99, 101–110, 112–114, 119–123, 125, 126, 129–132, 206–211, 214, 215, 217, 218, 260–263, 271, 278, 280, 281, 285–287
Glycoproteomics 25, 36–38, 41, 169–172, 175–180
Glycosylation 169–180, 223, 224, 226, 228–252
Glycosylation pathways 75, 76
Glycosylation patterns 119–121, 123–132
GlycoWorkbench 59, 65

H

High mannose type 56
HILIC-UPLC 62, 64, 66
Homeostasis 1–3, 5, 6, 8–16, 224–226, 231–235, 237–241, 247, 249, 250
Host-pathogen interaction 119–123, 125–128, 130, 132, 260, 261, 271, 272, 283, 285–287
Hybrid type 56

I

Immune dysregulation 233
Immune homeostasis 233, 240
Immune modulation 70, 72, 74
Immune regulation 170, 173, 174, 178–180
Immunotherapy 169, 171, 172, 174–179
Infection 24–26, 32–34, 39, 48, 95, 97, 105–107
Infection progress 119, 122, 126, 127
Infectious diseases 120, 121, 123, 126, 128, 130, 132
International Union of Pure and Applied Chemistry (IUPAC) 185
Isomeric N-glycans 206, 217, 218

L

Lectin Microarray-based assays 206

M

M. pruriens 120–128, 130–132
MALDI-MS and mass spectrometry 206
MALDI-TOF MS 53, 54, 62, 65
Mass spectrometry 96, 109, 112, 114
Maternal glycobiology 142, 146
Maternal glycome 142
Maternal-foetal interface 147
Medicine strategies 170
Metabolomics 25, 29, 39
Microbial glycomics 25, 32
Microbial invasion 119, 120, 130
Milk oligosaccharides 145, 146, 160

N

N- and O-glycobiomarkers 206
Neurodegenerative diseases 95, 97, 98, 104, 226, 236, 241, 245, 248, 249, 250, 252
N-glycans 53–60, 62–66
NMR 230

P

Pattern-recognition receptors 70
Paucimannosidic type 56
Placental glycobiology 147, 149, 150
Placental glycome 156
PNGase A 53, 57, 58
PNGase F 57, 58
PNGase H+ 53, 57, 58, 60, 62, 63, 66, 67
Portal 189–193
Precision medicine 24, 96, 98, 99, 101, 114
Pre-eclampsia 136, 142, 146, 147, 150, 153, 155, 156, 158, 160
Pregnancy 136, 137, 140–145, 147–157, 159, 160
Protein folding 1, 2, 4, 7, 10, 12, 13, 17

R

Receptor modulation 8
Repository 183, 188, 191, 193

S

Semantic web 184, 186, 188–193, 195, 198
Solid-phase extraction 62, 64
Symbol Nomenclature for Glycans (SNFG) 185, 190

T

Therapeutic strategies 120, 121, 123, 125, 126, 128, 129, 132
Therapeutic targets 25, 26, 71, 78, 79
Trophoblast invasion 136, 142, 145, 149, 156, 160

W

Web3 Unique Representation of Carbohydrate Structures (WURCS) 185

For Product Safety Concerns and Information please contact our EU representative GPSR@taylorandfrancis.com Taylor & Francis Verlag GmbH, Kaufingerstraße 24, 80331 München, Germany